Atoms and d^n – Ions in Perturbation Fields

Atoms and d^n – Ions In Perturbation Fields

(Theory and calculation basics of Atomic and Crystal Field States and Relative Energies)

Sisir Chandra Rakshit, Ph.D.

Formerly, Professor of Chemistry
The University of Burdwan

LEVANT

Levant Books

Kolkata, India

Atoms and d^n – Ions in Perturbation Fields

First Published, 2010

Published by
Levant Books
27C, Creek Row
Kolkata – 700 014

And exclusively distributed by
Sarat Book Distributors
18B, Shyama Charan Dey Street
Kolkata – 700 073, India
E-mail: books@saratbookhouse.com

ISBN: 978-93-80663-04-3

Typeset, Printed and bounded at
Sarat Impressions Pvt. Ltd.
18B, Shyama Charan Dey Street, Kolkata – 700 073, India

To my lovely little granddaughter,

Elora Basu

who would like to streak the pages with crayons or,
in a joyful spurt of activity, tear these into pieces.

....

Acknowledgement:

I express my sincere thanks and indebtedness to

i) Soma Basu and Amitava Basu for word-processing the entire manuscript on a PC and a laser print of the same,

ii) Dr. Manas Banerjee, my ex-professional colleague, Chemistry Department, The University of Burdwan, for suggesting rearrangement and improvement, and finally,

iii) Dr. Biren Hazra, my ex-research colleague, for helping me with some drawings and diagrams.

– Author

I am specially thankful to Sri Pranjit Roy for preparing the new-impression of the manuscript in press-ready form.

– Author

CONTENTS

INTRODUCTION

To understand the magnetic and the special properties of the transition metal complexes, essential for revealing their structures, one ought to know the origin, nature and classification of the spectral lines of atoms and the spectra of complex ions. In essence, the reader should be familiar with the atomic states, the crystal field states, Zeeman levels and other types of split levels. This useful "behind the scene" information of the spectra concerned is the prime goal of our study here.

The states, mentioned above, evidently demand of us a knowledge of the associated eigenfunctions and energy eigenvalues. The classification would then naturally follow from the symmetry types of the eigenfunctions or a reverse order may also be feasible. It is amazing to think how wonderfully one can have access to this "behind the scene" information if he is adequately conversant with a number of disciplines. These are
(i) the commutation properties of angular momenta
(ii) their coupling
(iii) the formality of expressing wavefunctions in the form of kets and energies in forms involving parameters
(iv) the theory of groups and double groups and
(v) the perturbation theory.

One needs to know the symbolic ket forms of eigenfunctions and only the parameter-laden energy values instead of the actual full functional forms of the eigenfunctions and the numerical magnitudes of energies.

The "vector model of an atom" method and the perturbation method are the two avenues leading to this goal. But while the "vector model" method involves concepts many of which are not acceptable in quantum mechanics, perturbation methods for deriving the states and energies never transgress the confines of quantum mechanics and are always in harmony with it. Additionally, the crystal field states, which lie within the reach and ambit of the perturbation theory, are beyond the pale of "vector model" theory.

The basic approach will then be to enlist the internal and external perturbative interactions, form the corresponding hamiltonians, and to successively apply the perturbations in order of their relative dominance. While the relative dominance will depend on situations and systems, successive applications of perturbations would entail successive building-up and solutions of perturbation matrices of the respective hamiltonians. The solutions are in the forms of eigenvectors (leading to eigenfunctions) and energy eigenvalues. In many instances, however, alternatives to perturbation matrix – based approach are feasible and are resorted to, if found less laborious and more elegant.

No knowledge of any higher level topic in chemistry is complete unless its relevant mathematical basics are well understood and mastered. It is with this aim in view that the present treaties sets out to derive the quantitative formulation from the viewpoint of theory. The applications, dealt with in other books, have mostly been excluded except in the forms of a few worked out examples and related exercises.

The chapterwise distribution of materials in the present book runs as follows:

The two initial chapters (1 and 2) open with a brief sum-up of the essential aspects of the commutation principles of angular momentum, their couplings in atoms, spherical harmonics, central field and different perturbative field hamiltonians, quantum mechanical perturbation theory and specific theorems and results of group theory. Special attention has been paid to the mathematical genesis of the different perturbation hamiltonians in polyelectron atoms and to the matrix elements of these hamiltonians between the determinantal wavefunctions of the multielectron atoms.

Chapters 3, 4, 5 accommodate the calculations of eigenfunctions and energy eigenvalues of the split levels under different in-built and external perturbations of atoms in which the couplings may be either LS or JJ type. This discussion on free atom eigenstates percolates into chapter 6 with emphasis on electrostatic perturbation energies in the higher d^n systems and on the choice of energy parameters.

The crystal field eigenstates, their symmetries, energy values and nature of their splittings including their quantitative calculations in different symmetry environments for d^n complex ion form the contents of chapters 7 to 11. Once again emphasis is laid in chapter 7 on the structures of the crystal field perturbation hamiltonians and the matrix elements in different environmental situation for a central d^n ion. These are then put to use for quantitative calculations in chapters 8, 9 and 10 with a spill-over to chapter 11 terminating in a short and incisive treatment of magnetic susceptibility and moment of d^n systems in crystal fields.

There has been a liberal use of conventional abbreviations for extended expressions and words including mathematical formulations. One unconventional brevity that occurs frequently is the word "det" for determinantal wave function. Finally, it is emphasised that the book deals only with the quantitative calculational basis in free and crystal field environments and generally steers clear of many of their thumb-rule and qualitative applications.

The book is intended for the serious and diligent post-graduate and pre-doctoral level chemistry students and to an extent, for the physics students as well, who want to furrow deep and learn the subject at its very depth. It is also hoped that the book will help the teachers in planning the contour, pruning or adding certain elements and stressing on certain refinements of their class lectures. The author will have his labour amply rewarded if the book meets the recommendations of the teachers and the expectations of the taught.

Chapter One

Angular Momentum, Spherical Harmonics and Couplings

General:

The different mosaics, each composed of a judicious mix of various theorems, formulations and properties of special functions, angular momentum, wave mechanics and group theory, underlie all the calculations pertaining to atoms or ions in free, crystal field or other environments. A fairly deep knowledge of these relevant topics is, therefore, a prerequisite for a clear understanding of the basic techniques of calculations in such fields. While it will be assumed that the readers are well conversant with the above, we shall, nevertheless, mention in a very general manner the pertinent features of the foregoing topics before plunging into calculational techniques. These will generally comprise listings only, unattended with proofs. Only in cases of the angular momentum and the hamiltonian operators will the treatments be fairly more than marginal and distributed over more than one location.

In the present chapter, a cursory glance will be cast at the commutation properties of the angular momentum. Legendre's and the associated Legendre's polynomials, spherical harmonics, states of atoms with one or more optical electrons, i.e., the Russell Saunders' and j-j couplings.

The relevant features of the form of hamiltonian. perturbation formulae, perturbation matrices, the necessary aspects of group theory, symbolic ket notation and the conventional representation of the polyelectronic determinantal wave functions of atoms will be described in the chapter to follow. One has to wriggle his way out of this labrynth of lanes and bylanes of the preliminaries, bordering on sophistication at some points, before he emerges on to a field providing the take-off stage for our intended calculations from Chapter 3 onwards.

1.1. Angular Momentum Commutation Rules, Step Up, Step down Operators and Properties:

The classical definition of angular momentum of a moving particle about a point is $l^* = r \times p$. where r is the radius vector from the point to the particle and p. its linear momentum. As the definition involves a crossproduct, it is evident that the angular momentum will be zero if r and p are collinear, while for other relative dispositions, the value is different from zero. Since in quantum mechanics one cannot allow for the concept of particle track, any zero value of the angular momentum of a microsystem (e.g., an electron in an atom) is some time-average value of the mechanical quantity without any pictorial significance of collinearity of r with the velocity track.

Expressing $r \times p$ in determinantal form, it is seen that the cartesian components of angular momentum l^* are

$$l_x^* = i\left(yp_z - zp_y\right), \quad l_y^* = j\left(zp_x - xp_z\right), \quad l_z^* = k\left(xp_y - yp_x\right) \qquad \ldots\ldots(1.1)$$

The operator forms of angular momentum in cartesian coordinates are, therefore,

$$\left.\begin{aligned} l_x &= (\hbar/i)\{y(\partial/\partial z) - z(\partial/\partial y)\} \\ l_y &= (\hbar/i)\{z(\partial/\partial x) - x(\partial/\partial z)\} \\ l_z &= (\hbar/i)\{x(\partial/\partial y) - y(\partial/\partial x)\} \end{aligned}\right\} \qquad \ldots\ldots(1.2)$$

Expressed in polar coordinates, the forms of the operators are

$$\left.\begin{aligned} l_x &= (\hbar/i)\{-sin\phi(\partial/\partial\theta) - cot\,\theta(\partial/\partial\phi)\} \\ l_y &= (\hbar/i)\{cos\,\phi(\partial/\partial\theta) - cot\,\theta\, sin\,\phi(\partial/\partial\phi)\} \\ l_z &= (\hbar/i)(\partial/\partial\phi) \\ \text{and } l^2 &= -\hbar^2\{(1/\sin\theta)(\partial/\partial\theta)(\sin\theta(\partial/\partial\theta)) + (1/\sin^2\theta)(\partial^2/\partial\phi^2)\} \end{aligned}\right\} \qquad \ldots\ldots(1.3)$$

The ranges of θ and ϕ being $0 \le \theta \le \pi$ and $0 \le \phi \le 2\pi$ respectively.

The forms (1.3) can simply be arrived at from first principles by noting that $x = rsin\theta\ cos\phi$, $y = rsin\theta\, sin\phi$ and $z = rcos\,\theta$.

Two other operators of great importance, called step up and step down operators, are defined and finally expressed in polar coordinates:

$$\left.\begin{aligned} l_+ &= l_x + il_y &= e^{i\theta}\{\hbar(\partial/\partial\theta) - cot\theta l_z\} \\ l_- &= l_x - il_y &= e^{-i\theta}\{-\hbar(\partial/\partial\theta) - cot\theta l_z\} \end{aligned}\right\} \qquad \ldots\ldots(1.4)$$

where l_z operator is given by (1.3)

The significance of the terms step up and step down (also called raising and lowering) operators will be clear in Sec 1.4 where we discuss the spherical harmonics and angular momentum. For a single election atom l^*, arising from its motion in space, is called orbital angular momentum of the electron. The following commutation, rules, as may be proved by considering relations (1.1, 1.2, 1.3, 1.4), are obeyed by the operators.

$$\left[l^2, l_z\right] = \left[l^2, l_x\right] = \left[l^2, l_y\right] = 0 \qquad \ldots\ldots(1.5)$$

$$\left.\begin{aligned} \left[l_x, l_y\right] &= \left(l_x l_y - l_y l_x\right) = i\hbar l_z \\ \left[l_y, l_z\right] &= i\hbar l_x \\ \left[l_z, l_x\right] &= i\hbar l_y \end{aligned}\right\} \qquad \ldots\ldots(1.6)$$

Also

$$\left.\begin{aligned} l_z(l_x + il_y) &= (l_x + il_y)(l_z + \hbar) \\ l_z(l_x + il_y)^n &= (l_x + il_y)^n (l_z + n\hbar) \\ l_z(l_x - il_y) &= (l_x - il_y)(l_z - \hbar) \\ l_z(l_x - il_y)^n &= (l_x - il_y)^n (l_z - n\hbar) \end{aligned}\right\} \quad \ldots\ldots(1.7)$$

where n is some positive integer.

Experimental observation of fine structures in the s.p.d. *f* series of atomic spectral lines led to the assumption (Uhlenbeck and Goudsmit) of an additional intrinsic angular momentum of an electron. This additional momentum introduced into atomic theory, was named spin angular momentum. s^* though it has got nothing to do with the spin motion, if any be there, of an electron. However, the name has stuck a deep root and is now irreplaceable. For an electron the spin moment has the invariable value, $s^* = \sqrt{s(s+1)}\hbar = \sqrt{1/2(1/2+1)}\ \hbar$ where s = spin quantum number = 1/2 and $s_z^* = m_s\,\hbar$ where $m_s = \pm 1/2 =$ spin quantum number of the *Z* directional angular momentum s_z^*. Like *l*, we can conceive of operators s_x, s_y, s_z, s^2, s_+ and s_- which are assumed to obey commutation rules similar to those of (1.5), (1.6) and (1.7). Thus

$$[s^2, s_z] = [s^2, s_x] = [s^2, s_y] = 0 \quad \ldots\ldots(1.8)$$

$$[s_x, s_y] = i\hbar s_z \qquad \text{etc.} \quad \ldots\ldots(1.9)$$

$$\left.\begin{aligned} s_z(s_x + is_y) &= (s_x + is_y)(s_z + \hbar) \\ &\text{etc.....} \end{aligned}\right] \quad \ldots\ldots(1.10)$$

It should, however, be noted that relations analogous to (1.1 -1.4) cannot be written down for the spin operators due to our lack of knowledge of the operator forms of spins in spin coordinates in spin spice.

However, since the space and spin coordinates are mutually independent, l^2 and each of l_i 's ($i = x, y, z$) commute with s^2 and all s_i 's separately.

We can, therefore, now conceive of a total mechanical momentum j^* of an electron which is the vector sum of l^* and s^*. Now j^2 and the component operators j_x, j_y, j_z obey commutation rules, similar to (1.5) to (1.7).

$$\left[j^2, j_z\right] = \left[j^2, j_x\right] = \left[j^2, j_y\right] = 0 \quad \text{......(1.11)}$$

$$\left[j_x, j_y\right] = i\hbar\, j_z \qquad \text{etc.} \quad \text{......(1.12)}$$

$$j_z\left(j_x + ij_y\right) = \left(j_x + ij_y\right)\left(j_z + \hbar\right) \text{ etc.} \quad \text{......(1.13)}$$

For a polyelectronic atom, all electrons have their individual l_i's and s_i's (i = 1,2,3) Depending on the nature of dominating interaction forces, the final coupled states of vectors arise giving rise to Russell Saunders' or the *j-j* coupled states of atoms in two extreme coupling conditions. In the Russell Saunders' or LS coupling the preliminary coupled vectors are $L^* = l_1^* + l_2^* + l_3^* + ...$ and $S^* = s_1^* + s_2^* + s_3^* + ...$ which then undergo a secondary coupling giving rise to a resultant vector $J^* = L^* + S^* = \Sigma\, j_i^*$. The primary couplings are essentially electrostatic which are followed by an overall secondary magnetic interaction coupling. It can be shown that all the commutation relations (1.5) through (1.13) are valid for such multielectron atom, the only modification being the replacement of the small case letters *l.s.j* by the upper ones i.e., *L.S.J.*

Prototypes of Eqns. (1.4) to (1.13) with uppercase L.S.J	......(1.4′) ......(1.5′) ....(1.13′)

Finally it can be shown :

$$L_+L_- = \left(L_x + iL_y\right)\left(L_x - iL_y\right) = L^2 - L_z^2 + \hbar L_z \quad \text{......(1.14)}$$

$$L_-L_+ = \left(L_x - iL_y\right)\left(L_x + iL_y\right) = L^2 - L_z^2 - \hbar L_z \quad \text{......(1.15)}$$

and $$L^2 = \Sigma l_i^2 + 2\Sigma_{i<j}\, l_{iz}\, l_{jz} + \Sigma_{i<j}\left(l_{i+}\, l_{j-} + l_{i-} l_{j+}\right) \quad \text{......(1.16)}$$

where i = 1,2,3 ......

We are not yet through with our discussion of angular momentum. Once again we shall recall these relations after we deal with some special polynomials in the next section.

1.2 Legendre's Polynomials and Associated Legendre's Polynomials

These polynomials occur in many important segments of mathematical physics, quantum mechanics, electrostatics, electrochemistry and group theory. Legendre's polynomials have been introduced in many ways, e.g., i) Rodrigue's function, ii) Generating function method iii) the polynomial solutions of special differential equation of the second order, known as Legendre's equation. The last equation referred to is

$$\left(1-x^2\right)d^2z/dx^2 - 2xdz/dx + n(n+1)z = 0 \qquad \ldots\ldots(1.17)$$

The solution is a polynomial in x given by

$$z = P_n(x) = \left(1/2^n n!\right)\left(d^n/dx^n\right)\left(x^2-1\right)^n \qquad \ldots\ldots(1.18)$$

i) This solution is valid only for real variable x and for integral positive value of the parameter n including zero.

ii) Any two such polynomial solutions of degree n and l, viz. $P_n(x)$, $P_l(x)$ are mutually orthogonal and each is individually normalized within the range $-1 \le x \le 1$

i.e. $\int_{-1}^{+1} P_n(x), P_l(x)dx = \delta_{n,l}$

iii) Other solution with nonintegral or nonpositive value of n in Eq. (1.17) may also occur, but such cases do not arise in physical problems dealt with by quantum mechanics or in problems of electrostatics.

iv) Very often, in problems of interest, the variable $x = cos\,\theta$, where θ is the polar angle in the spherical polar coordinate system.

In such cases, (1.18) transforms to

$$P_n(cos\,\theta) = \left\{(-1)^n/2^n n!\right\}\left\{d^n/(d\,cos\,\theta)^n\right\}\left(sin^{2n}\,\theta\right) \qquad \ldots\ldots(1.19)$$

The integral properties mentioned under (ii) are also retained by this function (1.19) the limits for θ being 0 to $+\pi$.

Associated Legendre's Polynomials $P_n^m(x)$

As in the previous case, the polynomial solutions that occur from the differential equation

$$\left(1-x^2\right)d^2u/dx^2 - 2xdu/dx + \left\{n(n+1) - m^2/\left(1-x^2\right)\right\}u = 0 \qquad \ldots\ldots(1.20)$$

are $u = P_n^m(x) = \left(1-x^2\right)^{|m/2|}\left(d^m/dx^m\right)P_n(x)$

$$= \left(1-x^2\right)^{|m/2|}\left(d^{n+|m|}/dx^{n+|m|}\right)\left\{\left(1/2^n n!\right)\left(x^2-1\right)^n\right\} \qquad \ldots\ldots(1.21)$$

subject to the restriction n= positive integer including zero and m, capable of taking both positive and negative integral values including zero in the range $-n \le m \le +n$

i) The polynomial function of degree n and order m, viz., $P_n^m(x)$ as given by Eq (1.21) is unnormalized.

ii) A normalized $P_n^m(x)$ within the limits –1 to +1 , is given a new symbol $\Theta_{n,m}(x)$ which is

$$\Theta_{n,m}(x) = \sqrt{\{(2n+1)/2\}\{(n-|m|)!/(n+|m|)!\}}\, P_n^m(x) \qquad \ldots\ldots(1.22)$$

$$= \sqrt{\{(2n+1)/2\}\{(n-|m|)!/(n+|m|)!\}}\times$$

$$(1/2^n n!)(1-x^2)^{|m/2|}\left(d^{n+|m|}/dx^{n+|m|}\right)(x^2-1)^n \qquad \ldots\ldots(1.23)$$

iii) Two such functions $\Theta_{n,m}(x), \Theta_{n,m'}(x)$ are mutually orthogonal within the aforesaid interval range. So are $\Theta_{n,m}(x)$ and $\Theta_{l,m}$ within the same range.

iv) In almost all the problems of physical interest where the Associated Legendre's polynomicals occur, the variable x is $\cos\theta$, where θ is the polar angle. Eqn. (1.20) and its normalized polynomial solutions (1.23) can then be expressed as (writing $\Theta_{n,m}$ for normalized u)

$$(1/\sin\theta)\,(d/d\theta)\{(\sin\theta)(d\Theta_{n,m}/d\theta)\} + \{n(n+1) - m^2/\sin^2\theta\}\Theta_{n,m} = 0 \quad \ldots(1.24)$$

and

$$\Theta_{n,m} = \{(-1)^{n+m}/2^n n!\}\sqrt{\{(2n+1)/2\}\{(n-|m|)!/(n+|m|)!\}}\left(\sin^{|m|}\theta\right)$$

$$\left\{d^{n+|m|}/(d\cos\theta)^{n+|m|}\right\}\sin^{2n}\theta \qquad \ldots..(1.25)$$

The $(-1)^m$ factor is introduced in consonance with the phase convention of Condon and Shortley [2].

v) It may be noted that Legendre's equation (1.17) and the corresponding Legendre's polynomial (1.18) are just respectively the special forms of the more general Associated Legendre's equation and of the corresponding polynomial with the parameter value $m = 0$.

1.2.1 Legendre's Polynomials and Distance

Besides the few important properties of Legendre's polynomials mentioned in the earlier section, there is one of special importance which correlates the distance between the tips of two vectors r_i, r_j in space in terms of Legendre's polynomials in $\cos\omega$, where ω is the intervector angle between r_i and r_j.

Consider, as in Fig.1.1, the two vectors r_i, r_j with angular orientations (θ_i,ϕ_i) and (θ_j,ϕ_j) respective shooting out from the origin O and having common polar axis OZ. Let ω be the intervector angle and r_{ij} the distance vectors between the points i and j

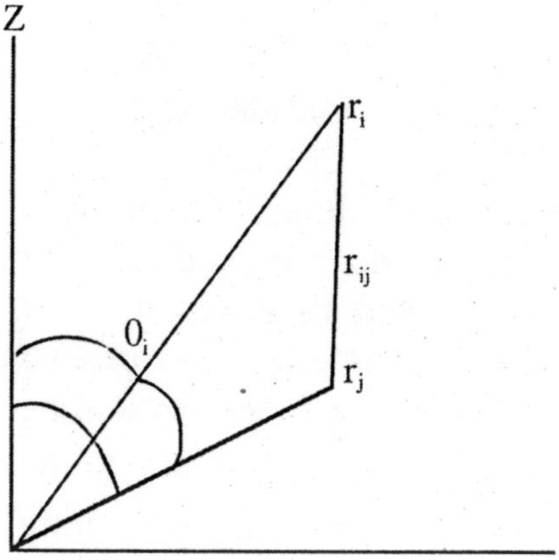

Fig. 1.1

It is easily seen that the scalar magnitude

$$1/r_{ij} = 1/\left(r_i^2 + r_j^2 - 2r_ir_j.\cos\omega\right)^{1/2}$$

$$= \frac{1}{r_>}\left(\frac{1}{\sqrt{1+z^2-2z\cos\omega}}\right)$$

where $z = r_< / r_>$ and $r_>$ is the larger of the two vector r_i and r_j

The method of building up Legendre's polynomials via generating function process, mentioned but not detailed earlier, shows that $\left(1+z^2-2z\cos\omega\right)^{-1/2}$ is just $\Sigma_n \left(r_< / r_>\right)^n P_n(\cos\theta)$ and hence

$$1/r_{ij} = \Sigma_n\left(r_<^n / r_>^{n+l}\right)P_n(\cos\omega) \qquad \text{........(1.26)}$$

It will be seen later (Sec 1.3) that $P_n(\cos\omega)$ is expansible in terms of spherical harmonics involving Associated Legendre's polynomials two in different polar angles θ_i and θ_j of the two vectors r_i *and* r_j. It is in this latter form of r_{ij}, that the latter will occur in atomic and crystal field calculations. A table of normalized Associated Legendre's polynomials is provided below for future use (see Appendix)

1.3. Spherical Harmonics and Their Properties

If the normalized Associated Legendre's polymials. $\Theta_{n,m}(\cos\theta)$ be multiplied by $\left(1/\sqrt{2\pi}\right)e^{im\phi}$ notated by $\Phi_m(\phi)$, a normalized harmonic oscillatory function, one gets the very important class of functions, called the spherical harmonics and denoted by. $Y_{n,m}(\theta,\phi)$ Thus,

$$\begin{aligned} Y_{n,m}(\theta,\phi) &= \Theta_{n,m}(\theta)\Phi_m(\phi) \\ &= \left\{(-1)^{n+|m|}/2^n n!\right\}\sqrt{\{(2n+1)/4\pi\}\{(n-|m|)!/(n+|m|)!\}}\left(\sin^m\theta\right) \\ &\quad \left\{d^{n+|m|}/(d\cos\theta)^{n+|m|}\right\}\sin^{2n}\theta e^{im\phi} \end{aligned} \qquad \text{........(1.27)}$$

where $(-1)^m$ is phase convention factor. In the above relations the parameters n and m carry the same restrictions as mentioned before.

Some physical or pictorial ideas of these spherical harmonics can be roughly obtained in the following manner. We know if we have a stretched string attached to fixed points at its two ends, one dimensional stationary vibrational waves of suitable frequencies can be generated in it. Similarly one can generate two dimensional stationary harmonic waves on a plain thin metal sheet. If we consider a flooded planet, or alternatively, a spherical planet composed of only liquid water of uniform density at all points and try to generate three dimensional stationary waves in it having its source at the centre of the planet, what shapes do we predict? The expectation is something like this. With *l* (written for *n*) and m retaining their restrictions, if we build real spherical harmonic functions by superposition of two suitable $Y_{l,m}(\theta,\phi)$ *'s* with the same *l* and different *m's*, the resultant functions will be depicted by real three dimensional stationary waves on this hypothetical planet. Some of the $Y_{l,m}$ *'s* with $m=0$ are themselves real and need no mixing in generating spherical harmonic waves. Others need mixing in various ways to generate all other forms of real stationary waves. Readers will thus realize the basics of the pictorial (graphical) forms of the angular parts of *s, p, d, f* atomic orbitals. The spherical harmonic $Y_{l,m}(\theta,\phi)$ *'s* are thus the ingredients for the construction of real three dimensional harmonic waves.

1.3.1 Properties of Spherical Harmonics

We shall mention here some of the properties of spherical harmonics. Only such properties as will be relevant for our objective in this book will be mentioned.

i) The Spherical harmonics (1.27) are individually normalized and mutually orthogonal within the range of $0 \le \theta \le \pi$ and $0 \le \phi \le 2\pi$.

ii) The $Y_{l,m}$ *'s* form a complete set and any arbitrary function, obeying the boundary conditions of the functions of the set, can be expanded in terms of $Y_{l,m}$ *'s* with all possible values of *l*'s and the corresponding *m's*.

iii) The $Y_{l,m}$ *'s* constitute the rotational eigenfunctions of Schrodinger equation for rotations of rigid rotator with free axis. They also form the angular parts of the wavefunctions of single electron atom.

iv) The single-electron atomic orbital approach for solutions of multielectron problem starts with a product function of a radial part (e.g. Slater's radial part) and a spherical harmonic. Thus arise finally the s, p_1, p_o, p_{-1}, d_2, d_1, d_0, d_{-1}, d_{-2}, and other single electron atomic orbitals.

v) The $Y_{l,m}$ *'s* constitute the standard base functions for the representations of different molecular point groups. The functions $Y_{l,m}$ *'s* for any particular value of *l* divide themselves into groups forming the basis sets of one or/and other

IR's of point group. In case of completely spherical groups (such as an atom or ion with intra electronic repulsions) the set of $Y_{l,m}$ *'s* with a particular L and different *M's* constitute the basis set for an L-dimensional representation of a degenerate *IR* of the spherical group.

vi) The immensely useful relations [3,4] obtained by applying angular momentum operators (section 1.1) on spherical harmonics are assembled in the next section (see 1.4).

vii) Taking advantage of property ii) above and subjecting relations (1.26) to an expansion procedure [5,6] one arrives after a number of steps at the final result, viz.

$$1/r_{ij} = \Sigma_{n=0}^{\alpha} \Sigma_{m=-n}^{n} \left(r_<^n / r_>^{n+1}\right) 4\pi/(2n+1) Y_{n,m}(\theta_i \phi_i) Y_{n,m}^*(\theta_j \phi_j) \qquad \text{.......(1.28)}$$

In arriving at (1.28) one needs to express $P_n(\cos\omega)$ of (1.26) in terms of spherical harmonics containing Associated Legendre's Polynomials in $\theta_i\theta_j$ of the two vectors r_i and r_j Relation (1.28) will play a prime role in our forthcoming calculations.

1.4. Angular Momentum Operators and Spherical Harmonics

A theorem in wave mechanics ensures the existence of a simultaneous set of eigenfunctions for any pair or a multiple number of commuting operators. Since L^2 and L_z commute, they should have simultaneous set of eigenfinctions with their respective eigenvalues. The fairly common knowledge that L^2 and L_z operators have their common rotational eigenfunctions with eigenvalues $L(L+1)\hbar^2$ and $M_L\hbar$ follows from solving Schrodinger equation for a rigid rotator with a free axis.

Wave mechanics apart, this entire capsule of information and additionally something more can simply be obtained as consequences flowing from the commutation relations (Sec 1.1) of angular momentum operators. Because of their intrinsic importance in the theory of atoms, we shall discuss, in some detail, some of the results accruing from operator principles.

i) Let us take ψ the functional form of which is yet to be known to be a member of the common eigenfunction set of the commuting operators L^2 and L_z. Assume ψ to be normalized and it is a common eigenfunction it should satisfy the relations.

$$L^2\Psi = \alpha\psi,\ \alpha \text{ in units of } \hbar^2 \qquad \text{.......(1.29a)}$$

$$L_z\Psi = k\psi,\ k \text{ in units of } \hbar \qquad \text{.......(1.29b)}$$

where α and k are the two eigenvalues. Now using equation (1.7') we can write

$$L_z(L_x + iL_y)\psi = (L_x + iL_y)(L_z + \hbar)\psi = (k + \hbar)(L_x + iL_y)\psi \qquad \text{.........(1.30)}$$

and also

$$L^2(Lx + iL_y)\psi = (L_x + iL_y)L^2\psi = \alpha(L_x + iL_y)\psi \qquad \text{......(1.31)}$$

It thus transpires that if ψ be a member of the common eigenfunction set, then we can construct another eigenfunction member, viz. $\{(L_x + i\,L_y)\psi\}$ with eigenvalues, one stepped upto $(k + \hbar)$ for L_z and another remaining unchanged, viz. α for L^2 with the step up operator $(1.7')$ acting on ψ. Applying this operator repeatedly, we are led to a series of eigenfunctions and eigenvalues as follows:

Common eigenfunctions ψ, $(L_x + iL_y)\psi, (L_x + i\,L_y)^2\psi$............$(L_x + i\,L_y)^n\psi$

Eigenvalues of L_z : k, $(k + \hbar)$, $(k + 2\hbar)$, $(k + n\hbar)$ .....(1.32)

Eigenvalues of L^2 α, α, α, α

If we carry out a similar set of operations on Ψ with the step-down operator $(L_x - i\,L_y)$ successively we can similarly find newer and newer common eigenfunctions with eigenvalues stepping down to $(k - \hbar)$, $(k - 2\hbar)$.......$(k - n\hbar)$ for L_z and those L^2 remaining unaltered, viz. α. Thus we can write, like (1.32).

Common eigenfunctions ψ, $(L_x - iL_y)\psi, (L_x - i\,L_y)^2\psi$............$(L_x - i\,L_y)^n\psi$

Eigenvalues of L_z : k, $(k - \hbar)$, $(k - 2\hbar)$, $(k - n\hbar)$ .....(1.33)

Eigenvalues of L^2 α, α, α, α

It needs scarcely be mentioned that k and α in (1.32) and (1.33) are some numerical multipliers of $\hbar$ and $\hbar^2$ respectively. We shall shortly find their actual natures. Combining the implications of (1.32) and (1.33) we conclude that starting from an arbitrary member ψ of the common eigenfunction set, one can build up, with the help of step-down and step-up operators, an eigenfunction ladder consisting of ...... $(L_x - iL_y)^n\psi$, $(L_x - iL_y)^{n-1}\psi$,...$(L_x - iL_y)\psi$...$(L_x - iL_y)^y\psi$... with changing eigenvalues of L_z and unchanged eigenvalue of L^2.

ii) It may now be remarked that the new eigenfunction constructed from normalized ψ by operating with L_+ or L_- are not normalized, but have to be done so individually.

Let $(L_x + iL_y)\psi = \psi' = \sqrt{N_1}\psi_u$ where ψ_u is an upper rung normalized eigenfunction and similarly $(L_x - iL_y)\psi = \psi'' = \sqrt{N_2}\psi_1$ where ψ_1 is a lower rung normalized eigenfunction. For convenience and that too without impairing generalities we can take both the factor, N_1 and N_2, to be real positive quantities. We shall write down the explicit expressions for the normalizing factors (N_1, N_2) for the common eigenfunction set of spherical harmonics later.

iii) It can be shown that the eigenfunction ladder is not infinite, it terminates both ways. If ψ be a normalized member (not being the highest rung member in case the ladder be finite) then

$$<\psi|(L_x - iL_y)(L_x + iL_y)|\psi> = \sqrt{N_1 N_2} > 0$$

But if ψ be the uppermost rung eigenfunction (in case the ladder be finite), then

$$<\psi|(L_x - iL_y)(L_x + iL_y)|\psi> = 0$$

Hence in general $<\psi|(L_x - iL_y)(L_x + iL_y)|\psi> \geq 0$

Or using (1.15) $(\alpha - k^2 - k\hbar) \geq 0$ ....(1.34)

Arguing in the similar vein we can also write, using 1.14

$$<\psi|(L_x + iL_y)(L_x - iL_y)|\psi> \geq 0$$

i.e., $(\alpha - k^2 + k\hbar) \geq 0$ ....(1.35)

Adding (1.34) and (1.35), one can write $\alpha \geq k^2$ ....(1.36)

i.e., the eigenvalue α of L^2 should be greater than or at least equal to the square of eigenvalue of L_z pertaining to ψ of the eigenfunction set. Equations, like (1.36), exist for each member of the eigenfunction set, i.e., the general form will be

$$\alpha \geq (k + p\hbar)^2 \qquad \text{......(1.37)}$$

where p may be any of a set positive or negative integers including zero. It is clear that for (1.37) to be valid with an unchanged value of α and any admissible value of p, the eigenfunction ladder must terminate both ways restricting the total number of different L_z eigenvalues. Since, as has been mentioned before, $k = M\hbar$ and L_z eigenvalue the finite eigenvalue sets of L_z and L^2 are

$$(M - n')\hbar \text{.......} (M-1)\hbar, M\hbar, (M+1)\hbar \text{.......} (M+n)\hbar \quad \alpha \qquad \text{.........(1.38)}$$

iv) Nature of Quantized values and Interelations of Quantum Numbers:

Since the series terminates, let us take ψ^+ and ψ^- to be the two terminal eigenstates with L_z eigenvalues β^+ and β^- respectively. It is evident that $\beta^+ = (M+n)\hbar \text{ and } \beta^- = (M-n')\hbar$ Now from the know pair of integrals

$$< \psi^{+} |(L_x - iL_y)(L_x + iL_y)| \psi^{+} >= 0$$

$$< \psi^{-} |(L_x + iL_y)(L_x - iL_y)| \psi^{-} >= 0 \quad \text{.........(1.39)}$$

One can scoop the following pair of relations, using (1.14) and (1.15)

$$\alpha - \beta^{+2} - \beta^{+}\hbar = 0 \quad \text{........(1.40)}$$

$$\alpha - \beta^{-2} + \beta^{-}\hbar = 0 \quad \text{........(1.41)}$$

which, on subtraction yields

$$(\beta^{+} + \beta^{-})(\beta^{+} - \beta^{-} + \hbar) = 0$$

Hence, either $\beta^{+} + \beta^{-} = 0$ ....(1.42a)

Or $(\beta^{+} - \beta^{-} + \hbar) = 0$ ....(1.42b)

Of these (1.42b) is inadmissible since it demands that the lowest eigenvalue of L_z, namely viz., β^{-} be greater that the highest, β^{+}.

From (1.42a) $\beta^{+} = -\beta^{-}$

i.e., $(M + n)\hbar = -(M - n')\hbar$ which means

$$L\hbar = -L'\hbar \quad \text{........(1.43)}$$

This leads to $L' = -L$ for the terminal quantum number associated with the eigenvalue of L_z. Writing the general quantum number M_L it is clear that

$$-L \le M_L \le L \quad \text{........(1.44)}$$

Substitution of (1.43) in (1.40) yields

$\alpha = L(L+1)\hbar^2$ as the eigenvalue of L^2 ....(1.45)

The difference between two successive eigenvalues of L_z is $\hbar$ (compare 1.38). The difference between the highest and the lowest eigenvalues of L_z should thus be a positive integral multiple of $\hbar$ and this is from (1.44) is

$$L\hbar - (-L\hbar) = 2L\hbar \quad \text{.......(1.46)}$$

whence it follows that *L* must be a positive integer or positive half integer, including zero, to be in keeping with the demand that *2L* = positive integer.

Again, since the terminal values of M_L are +*L* and –*L* respectively and the successive values of M_L differ from one another by ± 1 (1.44) and 1.38), it turns out that the value of M_L would be either integer (positive or negative including zero) or half integer (positive or negative) depending on the nature of *L* (always positive including zero if possible). Conversely the actual nature of L can be discerned from the nature of M_L.

v) Interlude: Interrelations in terms of J^2 and J_z, S^2 and S_z.

All the relations and conclusions drawn from Eqns. (1.29a to 1.46) are entirely based on the commutability of L^2 and L_z operators for a multi electron atom and their commutation properties. Nowhere has the specific nature of the common eigenfunctions played a role to affect or bear upon the validities of the equations or of the conclusions drawn therefrom.

Attention may now focused on the commutability of J^2 and J_z operators of a multielectron atom. Commutation relations of J^2, J_z, J_+ and J_- are entirely similar to those of L^2, L_z, L_+ and L_- respectively (See 1.1). It is thus clear that the equations and the conclusions derived and assembled in (iv) are also applicable to the nature and interrelations of the eigenvalues of J^2 and J_z. Thus calling the quantum numbers J and M_J associated with the operators J and J_z, we can write

$$J^* = \text{the total angular momentum} = \sqrt{J(J+1)}\hbar \qquad \text{..........(1.47)}$$

and $$J_z^* = \text{the Z component of } J^* = M_J \hbar \qquad \text{.........(1.48)}$$

Where M_J values are restricted to $-J \le M_J \le J$ and capable of taking up positive and negative half integral values or positive negative integral values including zero depending on positive half integral or positive integral (including zero) nature of J. Conversely also the actual nature of M_J will fix the nature of J.

Identical reasonings can be extended to S^2 and S_z operators leading to

$$S^* = \text{the total spin momentum} = \sqrt{S(S+1)}\hbar \qquad \text{........(1.49)}$$

and S_z^* the Z component of $S^* = M_S \hbar$ ....(1.50)

The natures of restrictions applicable to the quantum numbers S and M_S are similar to those of J and M_J.

vi) Normalized Eigenfunctions – Interconnections

If in a ladder we start from a normalized function $\psi^{(n)}$, (the superscript n is added to indicate normalized state) then in a step-up process, we can get a function, viz. $\{(L_x + iL_y)\psi^{(n)}_{L,M_L}\}$, which on testing, is found to be unnormalized. When this upper function is normalized [3,5], the resultant function is

$$\left\{\left(L_x + iL_y\right)\psi^{(n)}_{L,M_L}\right\} / \hbar\sqrt{L(L+1) - M_L(M_L+1)} = \psi^{(n)}_{L,M_L+1} \qquad \text{.......(1.51)}$$

$$\left\{\left(L_x + iL_y\right)\psi^{(n)}_{L,M_L}\right\} = \hbar\sqrt{L(L+1) - M_L(M_L+1)}\,\psi^{(n)}_{L,M_L+1} \qquad \text{......(1.52)}$$

Similarly in a step-down process from $\psi^{(n)}$ it is found

$$\left\{\left(L_x - iL_y\right)\psi^{(n)}_{L,M_L}\right\} = \hbar\sqrt{L(L+1) - M_L(M_L-1)}\,\psi^{(n)}_{L,M_L-1} \qquad \text{.......(1.53)}$$

Relations (1.52) and (1.53) are extremely important and will be frequently referred to later without, however, the superscript (*n*) although the normalization will be implied.

Since upto this moment, we have not brought in the specificities of the actual nature of wave functions, Y_{LM_L} relations like (1.52) and (1.53) can also be written down for J_+ and J_- operators or for spin step-up step-down operators S_+ and S_- operators. Thus writing the normalized eigenfunctions in ket notations, we have

$$\left\{\left(J_x + iJ_y\right)\left|J, M_J\right>\right\} = \hbar\sqrt{J(J+1) - M_j(M_J+1)}\,\left|J, M_J+1\right> \qquad \text{.......(1.54)}$$

$$\left\{\left(J_x - iJ_y\right)\left|J, M_J\right>\right\} = \hbar\sqrt{J(J+1) - M_j(M_J-1)}\,\left|J, M_J-1\right> \qquad \text{.......(1.55)}$$

and

$$\left\{\left(S_x + iS_y\right)\left|S, Ms\right>\right\} = \hbar\sqrt{S(S+1) - Ms(Ms+1)}\,\left|S, Ms+1\right> \qquad \text{......(1.56)}$$

$$\left\{\left(S_x - iS_y\right)\left|S, Ms\right>\right\} = \hbar\sqrt{S(S+1) - Ms(Ms-1)}\,\left|S, Ms-1\right> \qquad \text{.......(1.57)}$$

vii) Angular momentum eigenfunction sets and wavefunctions of multielectron atoms

From the commutation properties enlisted in See 1.1, it can be established that for a multielectron atom the set of operators

a) L^2, S^2, L_z, S_z, commute with each other.

b) L^2, S^2, J^2, J_z commute among themselves

We can thus hope to have two types of common set angular momentum eigenfunctions, which in ket notations are $\left|L, S, M_{L,} M_S\right>$ and $\left|L, S, J, M_J\right>$. Are the wavefunctions of the atom the same as or different from these two differently described common eigenfunction set a) or the set b) ? The answer depends on whether the atomic hamiltonian, H, commutes or not with each of the operators in set a) and of set b). It turns out that if we deal with hamiltonian, independent of spin (vide infra) then H commutes with all in set a) as well as with those of set b). The common set of eigenfunctions of H and of the angular momentum operators simply expressed in the form $|L,S,M_L,M_S>$ with the implicit understanding that the radial part has also been included in the form of a product, will themselves be the wavefunctions of the atom. Similarly will be the kets $|L,S,J,M_J>$'s, radial parts being included. Of course, we can express each ket of the latter type as a linear combination of the kets $|L,S,M_L,M_s>$. But this will engage our attention for the moment.

Let us now address ourselves to the task of discerning the specific nature (forms) of wavefunctions, i.e., the common eigenfunction set when the hamiltonian does not include spin interaction energies.

viii) Specific Nature of Wavefunctions of Atoms (Common Eigenfunction Set)

For a zero-order central field spinless hamiltonian in a multielectron atom, the general form of wavefunctions or a ket $|L,S,M_L,M_S>$ is $R(r)\Theta(\theta)\Phi(\phi)\{\alpha/\beta\}$ where α or β are spin wavefunction components independent of space coordinates. From angular momentum operator principles alone and the conclusions already assembled in the present section, it can be shown that the Φ dependent part of the wavefunction is $\Phi_{M_L} = (1/\sqrt{2\pi})e^{iM_L\phi}$. To ensure the singlevaluedness of Φ_{M_L}, it is imperative for M_L to assume only integral value (positive, negative including zero). This restricts the L values to the domain of positive integral values only including zero (*cf iv* of sec 1.4). Further applications of relations (1.52 and 1.53)[4] lead to the result that the Θ component of the wavefunction is the normalized Associated Legendre's polynomial (1.23, 1.24). Hence the total angular component of the wavefunction is a spherical harmonic

$$Y_{L,M_L}(\theta,\varphi) = \Theta_{L,M_L}(\theta)\Phi_{M_L}(\phi) \quad \text{.......(1.58)}$$

The total spacepart of the wavefunction in (L,S,M_L,M_s) scheme is

$$\psi(L,S,M_L,M_S) = R(r)Y_{L,M_L}(\theta,\varphi) \quad \text{........(1.59)}$$

which unfolds only partially its specific functional form, the radial part remaining unknown. Thus L and M_l play roles of good and effective quantum numbers when $\mathscr{H}$ does not include spin interactions. But when the hamiltonian includes spin interactions, it can still be shown once again from angular momentum principles that the Φ component of the wavefunction is $\Phi_{M_J} = (1/\sqrt{2\pi})e^{iM_J\phi}$, but the singlevalued character can be ensured either when both J and M_J are integers including zero, or when both J and M_J are half integers positive for J and positive or negative for M_J. In view of such alternatives, the total angular momentum component of the space part of the wavefunction cannot always be a spherical harmonic.

As to the quantum numbers S and M_s involved in the spin part of the wavefunction, - whether or not the hamiltonian is spin free or spin dependent –

nothing more precise than what is described under *v* of sec 1.4 can be stated. So depending on the spin orientation and the number of electrons, M_s will admit of either integral or half integral values (positive or negative including zero) restricting the total *S* to positive integral including zero or positive half integral values respectively.

1.5 Couplings and Atomic States: Russell Saunders' and *j-j* Couplings

The spectroscopic states of multielctron atoms for different electronic configurations, e.g. (*pd*), (*spd*). (*d.d*) (d^2) are usually obtained readily by either Russell Saunders' (also called *LS*) coupling scheme or by *j-j* coupling scheme. The thumbrules prescribed by the two schemes are too widely known [7,8] to merit any elaboration here. Each of the two schemes proposes an initial strong coupling of vectors (called primary coupling) capable of producing any of a possible set of resultants subject to certain quantum restrictions. The resultants then undergo a rather weak coupling (termed the secondary coupling) to produce the final atomic states. In the Russell Saunders' scheme the primary couplings (all l_i^* 's producing number of possible L^*'s and all s_i^* 's yielding a number of possible S^*'s) are caused by electrostatic effects and the secondary couplings originate from magnetic interaction of these L^*'s and S^*'s yielding the final atomic states. All the relatively lighter elements inclusive of those in the first transition group are covered by this scheme.

In the *j-j* scheme applicable to rather heavy elements, the primary couplings ($l_i^*.s_i^*$ interactions) are magnetic in nature. These are then followed by electrostatic$(j_i\, j_i)$ interactions constituting the secondary stage in the total coupling scheme. The resultant J^*'s carry some quantum restrictions in their final achievable magnitudes. However, to many elements in between the light and heavy, mixed coupling schemes apply.

Some illustrations of these two schemes will be presented in the next section. Attention of the readers is, however, drawn to the fact that the two coupling schemes were initially proposed by considering "a vector model of an atom" which in essence is a wrong and inadmissible concept. But the predictions on the basis of vector model theory for the final atomic states are correct and supported by the results obtained by applying quantum mechanics.

The correct methods of arriving at the possible atomic states of multielectron atoms are

i) Angular momentum commutation principles and the symmetry theory of spherical groups

ii) Angular momentum principles and the quantum mechanics.

Without delving into details of the method, we shall make some general remarks.

The question arises as to why the primary couplings should, at all, take place. It is because forces are operative within an atom that cause a mutual inter-dependence of the different vectors leading to the formation of any of the possible resultants. But what are the forces within the atom? These are (apart from the internuclear forces):

i) Electrostatic forces of attraction and repulsion

ii) Spin exchange forces

iii) Magnetic interaction forces (spin orbit, spin spin, orbit orbit)

iv) Centrifugal forces consequent on the kinetic motion of the electrons.

While the attractive forces balance the destabilizing centrifugal and the overall repulsive force, the electrostatic repulsion among the electrons will cause the first primary coupling of the l_i^* vectors of the atom in the Russell Saunders' scheme. For a spherically symmetric multielectron atom, group theory and angular momentum principles jointly permit the validities of the different conjectures on the magnitudes and the possible numerical values of L^*, L, L_z^* and M_L initially proposed empirically by Russell and Saunders.

The spin exchange forces ii) cause another primary coupling leading to the inter-dependence of spin vectors and hence, of final resultant S^*. But how to elucidate the spin exchange forces? The primary electrostatic repulsions as envisaged above will cause the electrons to move in a correlated manner so that the electrons tend to be "as far as possible form each other" during their motions around the nucleus, thus causing as low a value of repulsion energy as possible. This reduction in electrostatic repulsion is further aided (i.e., the movements become more correlated) if the spins of the electrons are oriented in parallel direction. But why should parallel orientation of spins cause a reduction in repulsion energy? The answer, though not clearly imaginable, is, however, indirectly provided by quantum mechanical treatment in which an extra negative to electronic energy, viz., exchange energy, comes from the assumption of indistinguishability of the electrons and of the anti-symmetry principle holding good under exchange of coordinates of a pair of electrons. The mathematical structure of this exchange integral conveys the physical idea that the coordinates of a pair or of pairs of electrons with parallel spins have been once or successively exchanged. Hence is the name exchange force leading to this exchange energy.

In finality, this means that the coulomb repulsion energy integral generated by inter-electron repulsions i) overestimates the destabilizing repulsion energy and that the exchange integral (with parallel spins) is a corrective term electrostatic in nature but with opposite sign. The exchange energy happens to be a by-product of the theoretical treatment of a multielectron atom based on single-electron orbital approach. The exchange force is thus not immediately apparent at the outset when one attempts a classification of the types of forces extant within an atom, but makes its presence felt in a quantal treatment on the basis of single-electron orbital approach.

In the *j-j* scheme, the primary coupling $\left(l_i^*.s_i^*\right)$ is due to dominant magnetic interaction: The electrostatic repulsions inclusive of the spin exchange consequences cause only a weak secondary coupling $\left(j_i^*.j_j^*\right)$. The quantum numbers *L* and *S,* associated with the strong primary couplings into resultants L^* and S^* in the Russell Saunders' scheme, are good quantum numbers since in the zero-order approximation this angular momentum eigenstate is fairly stable. The secondary weak coupling of L^* and S^* causes the formation of J^* with quantum number *J.* But these may be deemed as the first order perturbation of the zero-order eigenstates of L^* and S^2. Hence, the spectroscopic notation of an atomic state in which L^* S^* coupling has taken place, involves three quantum numbers *S, L* and *J.* The convention followed to represent an atomic state is

$^{2s+l}L_J$ where $2S+1$ = spin multiplicity

L = orbital quantum number = 0,1,2,3 ... = *S,P,D,F...*

and J . $= (L+S)........|(L-S)|$ in which each successive value of J differs from the previous one by unity.

As to the notations of the *j-j* coupled states, there is no universally common custom. But many authors prefer to express the state, say for a two-electron system, by three quantum numbers, two of which, written inside a parenthesis, represent the quantum number of the resultants of the primary couplings with the secondary weak coupling quantum member. *J,* as a subscript outside it. *Aj-j-* coupled state of a two-electron atom is thus expressed as.

$\left(j_i, j_j\right)_J$ Where *J* values range across $\left(j_i + j_j\right)........\left|\left(j_i - j_j\right)\right|$

1.5.1 (LS) and (j-j) states of some Electronic Configurations:

There are different, but perfectly equivalent ways of finding the *LS* terms, such as the vector diagram method, Breit's tabular scheme or the branching method, each of the three represents the implementation of the basic scheme presented in sec 1.5. The last two are fairly well known short cuts. We shall just confine ourselves to finding the *LS* lerms to two electronic configurations i) a (*pd*) configuration and ii) (*d.d*) configuration by the branching (proliferation of states) method.

The scheme is to start with the spectroscopic *LS* state of a single electron. Add another required electron to make up the corresponding electronic configuration and then observe the following two rules in the course of branching (proliferation) of the original *LS* state.

1. Spin multiplicity goes up and down by unity.
2. Original L goes up and down by *l* value of the added electron and additionally includes all the integral numbers in the range $(L+l)........|(L-l)| \geq 0$ to generate

the set of *L* values of the new states.

To each of the new *LS* terms so obtained finally, the J quantum number may be ascribed as a subscript, the values being in the range $(L+S)........|(L-S)|$ and never less than zero.

Illustration (a) LS coupling for (*pd*) configuration

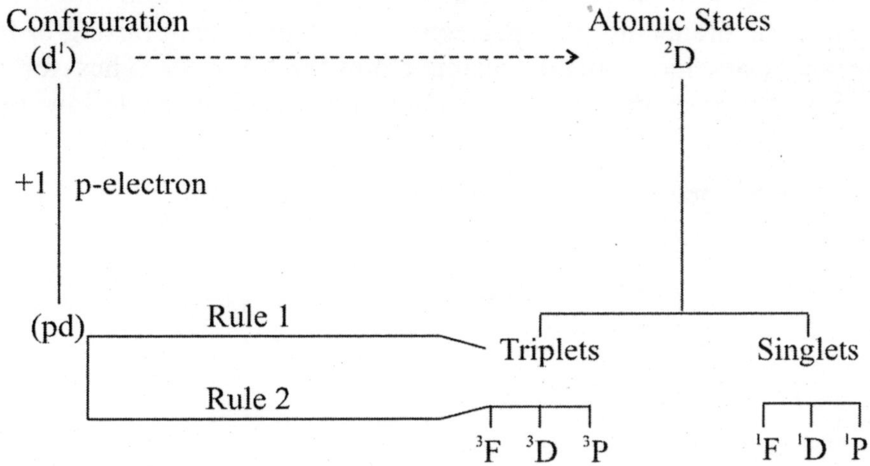

These then undergo weak magnetic interaction to give the final structures

$\left({}^3F_4, {}^3F_3, {}^3F_2\right), \left({}^3D_3, {}^3D_2, {}^3D_1\right), \left({}^3P_2, {}^3P_1, {}^3P_0\right), {}^1F_3, {}^1D_2, {}^1P_1$

Illustration (b) *LS* coupling for non equivalent (*d.d*) configuration

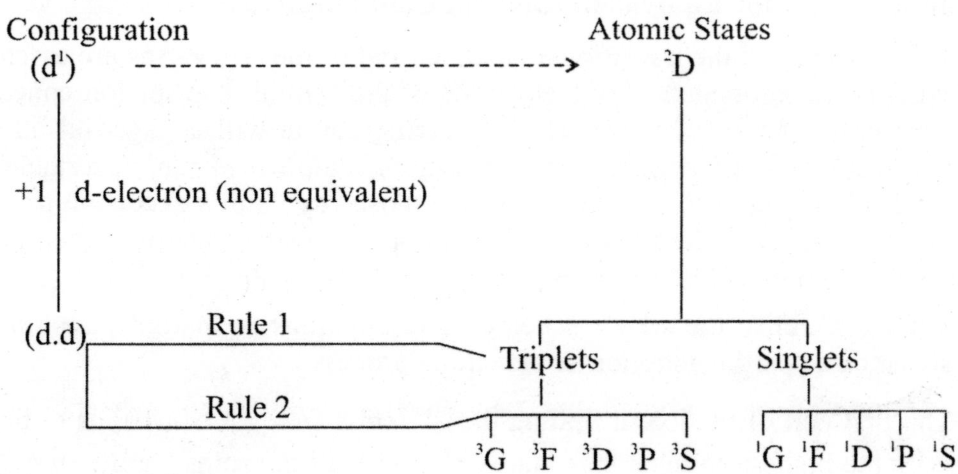

These are then split into fine structures by magnetic interactions between L^* *'s* and S^* *'s*.

Thus in the absence of any external field a (*pd*) configuration consists of ultimately 12 fine structure states and a (*d.d*) configuration, of 18 fine structure levels.

One can extend the branching technique to new electronic configurations covering three or more electrons, say (*pds*) or (*d.d ps*). For (*pds*), one make a start from each of the intermediate *LS* terms of the foregoing (*pd*) configuration, add one s electron to achieve new states by applying the branching rule technique.

The (*j-j*) coupled terms for any electronic configuration can also be quickly obtained by the same branching rule, neglecting the steps of spin raising or lowering and focusing one's attention on the quantum number *J's* only. Thus for the (*j-j*) coupled states of a (*pd*) configuration branching rules lead to the following atomic states:

Illustration (c) (*j-j*) coupled terms for (*pd*) configuration

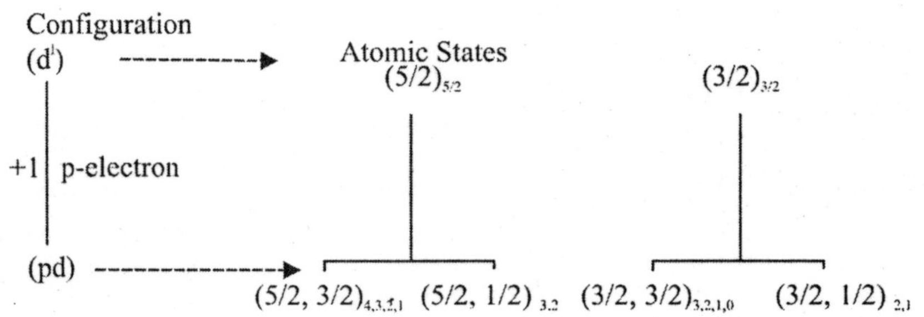

It may be noted that the number of fine structure levels in the above case is the same as of the (*LS*) states in (*pd*) configuration.

1.5.2. Atomic States for Equivalent Electron Configurations:

The problems of the transitional elements and of the rare earths are essentially concerned with the equivalent *d* or *f* electrons of the central atom or ion concerned. The atomic terms of the equivalent electron configuration will engage our attention here. Such calculation of terms, though a little more complicated can, nevertheless, be made by considering the additional restriction imposed by Pauli's exclusion principle. Since any two or more *d* electrons in the transition series elements have their *n* and *l* quantum numbers the same, these cannot have the same pair of (m_l, m_s) values. We can, therefore, simplify the whole scheme of calculations by partial adoption of a routine scheme and partial adoption of branching process.

The first part of the job in finding the d^n terms $(n = 2,3........10)$ is to find the term(s) with the highest possible spin states, i.e., with all the spins parallel $n = 2$ to 5. The terms, with n = 6 to 10.will follow later. These highest spin state terms can be inferred from the construction of a table of (m_l .m_s) values of an electron. This is followed by further construction of a second incomplete partial table of (M_l, M_s)

values of the d^n (n=2 to 5) system. Thus for any d electron the permissible pairs of (m_l, m_s) values are

m_s	1/2	1/2	1/2	1/2	1/2	-1/2	-1/2	-1/2	-1/2	-1/2
m_l	2	1	0	-1	-2	2	1	0	-1	-2
	a	b	b	d	e					

Now for the highest spin state of a d^n – system (n =2 to 5) we may only partially consider the values of the positive spin parts of the foregoing table and build up their corresponding (M_L, M_s) tables (partial) as follows:

System	(m_l, m_s) combination	(M_S, M_L) combination (partial)	Inference of highest spin state terms from partial (M_L, M_s) combination
i) d^2	*ab, bc, cd, de* *ac,bd,ce* *ad, be* *ae*	(1,3) (1,1)(1,-1)(1,-3) (1,2) (1,0), (1,-2) (1,1) (1,-1) (1,0)	^3F, ^3P
ii) d^3	*abc, bcd, cde* *abd, bce* *abe, bde* *acd* *ace* *abe*	(3/2,3) (3/2,0) (3/2,-3) (3/2,2) (3/2,-1) (3/2,1) (3/2,-2) (3/2,1) (3/2,0) (3/2,-1)	^4F, ^4P
iii) d^4	*abcd, bcde* *abce* *abde* *acde*	(2,2) (2,-2) (2,1) (2,0) (2,-1)	^5D
iv) d^5	*abcde*	(5/2,0)	^6S

The term types appearing in the last column are indicated emphatically by the partial collections of the possible component values of (M_L, M_s) occurring under column 3).

II) The second part of the job is to find lower spin states of these d^n (n=2 to 5) adoption of the previous branching process in which the spin multiplicity of the proliferating new state goes down by the required extent.

i) The lower spin states of d^2 system $(n = 2)$ are obtained by applying branching rule to the highest spin state or states of the previous d^{n-1} system.

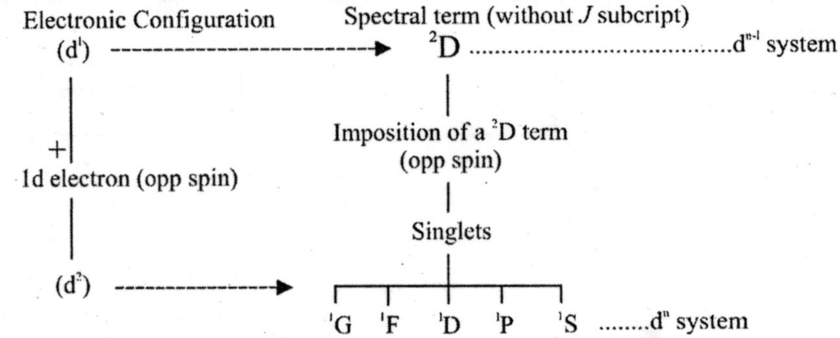

Of the singlets recorded above ^1F and ^1P are not independent terms, these two being the components of the highest spin state terms ^3F and ^3P respectively found earlier for the d^2 system. Hence the full complement of terms of (d^2) configuration comprises ^3F, ^3P, $^1G, ^1D$ and ^1S. These form only a part of the complete set two *d*- electron system when the electrons are non equivalent (sec 1.5.1)

ii) The lower spin states of d^3 system $(n=3)$ can be generated in a like manner from the highest spin states of the corresponding d^{n-1} systems. i.e., from ^3F and ^3P terms of the d^2 system.

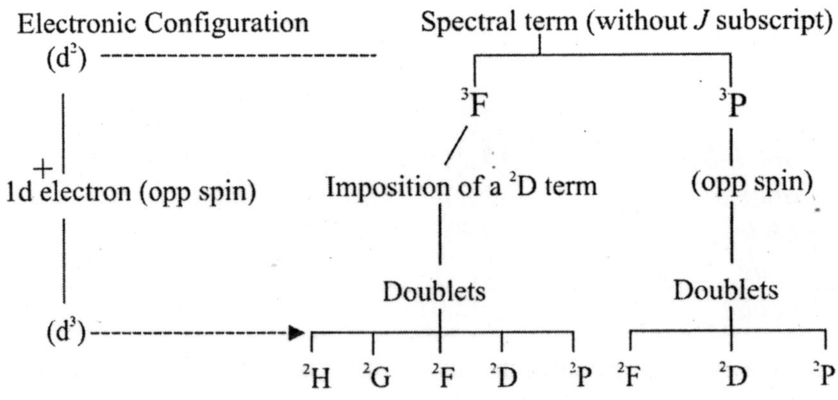

The highest spin states of d^3, as we already know are ^4F and ^4P. One ^2F and one ^2P of the above newly found states are just the components ^4F and ^4P . Hence, deleting these two, the total set of terms resulting from Russell Saunders' coupling of three equivalent d electrons comprises.

$^4F, ^4P$

$^2H, ^2G, ^2F, 2^2D, ^2P$

iii) The highest spin states (d^4) is already known to be ^5D. The lower spin terms of (d^4) can be obtained by (a) adding one d electron with opposite spin to each of the highest spin state, viz., ${}^4F, {}^4P$ of (d^3) and b) by adding two d electrons of similar spins, i.e., by imposing a ${}^3F$ and a ${}^3P$ terms to the opposite highest spin terms of (d^2) system. Thus from (a) and (b) we have

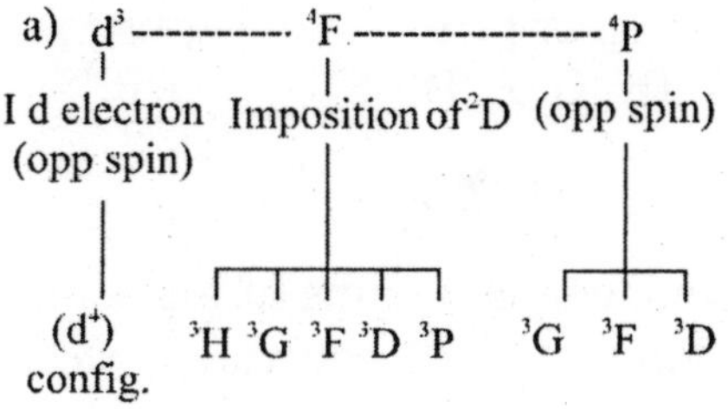

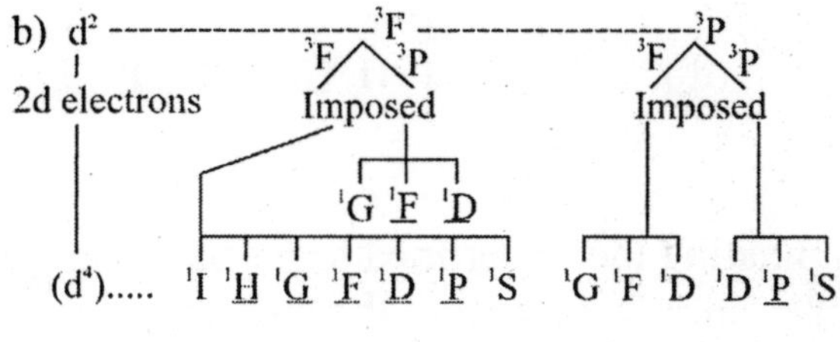

Now deleting lower spin component term(s), one each, for an admissible higher spin state term, the Russell Saunders' terms of four equivalent (d^4) configuration are: ${}^5D, {}^3H, {}^3G, 2{}^3F, {}^3D, 2{}^3P, {}^1I, 2{}^1G, {}^1F, 2{}^1D, 2{}^1S$

The underlined terms are omitted from the count as these form components of the higher spin state terms already included.

iv) D^5 system:

The distribution of five electron spins may be as follows:

a) All the five have similar spins yielding one ${}^6S$ state.
b) Four electrons have similar spins and one has opposite spin,

These lower spin state terms can be obtained by imposing ${}^2D$ state (opp spin) on ${}^5D$ state of (d^4) highest spin state. The spin multiplicities of the resulting states will diminish by unity.

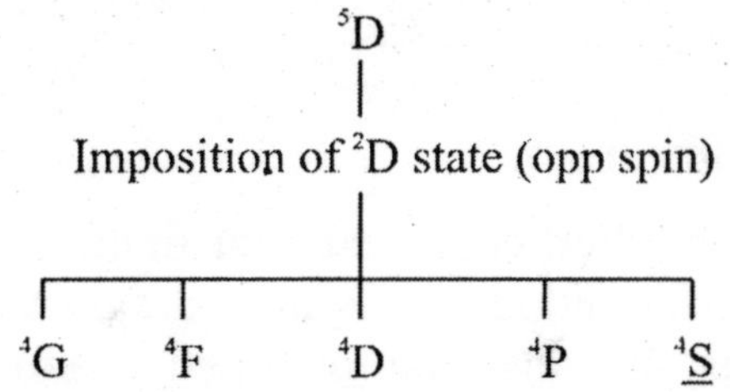

c) Three electrons of similar spins and two unlike spin:

The corresponding lower spin terms obtained by imposing a ${}^3F$ and ${}^3P$ terms separately on each of the highest spin states, viz., ${}^4F$ and ${}^4P$ of the (d^3) configuration. The spin multiplicities will reduce by two units. We can thus work out the terms :

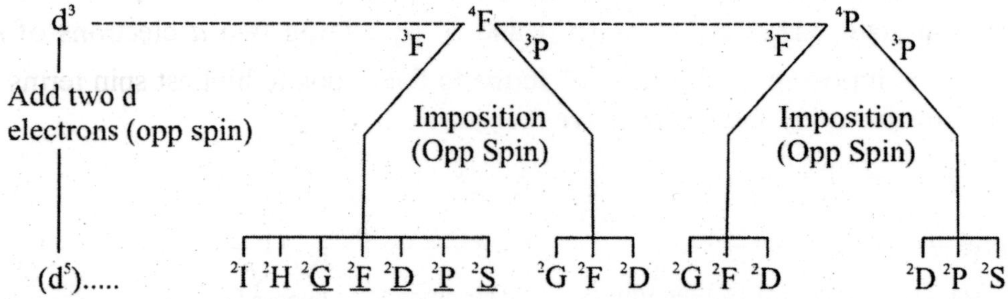

Eliminating the underlined terms as the lower component spin terms of the admissible higher spin state terms, the full complement of terms of (d^5) equivalent electronic configuration is ^6S, ^4G, ^4F, $^4D, ^4P, ^2I, ^2H, 2^2G, 2^2F, 3^2D, ^2P$ and ^2S.

The types of terms as well as the spin multiplicities of d^6 to d^{10} equivalent electron configurations are the same as d^4 to d^0 (i.e. d^{10-6} *to* d^{10-10}) electron configurations respectively. These follow from the hole-electron formalism discussed qualitatively in chapter 6 (sec 6.1). The term types of d^6 electron system are the same as d^4 hole systems. But since the electrons and the positrons (holes) electrostatically repel among themselves quantitatively in the same way, the term types of d^4 hole system will be identical with those of the d^4 electron system.

Terms of	(d^0)	are the same as of	(d^{10})
”	(d^1)	”	(d^9)
”	(d^2)	”	(d^8)
”	(d^3)	”	(d^7)
”	(d^4)	”	(d^6)

Terms of d^5 are just a different set.

The discussion of the short cut scheme for finding the Russell Saunders' terms for equivalent electron configurations necessarily drags in the ancillary problem in (*j*-*j*) coupling. These terms can also be obtained readily for equivalent electron configurations by drawing up a suitable abridged scheme parallel to the one described in sec 1.5.2 for the (*LS*) coupling. Only one has to use the weak field quantum number (n, l, j, m) instead of the strong field ones $(n, l, m_l\, m_s)$ while satisfying Pauli's exclusion principle in the (*j*-*j*) coupling process.

In the forthcoming chapters we shall deal with crystal field states of complexes of the d electron transition metal ions, where (*LS*) coupling is generally dominant. With the principles remaining the same, we omit here discussion of either *LS* or *j-j* couplings of the equivalent (f^n) electron system of the rare earth elements. A table is provided here for the different atomic terms that originate under *LS* coupling of equivalent electronic configurations.

Table T1.2 : Spectral Terms of Equivalent Electron Systems

Configuration	Terms
(p^1)	^2P
(p^2)	$^3P,^1D^1S$
(p^3)	$^4S^2P^2D$
$(p^4)=(p^2)$	$^3P\,^1D\,^1S$
$(p^5)=(p^1)$	^2P
$(p^6)=(p^0)$	^1S
$(d^1)=(d^9)$	^2D
$(d^2)=(d^8)$	$^3F\,^3P\,^1D\,^1G\,^1S$
$(d^3)=(d^7)$	$^4F\,^4P\,^2P\,^2D\,^2F\,^2G\,^2H^2D$
$(d^4)=(d^6)$	$^5D^3P^3D^3F^3G^3H\,^3P^3F\,^1S^1D^1F^1G^1I^1S^1D^1G$
(d^5)	$^6S^4P^4D^4F^4G\,^2S^2P\,^2D^2F^2G^2H^2I^2D^2F^2G^2D$
(d^{10})	^1S

Exercises

1. Using the abridged scheme of Sec 1.5.2, write down the Russell Saunders' terms of (f^2) equivalent electronic configuration.
2. Draw up a branching scheme for a ready generation (*j-j*) coupled terms in (p^2) and (p^3) electronic configurations.
3. Transform the cartesian forms of l_x and l_y operators in spherical polar coordinates.
4. Evaluate $\int_o^\pi \Theta_{lm}^*(cos\,\theta)\Theta_{lm'}(cos\,\theta)sin\,\theta\,d\theta$ to test the orthogonality of the functions.
5. Consider the integral $\int_o^{2\pi} \Phi_m^*(\varphi)\Phi_k(\varphi)\Phi_{m'}(\varphi)d\varphi$ where φ_n's are of the type $(1/2\pi)^{1/2}\,e^{in\varphi}$ and each is a single valued function. What is the survival condition of the integral. Write down its value.

Reference :

1. G.E. Uhlenbeck and S. Goudsmit, *Naturwissenschaften* 13, 953, 1925.
2. E.U. Condon and GH. Shortley : *Theory of Atomic Spectra* (Cambridge University Press)
3. J.I. Powell and B. Crasemamn: *Quantum Mechanics* (Addison Westey)
4. L. Harris and A.L. Loeb : *Wave Mechanics* (Mc Graw-Hill)
5. H. Eyring, J. Walter and G.E. Kimball : *Quantum Chemistry* (John Wiley & Sons)
6. H. Margenau and G.M. Murphy : *Mathematics of Physics and Chemistry* (Van Nostrand)
7. H.E. White : *Introduction to Atomic Spectra* (Mc Graw-Hill)
8. Pauling and S. Goudsmit : *Structure of Line Spectra of Atoms* (Mc Graw-Hill)

Appendix

A. Normalised Associated Legendre Polynomials $\Theta_{lm}(\theta)$:

$$\Theta_{00} = 1/\sqrt{2};$$
$$\Theta_{10} = (3/2)^{1/2} cos\,\theta;$$
$$\Theta_{1\pm1} = (3/4)^{1/2} \sin\theta;$$
$$\Theta_{20} = (5/8)^{1/2}(3\cos^2\theta - l);$$
$$\Theta_{2\pm1} = (15/4)^{1/2} \sin\theta\cos\theta;$$
$$\Theta_{2\pm2} = (15/16)^{1/2} \sin^2\theta;$$
$$\Theta_{30} = (63/8)^{1/2}(5/3cos^3\theta - cos\,\theta)$$
$$\Theta_{3\pm1} = (21/32)^{1/2}(5cos^2\theta - 1)sin\,\theta;$$
$$\Theta_{3\pm2} = (105/16)^{1/2} sin^2\theta\, cos\,\theta;$$
$$\Theta_{3\pm3} = (35/32)^{1/2} sin^3\theta;$$
$$\Theta_{40} = (9/128)^{1/2}(35cos^4\theta - 30cos^2\theta + 3);$$
$$\Theta_{4\pm4} = (315/256)^{1/2} sin^4\theta$$

B. Normalized Spherical Harmonics, $Y_{lm}(\theta,\varphi)$ in spherical polar coordinates:

These are obtained by multiplying the above Associated Legendre polynomials by $(\varphi) = 1/(2\pi)^{1/2}\exp(\mathrm{i}m\varphi)$; i.e., $Y_{lm}(\theta,\varphi) = \Theta_{lm}(\theta)\Phi_m(\varphi) = 1/(2\pi)^{1/2}\,\Theta_{lm}(\theta)exp(im\varphi)$; with *m* taking both positive and negative values, including zero, as may be the case. Following the phase convention of Condon and Shortley, the final expressions of Y_{lm} is preceded by *a* – (minus) sign if m is an odd positive integer.

C. Normalized complex p- and d- wavefunctions (with Condon Shortley phase convention):

$$p_0 = f(r)(3/4\pi)^{1/2} \cos\theta \; ; \; p_{\pm1} = \mp f(r)(3/8\pi)^{1/2} \sin\theta \exp(\pm i\varphi);$$

$$d_0 = F(r)(5/16\pi)^{1/2}(3\cos^2\theta - 1); \; d_{\pm1} = \mp F(r)(15/8\pi)^{1/2} \sin\theta\cos\theta\exp(\pm i\varphi)$$

$$d_{\pm2} = F(r)(15/32\pi)^{1/2} \sin^2\theta \exp(\pm 2I\varphi);$$

Here, *f(r)* and *F(r)* are the normalized radial functions associated with the corresponding *p*- and *d*- electrons, respectively.

D. Normalized real from s of p- and d- orbitals:

$$p_z = p_0; \; p_x = 1/2^{1/2}(p_{-1} - p_1); \; p_y = i/2^{1/2}(p_{-1} + p_1)$$

$$d_{z^2} = d_0 = F(r)(5/4\pi)^{1/2}(3z^2 - 1)/2r^2;$$

$$d_{x^2-y^2} = 1/2^{1/2}(d_{-2} + d_2) = F(r)(5/4\pi)^{1/2}\sqrt{3}(x^2 - y^2)/2r^2;$$

$$d_{xy} = i/\sqrt{2}(d_{-2} - d_2) = F(r)(5/4\pi)^{1/2}\sqrt{3}xy/r^2;$$

$$d_{xz} = l/\sqrt{2}(d_{-1} - d_1) = F(r)(5/4\pi)^{1/2}\sqrt{3}xz/r^2;$$

$$d_{yz} = i/\sqrt{2}(d_{-1} + d_1) = F(r)(5/4\pi)^{1/2}\sqrt{3}yz/r^2;$$

E. Normalized complex single-electron f- function (with Condon Shortely phase convention)

$$f_{30} = F(r)(7/8\pi)^{1/2}(1/2)(5\cos^2\theta - 3\cos\theta)$$

$$f_{3\pm1} = \mp F(r)(7/8\pi)^{1/2}(3/8)^{1/2}(5\cos^2\theta - 1)\sin\theta\exp(\pm i\varphi)$$

$$f_{3\pm2} = F(r)(7/8\pi)^{1/2}(15/4)^{1/2}\sin^2\theta\cos\theta\exp(\pm i2\varphi)$$

$$f_{3\pm3} = \mp F(r)(7/8\pi)^{1/2}(5/8)^{1/2}\sin^3\theta\exp(\pm i3\varphi)$$

F(r) is the normalized radial function.

F. Normalized real forms of f-orbitals:

$$f_{z(5z2-3r2)} = f_0 = F(r)(7/16\pi)^{1/2}(5z^2 - 3r^2)/r^3$$

$$f_{x(5x2-3r2)} = 1/\sqrt{2}\{(af_1 + bf_{-3}) - (af_{-1} + bf_3)\} = F(r)(7/16\pi)^{1/2}(5x^2 - 3r^2)/r^3$$

$$f_{y(5y2-3r2)} = 1/\sqrt{2}i\{(af_1 + bf_{-3}) + (af_{-1} + bf_3)\} = F(r)(7/16\pi)^{1/2}(5y^2 - 3r^2)/r^3$$

$$f_{y(z2-x2)} = i/\sqrt{2}\{(bf_1 + af_{-3}) + (bf_{-1} + af_3)\} = F(r)(105/16\pi)^{1/2} y(z^2 - x^2)/r^3$$

$$f_{x(y2-z2)} = 1/\sqrt{2}\{(bf_1 - af_{-3}) - (bf_{-1} - af_3)\} = F(r)(105/16\pi)^{1/2} x(y^2 - z^2)/r^3$$

$$f_{z(x2-y2)} = 1/\sqrt{2}(f_2 + f_{-2}) = F(r)(105/16\pi)^{1/2} z(x^2 - y^2)/r^3$$

$$f_{xyz} = 1/\sqrt{2}i(f_2 + f_{-2}) = F(r)2(105/16\pi)^{1/2} xyz/r^3$$

where a = $(3/8)^{1/2}; b = (5/8)^{1/2}$

Chapter Two

Hamiltonians, Perturbations, Wavefunctions, Matrix Elements and Some Group Theory Relations

2.1 Hamiltonians

The hamiltonian of a microsystem, viz., a multielectron atom, should, in principle, contain operator fragments characteristic of all types of energy interactions prevalent within the system. It is, therefore, necessary to analyze in the first place the nature of each type of energy and apportion an appropriate operator to it. The types of forces and energies present in a multielectron atom (save those inside the nucleus) have already been mentioned in Sec. 1.5. Since the exchange energy constitutes a correction term for the electrostatic correlation effect, we need to focus our attention on three types, viz., kinetic energy, electrostatic (both attractive and repulsive) energy and magnetic interactions due to magnetic fields and magnetic moments. The latter are so varied and numerous that the hamiltonian inclusive of all these becomes unwieldy and impractical for use. And add to these the additional interaction energies incidentally causing an enlargement of the hamiltonian structure, when the atom is placed in one or another external perturbing field. What do we do under the circumstances? We choose the components contributing to the major slice of the total energy and only such perturbation components to the Hamiltonians as are directly relevant for investigating the properties under review. Thus many parts of the total hamiltonian naturally drop out in any actual treatment.

The present discussion on the hamiltonian will be split into three parts:

a) Forms of perturbation hamiltonian in Sec. 2.1.1,

b) Genesis of mathematical forms in Sec. 2.2, and

c) Recasting some perturbation hamiltonian in more useful forms in Sec. 2.3.

2.11 Forms of Perturbation Hamiltonian

A. Atom in free-state

Neglecting the intraperturbations (Sec. B), the major and the important parts of the hamiltonian are

$$\mathscr{H} = -(\hbar^2/2m)\Sigma_i \nabla_i^2 + \Sigma_{i<j} e^2/r_{ij} - \Sigma Ze^2/r_i \quad \text{......(2.1)}$$

All the terms on the right hand side are too well known to merit any elaboration.

B.

The intraperturbations of an atom in field-free state:

In (2.1) we notice that the electrostatic form of intraperturbations, viz., $\sum_{i<j} e^2/r_{ij}$ has already been included. The other intraperturbations that inherently exist within an atom are mostly magnetic interactions depending on various spin-spin, spin-orbit and orbit-orbit interactions. Neglecting magnetic orbit-orbit interactions, which are rather small, the hamiltonian for the other intraperturbations is (cf. Hameka[1])

$$\mathscr{H}_{intra} = \mathscr{H}_{el} + \mathscr{H}_{el,n} + \mathscr{H}_n \qquad \ldots\ldots (2.2)$$

$$= (\mathscr{H}_{el(1)} + \mathscr{H}_{el(2)} + \mathscr{H}_{el(3)}) + \mathscr{H}_{el,n} + \mathscr{H}_n \qquad \ldots\ldots (2.3)$$

where $\mathscr{H}_{el(1)}$ =hamiltonian for the sum of interactions of the electron spin magnetic moments with the orbital moments (same electron).

$\mathscr{H}_{el(2)}$ =hamiltonian for the sum of interactions of the electron spin magnetic moments with the orbit moments of the other electrons.

$\mathscr{H}_{el(3)}$ =hamiltonian for the sum of interactions of the electron spin magnetic moments with the other electron spin moments.

$\mathscr{H}_{el,n}$= hamiltonian for the sum of all the electron spin moment interactions with nuclear spin magnetic moment.

$\mathscr{H}_n$ = hamiltonian for the sum of all the electron orbital moment interactions with nuclear spin magnetic moment.

The concrete expressions for such intraperturbation hamiltonian parts are given below. The mathematical genesis of such expressions will be indicated in the next section, without, however, the full derivation, to satisfy the readers' curiosity.

$$\mathscr{H}_{el,}(1) = (e/mc)\, \Sigma_i\, s_i \,.\, (1/2mc)(F_i \, x \, p_i) \qquad \ldots\ldots (2.4)$$

where F_I = electric field at the i th electron due to the apparent motion of the nucleus and $-(e/mc)s_i$, the spin magnetic moment of the i th electron.

$$\mathscr{H}_{el,}(2) = (e/mc)\Sigma_{i<j}\, s_i \cdot (e/mc)[(p_j - (1/2)p_i) x r_{ij}]/r_{ij}^3 \qquad \ldots\ldots (2.5)$$

where the latter factor in (2.5) denotes the magnetic field felt at the i th electron due to the relative orbital motion of the j th electron around it.

$$\mathscr{H}_{el,}(3) = (1/2)\Sigma_i \Sigma_{j\neq i}(e^2/m^2c^2)\{(s_i \cdot s_j) r_{ij}^2 - 3(r_{ij}\cdot s_j)\}/r_{ij}^5 \qquad \ldots\ldots (2.6)$$

where r_{ij}^2 and r_{ij}^5 are scalars and the other r_{ij}'s are vectors. The validity of (2.6) will be justified in the next section.

$\mathscr{H}_{el,\ n}$, being the interaction between the electron spins and the nucleus spin, will have a look similar to (2.6) with the addition of one extra term called Fermi contact potential[2]

$$\mathscr{H}_{el,n} = (e\gamma_n/mc)\Sigma\left[\{(s_i \cdot I_n)r_i^2 - 3(r_i \cdot s_i)(r_i \cdot I_n)\}/r_i^5 - 8\pi/3(s_i \cdot I_n)\delta_{i,n}\right] \quad \text{...... (2.7)}$$

I_n denotes the nuclear spin, γ_n its gyromagnetic ratio and r_i distance vector of the i th electron starting from the nucleus. The origin of Fermipotential will be explained in the next section.

$\mathscr{H}_n$, the last intraperturbation term arising from interaction of the electron orbital moments with the nuclear spinmagnetic moments is due to an effective magnetic field H' set up at the nucleus due to orbital motions of all the electrons.

$$\mathscr{H}_n = -\gamma_n I_n \cdot H' \quad \text{....... (2.8)}$$

C. Another type of intraperturbation may arise in a transition metal atom or ion placed in an environment of ligands of different symmetries. The crystal field theory visualizes the metal atom/ion as under electrostatic perturbations of the point charges or the point dipoles serving as proxies for the ligands. This intraperturbation, technically called crystal field hamiltonian, will be treated in detail in Chapter seven onwards where crystal field calculations are presented.

D. External Perturbation

The most common form of external perturbation playing on an atom is its exposure to electromagnetic radiation.

(a) Electromagnetic field perturbation

An electromagnetic field, defined through E and H, is equally well represented by a couple of potentials, a vector potential A and a scalar potential φ. There is a latitude of freedom in choosing the pair of terms A and φ allowed by a suitable choice of potential gauge. The selection of gauge may be such as to permit one to work within the framework of either of the following two relations holding good.

$$\nabla.A = 0 \quad \text{(Coulomb's gauge)} \quad \text{....... (i)}$$

$$\nabla.A + (1/c)\,\delta\phi/\delta t = 0 \quad \text{(Lorentz gauge)} \quad \text{....... (ii)}$$

The readers' familiarity with such preliminaries and with Maxwells' field equations[3,4] is assumed including acquaintance with the pair of relations

$$\left.\begin{aligned} E &= -(1/c)\,\delta\phi/\delta t - \nabla\varphi \\ H &= \nabla \times A \end{aligned}\right\} \quad \text{....... (2.9)}$$

The varying E and H of the electromagnetic field provide magnetic fields each at the orbiting and the spinning electron j. The orbital magnetic moment of the j th electron and its spin magnetic moment interact with the magnetic fields provided by E_j and H_j

and generate two perturbation Hamiltonians which, summed over all the electrons, are represented by $\mathscr{H}_1$ and $\mathscr{H}_2$. These external perturbations are

$$\mathscr{H}_1 = \Sigma_j \{(e/mc)(\hbar/i)(A_j.\nabla_j)+(e^2/2mc^2)A_j^2 - e\varphi_j\} \quad \ldots\ldots (2.10)$$

where e is the absolute magnitude of electronic charge and E_j and Hj occur covertly through the guise of A_j and φ_j.

$$\mathscr{H}_2 = \Sigma_j (e/mc)s_j.[(1/2mc)(E_j \times P_j) + H_j] \quad \ldots\ldots (2.11)$$

The remaining part of the external field perturbation is the interaction of the external magnetic field with the spin magnetic moment of the nucleus. The corresponding hamiltonian is

$$\mathscr{H}_3 = -\gamma_n 1_n.H \quad \ldots\ldots (2.12)$$

(ii) Static Magnetic and Electric Fields

The expressions for the perturbation hamiltonian in the above mentioned fields can be obtained from those of the electromagnetic field subject to certain specific conditions. For an electromagnetic field, φ (Eq 2.10) may be set equal to zero[5] for situations (which are quite common) when ρ (charge density) = 0. When static magnetic field and external static electrical field are there, vector potential A and the electrostatic potential φ will be independent of t. Thus electric field E will be given by- $\nabla\varphi$ and the magnetic field H by $\nabla \times A$, where the vector potential A should be of suitable form. Thus, for static electric and magnetic fields interacting with orbital motion

$$\mathscr{H}_{1(\text{static})} = \Sigma_j \{(e/mc)\hbar/i(A_j.\nabla_j) + e^2/2mc^2 A_j^2 - e\varphi_j\} \quad \ldots\ldots (2.13)$$

One should read the distinctions between (2.10) and (2.13) although outwardly these two appear to be the same. The perturbation of the static constant magnetic field (A = suitable form) with the spin moment of the electron is

$$\mathscr{H}_{2(\text{static})} = \Sigma_j (e/mc)s_j.H_j \quad \ldots\ldots (2.14)$$

Under constant external electrostatic field, as prevails in situations for Stark effect, (z direction)

$$\mathscr{H}_{1(\text{elec. const})} = \Sigma - e\varphi_j Z_j \quad \ldots\ldots (2.15)$$

and for constant external magnetic field, one is led to form (2.13) and (2.14) to the perturbation hamiltonian.

$$\mathscr{H}_{1(\text{mag. const})} = \mathscr{H}_{1(\text{stat})} + \mathscr{H}_{2(\text{stat})} \text{ with } \varphi_j \text{ dropped from (2.13)} \quad \ldots\ldots (2.16)$$

The conversion of (2.16) to a form more useful for application will be made later in Sec. 2.3. Besides, the constant external magnetic field will interact with the nuclear magnetic moment generating a perturbation similar to (2.12).

E. Symmetry Perturbation

If the symmetry of the environment of an atom/ion (as in a complex) be changed resulting in, say, a descent in symmetry, a perturbational energy will be involved. Such cases of perturbation may be treated by forming a perturbation hamiltonian independently for the new environment or, alternatively by resorting to symmetry arguments on branching of an original IR into new IR's in the lower symmetry environment. The readers should not fail to note the thin difference between the ideas of *C* and *E*. *E* presupposes *C* to be the unperturbed state and change in environment of *C* causes the perturbation.

2.2 Genesis of Mathematical Forms of Perturbation Hamiltonians

(a) Eq (2.1)

The components of this hamiltonian are all central field terms except $\Sigma_{i<j}e^2/r_{jj}$. If we separate the valence or optical electrons (*k*, *l*) from the core electrons (*m*, *n*), the intraperturbation terms $\Sigma_{i<j}e^2/r_{ij}$ can be split into three parts, viz.,

$$\Sigma_{i<j}e^2/r_{ij} = \Sigma_{m<n}e^2/r_{mn} + \sum_m \Sigma_k e^2/r_{mk} + \Sigma_{k<1}e^2 r_{k1} \qquad \ldots\ldots (2.17)$$

We shall further amend this intraperturbation form in Sec. 2.3.

(b) Eqs (2.4), (2.5), (2.6) and (2.7)

(i) Physics tells us that if a magnetic dipole μ be present in a magnetic field *H*, the interaction energy is

$$\varepsilon = -\mu.H \qquad \ldots\ldots (2.18)$$

(ii) The magnetic field[6] produced at a point with respect to which a charge is in relative motion is given by Biot Savart law

$dH = \frac{1}{c}\left(idl \times {}^{r}\!/_{r^3}\right) = \left(\frac{1}{c}\right)ev \times {}^{r}\!/_{r^3}$, where all the quantities are in cgs electrostatic units. *e* is the charge in e s u and r is the vector drawn from the current element, id 1, to the point at which the magnetic field at that moment is sought. The charge e includes its sign and has a relative velocity v at that moment. An electron i, moving with a velocity v_i will cause an apparent orbital rotation of the nucleus (charge Ze expressed in e s u) with a velocity $-v_i$ around the former. Thus a varying electric field $Zer_i/r_i^3 = F_i$ (Gaussian unit) will act on the electron making it feel a magnetic field (Biot Savart law) at that moment

$$H_i = \left(Ze/cr_i^3\right)(-v_i)\times r_i = (Ze/c)\left\{\left(r_i/r_i^3\right)\times v_i\right\} = \begin{cases}(1/mc)(F_i\times p_i) & \text{.... (2.20)}\\ \left(Ze/mcr^3\right)\left(r_j\times p_i\right) & \end{cases}$$

$$= Ze/mc\left(1/r^3\right)l_i^* \qquad \text{.... (2.19)}$$

where l_i^* is the orbital angular momentum of the electron and $\left(1/r^3\right)$ is the average of the reciprocal of the cube of the electron distance from the nucleus.

These expressions for the magnetic field at the electron are modified by the relativity effect[7] of the spinning electron undergoing orbital motion through the introduction of a multiplying factor ½ (Thomas precession) giving

$$H_i = (1/2mc)(F_i\times P_i) = (1/2mc)\left(Ze/r_i^3\right)l_i^* \qquad \text{....... (2.21)}$$

(iii) The spin magnetic moment μ_s and the moment s_{i^*} of an electron are related to one another as

$$\mu_s = -(e/mc)s_i^* \qquad \text{....... (2.22)}$$

(iv) If there be two point dipoles μ_p and μ_q where r_{pq} is the vector distance from μ_p to μ_q, then the magnetic field at μ_q due to μ_p is given by [8]

$$H = \left\{3n\left(n.\mu_p\right)-\mu_p\right\}/r_{pq}^3 \qquad \text{....... (2.23)}$$

where $n = r_{pq}/r_{pq}$ =unit vector along r_{pq}. Substitution of this in (2.23) leads to

$$H = \left\{3r_{pq}\left(r_{pq}.\mu_p\right)-r_{pq}^2\mu_p\right\}/r_{pq}^5 \qquad \text{....... (2.24)}$$

Perturbation hamiltonians, given by Eqs. (2.4 and 2.5), can be obtained by employing Eqs (2.18), (2.21) and (2.22). Care should be taken in resurrecting Eq. (2.5). Eqs. (2.6) and (2.7) sans the Fermi contact potential term are easily recouped through the aids of Eqs. (2.22) and (2.24).

The Fermi contact potential in (2.7) arises from the fact that there exists for an s-electron a definite probability of being within the nucleus. A spinning nucleus with a magnetic moment, $\gamma 1_n$, produces, for a spinning s-electron when outside the nucleus, a magnetic field different from the one produced by magnetic induction within the nucleus when the electron is inside it. Thus there occur different contributions to the perturbation hamiltonian by an s-electron depending on its situations. This is taken care of by Dirac delta function in Eq. (2.7).

C) Eqs. (2.10) and (2.11)

Eq. (2.4), already recouped in the foregoing part, together with Eqs. (2.18) and (2.22) indicate how Eq. (2.11) can be made to appear. However, the perturbation represented by (2.10) is not that easy to derive. It demands finding the Lagrangian and the Hamiltonian functions for an electron moving in an electromagnetic field prior to setting up the hamiltonian operator. For derivation the reader may refer to the texts 1, or 5.

2.3 Redressing of some Perturbation Hamiltonians in Forms Adapted for Applications

a) Recasting of Eqs. (2.1) and (2.17)

The form of electrostatic intraperturbation given by (2.17) is recast in a form to constitute ultimately a central field hamiltonian. The core electrons moving in filled-up shells and subshells provide a spherical charge distribution generating a central field potential. The second term of RHS of (2.17) is not spherically symmetric. If this component is substituted by a central field term $Z'e^2/R_i$ with such a choice of Z' so as to make it effectively equivalent[9] to the former, then it is readily seen

$$\mathscr{H} = \left[\Sigma_j -\left(\hbar^2/2m\right)\nabla_j^2 - \Sigma Ze^2/r_j + \Sigma_{m<n} e^2/r_{mn} + Z'e^2/R\right] + \Sigma_{k<1} e^2/r_{k1}$$

$$= \mathscr{H}_{cent} + \Sigma_{k<1} e^2/r_{k1} \qquad \text{....... (2.25)}$$

When H is written in this form, it is realized that the electrostatic interaction amongst the valence/optical, electrons only constitutes the intraperturbation hamiltonian. The central field hamiltonian, $\mathscr{H}_{cent}$, with the major part of the electrostatic repulsion energy still forming a part of it, is the initial zero-order unperturbed hamiltonian.

b) Recasting of Eqs. (2.4) and (2.5)

Of these two spin-orbit interaction hamiltonians for the same and for the other orbits respectively, the second interaction is far too small to merit any further treatment. Eq. (2.4) is recast in a more useful form by utilizing the form of the magnetic field (2.21) and of the spin moment (2.22) of a single electron. Thus (cf. Eq. 2.4)

$$H_{el(1)} = \left(e/2m^2c^2\right)\left(Ze/r^3\right)ls \qquad \text{....... (2.26)}$$

$$= \xi(r)l.s \qquad \text{....... (2.27)}$$

where l and s should be in the operator forms of l^* and s^*.

Whenever the operator is applied we need the matrix element of the operator between the basis states each having a radial component and an angular part. In the applied form the product of the radial function and of $\xi(r)$, on integration, yields a result, say $\zeta/\hbar^2$, leaving us with the task of finding the matrix element of l.s operator between the angular parts of the basis states. The net result is that instead of $h_{el(1)}$ of (2.27), we use the spin orbit hamiltonian (in the applied form) as

$$h_{el(1)}^{(eff)} = (\zeta/\hbar^2)l.s = (\zeta/\hbar^2)^{1/2}\{j^2 - s^2 - l^2\} \qquad \text{....... (2.28)}$$

where $l.s$ operates on the angular parts of the base functions and

$$(\zeta/\hbar^2) = \xi' = < R^2\xi(r)r^2 >$$

Apart from (2.28) another useful form of the effective spin-orbit hamiltonian in the matrix element integral may be written when ($l.s$) is recast in the following form

$$l.s = (l_x s_x + l_y s_y + l_z s_z)$$

$$= \{l_z s_z + 1/2\,(l_+ s_- + l_- s_+)\} \qquad \text{....... (2.29)}$$

Thus (2.28), via (2.29), assumes the form

$$h_{el(1)}^{eff} = (\zeta/\hbar^2)[l_z s_z + 1/2(l_+ s_- + l_- s_+)] \qquad \text{....... (2.30)}$$

For a multielectron problem, the effective spin-orbit hamiltonian to be used in the evaluation of the matrix element is

$$\mathscr{H}_{el(1)}^{eff} = \Sigma_i\, \xi(r_i) l_{i.} s_i$$

$$= \Sigma_i\, \xi(r_i)\,[l_{zi} s_{zi} + 1/2(l_{i+} s_{i-} + l_{i-} s_{i+})] \qquad \text{....... (2.31)}$$

In some applications, the spin-orbit interaction matrix elements have to be formed directly in the basis set of the polylectron kets written in the schemes (L S $M_L M_S$) or (L S J M_J) without any prior decomposition of the former kets into dets (determinants) comprising single electron microkets. The operator to be used in such matrix elements of which the radial parts of the multielectron atom have already been intregated out becomes[10]

$$\mathscr{H}_{el(1)}^{eff} = \lambda/\hbar^2(L.S) = (\lambda/\hbar^2)1/2(J^2 - L^2 - S^2) \qquad \text{....... (2.32)}$$

leading to interaction energy

$$\epsilon_{sp-orb} = \lambda/2[J(J+1) - L(L+1) - S(S+1)] \qquad \text{....... (2.32')}$$

when the operand kets are eigenfunctions of the operators in (2.32).

c) Recasting of Eq. (2.16) in a constant magnetic field.

The hamiltonian (2.16), when $\varphi = 0$ is

$$\mathscr{H}_{mag} = \Sigma_j\{e/mc(\hbar/i)(A_j.\nabla_j) + e^2/2mc^2 A_j^2\} + \Sigma_j\, e/mc(s_j.H_j) \qquad \text{....... (2.33)}$$

Without the loss generality, the direction of the constant magnetic field may be assumed to be along the z-axis with H_x, H_y components equal to zero each. For such a field the vector potential A may be chosen as A= ½ $(H \times r)$ .... (2.34)

Hence for the jth electron, since $H = H_2$

$$A_j = 1/2(H_z \times r_j) \qquad \text{....... (2.35)}$$

From (2.35)

$$A_{xj} = -1/2 H_{zj} y_j ; A_{yj} = 1/2 H_{zj} x_j ; A_{zj} = 0 \qquad \text{....... (2.36)}$$

Introducing (2.36) in (2.33)

$$\mathscr{H}_{mag} = \Sigma_j \{ e/mc\, \hbar/i\, (A_{xj} \partial/\partial x_j + A_{yj}\, \partial/\partial y_j) + e^2/2mc^2 (A_{xj}{}^2 + A_{yj}{}^2) \}$$
$$+ \Sigma_j\, e/mc (s_{zj}.H_{zj})$$
$$= \Sigma_j \{ (e/2mc)\, H_{zj} \hbar/i (x_j\, \partial/\partial y_j - y_j \partial/\partial x_j) + (e^2/2mc^2) H_{zj}{}^2\, (x_j^2 + y_j^2)/4 \}$$
$$+ \Sigma_j\, e/mc (s_{zj}.H_{zj})$$
$$= \Sigma_j \{ (e/2mc) H_z (1_{zj} + 2s_{zj}) + (e^2/8mc^2) H_z^2 (x_j^2 + y_j^2) \} \qquad \text{....... (2.37)}$$

remembering that $H_z = H_{zj}$ and e = absolute magnitude of electronic charge

2.4 Perturbations – Principles and Formulations

There are a very large number of atomic and molecular state problems in physics and chemistry for which the schrodinger equation cannot be solved exactly. One adopts approximate methods of solving for wavefunctions and energies of such systems. Of the different possible approximate methods, perturbation method constitutes an important and often- used technique. Perturbation theories have developed formulations for first, second and other order corrections in energies and wavefunctions which are to be added to the energy and the wavefunction respectively of the unperturbed state to yield results of the perturbed state.

The principles of perturbation apply when the system under study is slightly distorted or deviated from the ideal or assumed undistorted state for which the wavefunciton and the energy are known. The perturbation theories and formulations differ depending on

i) Time independence or time dependence of the nature of perturbation;

ii) The nondegenerate or the degenerate state of the unperturbed state on which the perturbation is applied producing the perturbed states.

Thus the perturbation problems divide themselves into the following groups.

I. Time independent nondegenerate perturbation problems.

II. Time independent degenerate perturbation problems.

III. Time dependent perturbation problems.

Some specific example of the above three types are

I. H-atom (ground state) exposed to constant external magnetic field.

II. A p-electron of an atom (having different possible m_l m_s values) in a degenerate ^{2}p state. The allowance of the spin-orbit interaction on it constitutes a degenerate time independent perturbation problem.

III. An atom or a molecule exposed to electromagnetic field.

The total hamiltonian for the perturbed state is

$$\mathscr{H} = \mathscr{H}_0 + \mathscr{H}^{/} \qquad \text{...... (2.38)}$$

where $\mathscr{H}_0$ is the hamiltonian in the unperturbed state (also called the zero order approximation state) and $\mathscr{H}$ perturbation hamiltonian operator contributing additional energies (in positive or negative sense) arising from the perturbational cause. This $\mathscr{H}^{/}$should be small compared to $\mathscr{H}_0$ of the unperturbed state.

I. Time independent perturbation of nondegenerate state:

For such a state, the first and the second order corrections in energy due to $\mathscr{H}^{/}$ are

$$\in_n^{(1)} + \in_n^{(2)} = <\psi_n^0|\, \mathscr{H}^{/}\, |\psi_n^0> + \Sigma_k^{/} \left\{<\psi_k^0|\, \mathscr{H}^{/}\, |\psi_n^0> <\psi_n^0|\, \mathscr{H}^{/}\, |\, \psi_k^0\right\} / \left(E_n^0 - E_k^0\right) \qquad \text{... (2.39)}$$

$$= H_{nn}^{/} + \Sigma_k^{/} H_{nk}^{/} / \left(E_n^0 - E_k^0\right) \qquad \text{...... (2.40)}$$

where zero superscripts refer to terms in the unperturbed state $\Sigma_k^{/}$denotes a summation over all values of k except for k = n and ψ_n^0 is the unperturbed nondegenerate state on which the perturbation is applied. The first order correction in wavefunction is

$$\psi_n^{(1)} = \Sigma_k A_k \psi_k^0 = \left\{\Sigma_k^{/} H_{kn}^{/} H_{nk}^{/} / \left(E_n^0 - E_k^0\right)\right\} \psi_k^0 \qquad \text{...... (2.41)}$$

The second order correction in wavefunction can also be worked out but is seldom needed.

Sometime it may so happen that instead of the perturbation being lumped into one term, $\mathscr{H}^{/}$, as in Eq. (2.38), it may comprise a second smaller term (cf. Eq. 2.37- magnetic field perturbation), viz,

$$\mathscr{H} = \mathscr{H}_0 + \mathscr{H}^{/} + \mathscr{H}^{//} \qquad \text{...... (2.42)}$$

The energy correction terms are then,

$$\in_n^{(1)} + \in_n^{(2)} = <\psi_n^0|\mathscr{H}'|\psi_n^0> + \Sigma_k' <\{\psi_k^0|\mathscr{H}'|\psi_n^0><\psi_n^0|\ \mathscr{H}'\ |\psi_k^0\}/(E_n^0 - E_k^0)$$

$$+ <\psi_n^0|\ \mathscr{H}''\ |\psi_n^0>\} \qquad \text{......(2.43)}$$

$$= H'_{nn} + \{\Sigma_k' H'_{kn} H'_{nk} / (E_n^0 - E_k^0) + H''_{nn}\} \qquad \text{....... (2.44)}$$

The terms in the parenthesis make up the second order correction in energy.

II. Time independent perturbation of a degenerate state

In such cases for a time independent perturbation $\mathscr{H}'$, the perturbation hamiltonian is expressed in a form in the basis set of the unperturbed degenerate eigenfunctions. Thus if $\psi_1^0, \psi_2^0, ...\psi_m^0$ be the degenerate eigenfunction set of an m-fold degenerate energy, the perturbation Hamiltonian matrix is

$$\mathbf{H} = \begin{bmatrix} H'_{11} & H'_{12} .. & .. & ..H'_{1m} \\ .. & .. & .. & .. & .. \\ H'_{ml} & H'_{m2} .. & .. & ..H'_{nm} \end{bmatrix} \qquad \text{....... (2.45)}$$

where $H'_{mn} = <\Psi_k^0|\ \mathscr{H}'\ |\psi_m^0>$ etc. The matrix is worked upon for its eigenvalues and eigenvectors. From these, the corrections in energies and wavefunctions are written down.

Such types of degenerate problems often result in the lifting of degeneracy, either fully or partially, under the influence of perturbations. Further, the commutivity of the perturbation hamiltonian with the angular momentum operators and certain theorems on matrix elements of the matrix operator (2.45) often help to block diagonalise the matrix (2.45) facilitating easy solutions (chap 3).

Eq. (2.42) represents cases where two perturbation hamiltonians, $\mathscr{H}'$ and $\mathscr{H}''$ arise from certain physical cause, external or internal. We shall often come across cases where a number of physical causes of different degrees of importance contribute a perturbation hamiltonian each to the total hamiltonian of the perturbed system. For example, for a transition metal ion in a particular stereochemical environment, electrostatic repulsions, $\Sigma_{i,j} e^2 / r_{ij}$, crystal field interaction V_{crys} and spin-orbit interactions, $\Sigma_i \xi(r_i) 1_i . s_i$ simultaneously affect the unperturbed system contributing three perturbation hamiltonians. The problem of dealing with such system will be later treated in appropriate chapters.

III. Time dependent perturbations

There may arise various types of time dependent perturbations in different systems. One very important type is when an atomic/molecular system is exposed to electromagnetic radiation field. Various formulation covering different aspects of such cases have been theoretically developed. We shall, however, concern ourselves with the application of the transition moment integral which is measure of the probability of transition from one electronic state to another by absorption or emission of radiation. The transition moment integral expression is

$$T_{mn} = <\psi_m|\mathscr{H}'|\psi_n> \qquad \text{.......} (2.46)$$

where $\mathscr{H}'$ is the electric dipole moment operator which acting as a perturbation causes a transition from the nth quantum state to the mth quantum state.

2.5 Wavefuncitons of Polyelectronic Systems is Symbolic Ket Notations and Determinants (Dets)

The wavefunctions of a single electron in a central field potential, as in a hydrogenic system, is a product of a suitable radial function and a spherical harmonic- function of the angular coordinates, θ and φ.

$$\phi = R_{nl}(r)Y_{lm}(\theta,\varphi) = R_{nl}(r)\Theta_{1m}(\theta)\Phi_m(\varphi) \qquad \text{.......} (2.47)$$

This space orbital has to be multiplied by the electronic spin wavefunction, $\alpha \, or \, \beta$ of the spin coordinates, ω, give the net wavefunction $\phi\alpha$ or $\phi\beta$.

In the case of a multielectron system, two important principles have additionally to be taken care of in building the system wavefunciton on the basis of single electron orbital approach. The two principles are I) the indistinguishability of the electrons and ii) the antisymmetry of the system wavefunction under exchange of the coordinates (including the spin coordinates) of a pair of fermions, viz., electrons. Sticking to the single electron orbital approach for multielectron atom/ion, different methods are availed of, depending on the end in view, to build up the wavefunction and evaluate the energy.

i) In many cases, more or less accurate wavefunctions and good values of energies are needed for the problem on hand. In such cases we start with the requisite hamiltonian and an assumed form of an initial wavefunction (in determinantal form- vide infra) and through iterative technique continue to refine the wavefunction and modify the energy value till the attainment of self consistency within a desired limit.

ii) In other type of problems (such as Russell Saunders' coupling, electron spectra, Zeeman effect, spin-orbit interaction, crystal field effect) we are not interested in the overall total energy of the atom or ion, but rather in their departures from main (unknown) value of energy and in their corrections of the wavefunctions.

The whole hamiltonian is then suitably split into a central field part and a remaining perfurbational part. The energy and the wavefunction corresponding to $\mathscr{H}_{cent}$ are not solved. Instead, emphasis is laid on the perturbational part. In such cases also we have to start with an assumed from of wavefunction for the $\mathscr{H}_{cent}$, but we do not continuously maul or modify it as in the previous self consistent field case. We make do with it only to find corrections in energies and wavefunctions due to perturbations.

It thus turns out that, irrespective of the final objective, an assumed form of wavefunction for an n-electron system is the first step in the process. Each electron is described by a single electron spin orbital which is a product of a space orbital, ϕ and a spin wavefunction, α or β. Thus in an n-electron system, an electron with coordinates $r_1, \theta_1 \,\&\, \omega_1$, called electron 1, has the spin orbital a_1 where

$$a_1 = \phi_1\alpha = R_1(r_1)\Theta_1(\theta_1)\Phi_1(\varphi_1)\alpha(\omega_1)$$
$$= |l_1 s_1 m_{l1}\, m_{s1}> \qquad \ldots\ldots (2.48)$$

i.e., a ket in the (1 s, $m_l m_s$) scheme (cf. sec. 1.4 vii). The reader should note the different significances conveyed by ϕ, Φ and . For electrons 2,3 .... we may write the spin orbitals

$$a_2 = \phi_l\beta = R_1(r_2)\Theta_1(\theta_2)\Phi_1(\varphi_2)\beta(\omega_2) = |l_1 s_1 m_{l1}\, m_{s2}>$$
$$a_3 = \phi_2\alpha = R_2(r_3)\Theta_2(\theta_3)\Phi_2(\varphi_3)\alpha(\omega_3) = |l_2 s_2 m_{l2}\, m_{s1}> \qquad \ldots\ldots (2.49)$$
$$a_4 = \phi_2\beta = R_2(r_4)\Theta_2(\theta_4)\Phi_2(\varphi_4)\beta(\omega_4) = |l_2 s_2 m_{l2}\, m_{s2}>$$

where $m_{sl} = ½$, $m_{s2} = -½$ and ϕ_1, ϕ_2 ..... are called atomic orbitals. The net system wavefunction, i.e., the atomic wavefunction is a normalized antisymmetrised linear combination of the products of n spin orbital a_1, a_2, ... a_n. This antisymmetrised linear combination is mathematically equivalent to a Slater determinant wherein the individual spin orbitals serve as the elements of the determinant.

$$\psi = D = 1/\sqrt{n!} \begin{vmatrix} a_1(1) & a_1(2) \ldots\ldots a_1(n) \\ a_2(1) & a_2(2) \ldots\ldots a_2(n) \\ & \\ & \\ a_n(1) & a_n(2) \ldots\ldots a_n(n) \end{vmatrix} \qquad \ldots\ldots (2.50)$$

$1/\sqrt{n!}$ in (2.50) is the normalizing factor, and $a_n(k)$ represents an a_n spin orbital occupied by electron having the space and the spin coordinates r_k, θ_k, φ_k and ω_k. Any such electron is briefly called the kth electron. For economy in space, this determinantal wavefunction is written in an abbreviated form by writing only the diagonal product term (a_1, a_2, a_3 .... a_n) of the determinant within a parenthesis and that too without the explicit appearance of the normalizing factor or of the electron tag numbers. Thus, wavefunction (2.50) is indicated by

$$\psi = D = (a_1\ a_2\ a_3\ ......a_n) \qquad .... (2.51)$$

We should beware of the very large number of possible atomic states generated by the prevailing coupling scheme (say, Russell Saunders) for n electrons each having its l and s values (sec. 1.5). We assign to each atomic state determinantal forms of wavefunctions by selecting suitable sets of spin orbitals a_1, a_2, ...a_n for different atomic states. This will illustrated in some detail in Chapter 3. We suggest here, in advance, that for many such states the atomic wavefunction will be given not by a single determinant D_1 but by a linear combination of determinants with different sets of spin orbitals constituting the determinants.

$$\psi = \Sigma_j c_j D_j = c_1(a_1\ a_2\ a_3\ ....a_n) + c_2(b_1\ b_2\ b_3....b_n) + \qquad .... (2.52)$$

There is yet another angle of viewing the multielectron atomic system of n electrons each having its (1 s m_l m_s) values. The overall atomic state will have some resultant (L S M_L M_S) values as a result of Russell Saunders' coupling if that scheme prevails. The atomic state is then written in a ket notation of (L S M_L M_S) scheme.

$$\psi = |L\ S\ M_L\ M_s> = D = (a_1\ a_2\ a_3\ ....a_n) \qquad .... (2.53)$$

Thus the overall atomic ket is a combination of the products of the constituent electronic microkets such as a_1 a_2 a_3 ....a_n. Similar considerations apply if the coupling be of the (j-j) type, the microkets involved such cases being the types $|l_1\ s_1\ j_1\ m_{j1}>'s$

The ideas sought to be expressed in very general terms in the foregoing paragraphs are important enough to merit a little more elaboration by a concrete example. Of course details are put off until Chapter 3 as stated earlier. Suppose we have an equivalent electronic configuration (p^2) as in C-atom. The atomic states are sec. 1.5.2 Table $_{1.2}$ ^1D, ^3P and ^1S. The ^1D has the following overall ket forms :

$|2020>, |2010>, |2000>, |20-10>, |20-20>$ which are degenerate. Of these overall atomic kets; the $|2010>$atomic ket of ^1D is a linear combination of the products of microkets $|1^+>$and $|0^->$where only the m_l and m_s value are stated, the n, l, and s value of equivalent p electrons, being the same, are not stated, the + and – sign as superscripts denote m_s = + ½ and m_s = -½ respectively for the two electrons. Therefore, in determinantal from

$$\Psi(^1D, p^2) = |2020> = (1^+0^-) = 1/\sqrt{2!}\begin{vmatrix} 1^+(1) & 1^+(2) \\ 0^-(1) & 0^-(2) \end{vmatrix} \qquad .... (2.54)$$

But as will be shown later (chapter 4) ^{1}D ket, viz., $|2010>$ cannot be represented by a single det as in (2.54), but by a linear combination of dets (cf. Eq.2.62).

Such considerations also apply to the different degenerate atomic kets of the ^{3}P term, where depending on the nature of the kets, a single determinant or a multiple number of these may become necessary.

2.6 Matrix Forms of Hamiltonian and Other Operators

Suppose we have an eigenvalue eigenfunction equation, say, Schrodinger equation,

$$\mathcal{H}\psi = \in \psi \qquad \text{....... (2.55)}$$

Expanding the wavefunction Ψ in terms of an orthonormal set ϕ_j.s

$$\psi = \Sigma_j \, c_j \, \phi_j = \boldsymbol{\Phi}\, \boldsymbol{C} \qquad \text{....... (2.56)}$$

where ψ is the basis set of row matrix $\phi = (\phi_1 \; \phi_2 \; ...... \; \phi_n)$ and C, the coefficient column matrix. Eq. (2.35), mended through (2.56), changes to

$$\mathcal{H}\,\Psi = \in \Psi = \mathcal{H}\,\boldsymbol{\Phi C} \qquad \text{....... (2.57)}$$

i.e. $$\in \boldsymbol{\Phi C} = \mathcal{H}\,\boldsymbol{\Phi C} \qquad \text{....... (2.58)}$$

The verctor space language interpretation Eq. (2.56) is that the basis set functions, ϕ_j.s and the wavefunction Ψ, all belong to the same linear function space. The action of $\mathcal{H}$ operator on Ψ is to transform it into a new function, Ψ' of the same function space. This transformation operation by $\mathcal{H}$ can be alternatively described by the transformation of the old basis set ϕ_j.s to a new orthonormal basis set ϕ_j'.s, represented by the new matrix $\boldsymbol{\Phi}'$ with Ψ remaining unaffected. In the latter of sense of the operator $\mathcal{H}$ we have

$$\boldsymbol{\Phi}' = (\phi_1' \quad \phi_2' ...... \phi_n') = \mathcal{H}\boldsymbol{\Phi} = (\mathcal{H}\phi_1 \;\; \mathcal{H}\phi_2 \; .... \; \mathcal{H}\phi_n)$$

Each individual new basis function, ϕ_j', i.e., $\mathcal{H}\,\phi_j$ being a general function of the same linear space, can thus be written as a linear combination of the old basis set.

$$\left.\begin{array}{lllll}
\boldsymbol{\Phi}_1' & = \mathcal{H}\phi_1 & = & a_{11}\,\phi_1 + a_{21}\,\phi_2 + \;........ + a_{n1}\,.\phi_n \\
\boldsymbol{\Phi}_2' & = \mathcal{H}\phi_2 & = & a_{12}\,\phi_1 + a_{22}\,\phi_2 + \;........ + a_{n2}\,.\phi_n \\
... & ... & & ... \qquad ... \qquad\qquad ... \\
\boldsymbol{\Phi}_n' & = \mathcal{H}\phi_n & = & a_{1n}\,\phi_1 + a_{2n}\,\phi_2 + \;........ + a_{nn}\,.\phi_n
\end{array}\right\} \qquad \text{....... (2.59)}$$

Using the orthonormality of the basis sets, Φ and Φ', it is seen that

$$\left.\begin{aligned} a_{11} &= <\phi_1|\phi_1'> &= <\phi_1 \,|\, \mathrm{H}\,\phi_1> = h_{11} \\ a_{21} &= <\phi_2|\phi_1'> &= <\phi_2 \,|\, \mathrm{H}\,\phi_1> = h_{21} \\ \ldots & \quad \ldots & \ldots \quad\quad \ldots \\ a_{kj} &= <\phi_k|\phi_j'> &= <\phi_k \,|\, \mathrm{H}\,\phi_j> = h_{kj} \end{aligned}\right\} \quad \ldots\ldots (2.60)$$

In this way all the coefficients in (2.59) are know through (2.60) Eqs. (2.59) can thus be represented in matrix form

$$\Phi' = \mathscr{H}\,\Phi = (\phi_1\ \phi_2 \ldots\ldots \phi_n)\begin{bmatrix} h_{11} & h_{12} & \ldots\ldots h_{1n} \\ h_{21} & h_{22} & \ldots\ldots h_{2n} \\ h_{31} & h_{32} & \ldots\ldots h_{3n} \\ \ldots & \ldots & \ldots\ldots \ldots \\ h_{n1} & h_{n2} & \ldots\ldots h_{nn} \end{bmatrix} = \Phi\ \mathbf{H} \quad \ldots\ldots (2.61)$$

Eq. (2.6) tells that the operator $\mathscr{H}$ puts on the grab of a matrix , H, (in a certain basis set), i.e., the former becomes just a transformation matrix H in the formulations of equation. Now turning of Eqs. (2.57, 2.58 & 2.61) one may write

$$\mathscr{H}\,\Psi = \in \Psi = \mathscr{H}\Phi\mathbf{C}$$

i.e., $\quad \in \Phi\mathbf{C} = \mathscr{H}\Phi\mathbf{C} = \Phi'\,\mathbf{C} = \Phi\,\mathscr{H}\mathbf{C}$

$$\Phi \in \mathbf{C} = \Phi\,\mathscr{H}\mathbf{C}$$

$\therefore \quad \mathscr{H}\mathbf{C} = \in \mathbf{C} \quad \ldots\ldots (2.62)$

Eqs. (2.55 & 2.62) suggest that for each eigenvalue eigenfunction equation of an operator $\mathscr{H}$, there can be constructed an eigenvalue eigenvector equation of a corresponsding matrix transformation operator H in an implied basis set.

The considerations and conclusions made in respect of $\mathscr{H}$ operator can be extended to the operators, such as L^2, S^2, L_z, S_z as well as symmetry operators. In a suitable linear space each such operator may be represented as a matrix transformation operator and all such operators, having their respective eigenvalue eigenfunction equations, will also have the corresponding eigenvalue eigenvector equations in matrix forms. Thus the matrix form of L^2 operator, in the basis set of

$(2L+1)$ number of degenerate spherical harmonics, Y_{LMl} (Sec. 1.4) with $-L \le M_L \le L$ and constant eigenvalue $L(L+1)\hbar^2$ for L^2 operator, is a $(2L+1)$ dimensional square matrix, the diagonal elements being $L(L+1)\hbar^2$

$$L^2 = \begin{bmatrix} L_{11}^2 & L_{12}^2 \ldots\ldots L_{1n}^2 \\ L_{21}^2 & L_{22}^2 \ldots\ldots L_{2n}^2 \\ \ldots\ldots & \ldots\ldots\ldots\ldots \\ L_{n1}^2 & L_{n2}^2 \ldots\ldots L_{nn}^2 \end{bmatrix} = L(L+1)\hbar^2 \begin{bmatrix} 1 & \ldots\ldots\ldots & \\ . & 1\ldots\ldots\ldots & \\ . & . & \ldots\ldots \\ . & . & \ldots\ldots 1 \end{bmatrix} \quad \ldots\ldots (2.62a)$$

where the spherical harmonics $Y_{L-L}, Y_{L-(L-1)}, \ldots\ldots Y_{L(L-1)}, Y_{LL}$ have been sequentially numbered 1,2,3,....$2L,2L+1$. the orthogonality of the spherical harmonics with the same L and different $M_L's$ causes the disappearance of the offdiagonal elements.

Another illustration is taken up here to find the matrix form of the step up operator l_+ with the 3d single electron wavefunction $R_{32}\ d_2$, $R_{32}\ d_1$, $R_{32}\ d_0$, $R_{32}\ d_{-1}$, $R_{32}\ d_{-2}$ serving as the base functions. Remembering the operational effect of l_+ on normalized spherical harmonics (Eq. 1.52), we have

$$\begin{aligned} &l_+ d_2 = 0; \quad l_+ d_1 = \hbar\sqrt{\{2(2+1)-1.2\}}\, d_2; \\ &l_+ d_0 = \hbar\sqrt{\{2(2+1)}\, d_1; \quad l_+ d_{-1} = \hbar\sqrt{\{2(2+1)\}}\, d_2; \\ &l_+ d_{-2} = \hbar\sqrt{\{2(2+1)-2\}}\, d_{-1} \end{aligned} \quad \ldots\ldots (2.62b)$$

These form the key relations that enable one to write out the matrix form of the operator l_+ in the basis set of 3d wavefunctions.

We thus learn that an operator in an originally functional or differential form can always be substituted, in matrix formulation of the problem, by a matrix form in a basis set, $\phi_j.s$, spanning a linear function space. Since many basis sets may be chosen for the same space, it is evident that one can have different matrix representations of the same operator in any linear space. The eigenvalues of these different matrix forms of the same operator, say A, are the same and identical with the eigenvalue set of the operator A in its original eigenvalue eigenfunction equation (cf. Eq. 2.62). This identity of the eigenvalue sets means that the traces of the operator matrices of A, though structurally different, remain invariant as long as the basis sets of the same space are used to construct the matrix forms. Now a theorem in matrix[11] tells us that the eigenvalues of a matrix and consequently its trace remain unchanged under a similarity transformation (unitary transformation). Hence, we can divine that the structurally different any two matrix forms of the same operator, A, are related to each other through proper similarity transformation in the same linear space.

Thus if an operator A has the matrix forms $\mathbf{A}^f$ and $\mathbf{A}^g$ in the basis sets $\mathbf{f} = (f_1\ f_2\ f_3 \ldots\ldots f_n)$ and $\mathbf{g} = (g_1\ g_2\ g_3 \ldots\ldots\ldots g_n)$ respectively spanning the same linear space and $\mathbf{T}$ be the proper transformation connecting $\mathbf{A}^f$ and $\mathbf{A}^g$, i.e.,

$\mathbf{A}^g = \mathbf{T}^{-1}\ \mathbf{A}^f\ \mathbf{T}$ (a similarity transformation), then it can be shown that $\mathbf{T}$ connects the basis sets, viz., $\mathbf{g} = \mathbf{f}\,\mathbf{T}$. Conversely if this latter relation be assumed, the former relation follows.

2.7 Matrix Elements of Operators between Determinantal Wavefunction

All quantamechanical operators of multielectron atom/ion with which we shall normally deal are seen to consist of a sum of operator components which are functions of the coordinates of a single electron each and /or a sum of operator components each of which simultaneously contains the coordinates of a pair of electrons. Thus the total operator is the sum of single electron operators and /or the sum of two electron operator parts. To name a few, L^2, L_z, L_+, $\Sigma_i\ \xi(r_i)\ l_i.s_i$, are all sums of single-electron operators, while the total electrostatic repulsion operator, $\Sigma_{i<j}\, e^2/r_{ij}$, is a sum of two-electron operators, since the vector r_{ij} depends simultaneously on both the coordinates of the ith and the jth electrons. The hamiltonian operator, $\mathscr{H}$, (cf. Eq. 2.1) of a multielectron atom contains both one-electron and two-electron operator components.

In a central field the zero order atomic wavefunctions, as we know from sec. 2.5, are represented by single determinants or a linear combination of determinants (Eq. 2.52) consisting of microstate electronic kets, i.e., spin orbitals as their elements. Now any physically observable of an atom has thus to be obtained from the integral, $<\Psi|A|\Psi>$, where A is a operator corresponding to the observable. Since the wavefunction Ψ consist, in general, of a number of dets D_1, D_2, …. , the evaluation of the foregoing integral demands the calculation of diagonal and offdiagonal matrix elements of A between states represented by the dets D_1, D_2, etc., i.e., matrix elements like $<D_1|A|D_1>, <D_1|A|D_2>$ and such others. In terms of the description of nature of an operator, as given in the initial paragraph of the present section, evaluation of $<D_1|A|D_1>, <D_1|A|D_2>$ would involve handling the matrix elements of one and two-electron operators i.e., $<D_1|\Sigma_i f_i|D_1>$, $<D_1|\Sigma_i f_i|D_2>$, $<D_1|\Sigma_{i<j}\, g_{ij}|D_1>$, $<D_2|\Sigma_i f_i|D2>$, $<D_1|\Sigma_{i<j}\, g_{ij}|D_2>$, $<D_2|\Sigma_{i<j}\, g_{ij}|D_2>$ etc.,

where A $= \{\Sigma_i f_i + \Sigma_{i<j}\, g_{ij}\}$

A meandering through logic and arguments to appreciate the effect of permutation on the integral values with the possible rearrangement of the integrals in a new order leads to the general forms of the results for the matrix elements of the one-and the two-electron operators. We write down the general forms of the results recommending to the readers to look up suitable references [13,14,15] for the involved mathematical details and arguments.

a) Matrix elements of one-electron operators, $\Sigma_j F_j$ between dets.

i) Diagonal terms $<D_1|\Sigma_j f_j|D_1> = \Sigma_j <a_1|f_j|a_1> = \Sigma_j m_{fj}$ ……. (2.63) if the kets $|a_j>$'s are eigenkets of operators with eigenvalues m_{fj}'s for the operator f_j s, where $D_1 = (a_1\ a_2\ a_3\ \ldots..\ a_n)$.

Nondiagonal terms $<D_1|\Sigma_j\ f_j|D_2>$ with D_2 $(a_1\ a_2\ ..\ b_j\ ..\ a_n)$ where $|b_j>$is also an eigen ket of f_j. D_2 thus differs from D_1 in one electronic ket, $|b_j>$, the rest being the same.

$$<D_1|\Sigma_j\ f_j|D_2>=0 \qquad \ldots\ldots. (2.64)$$

The disappearance is caused by mutual orthogonality of $|a_j>$ and $|b_j>$. Thus the matrix form of the operators, $\Sigma_j\ F_j$, in the basis set of the above two dets is diagonal.

ii) A second alternative may also arise if $f_j\ |a_j>=c_j|b_j>$. Under such conditions, the matrix elements are all zeros.

$$<D_1|\Sigma_j\ f_j|D_1>=<D_2|\Sigma_j\ f_j|D_2>=0 \qquad \ldots\ldots. (2.65)$$

But the nondiagonal elements have non zero values.

$$<D_1|\Sigma_j\ f_j|D_2> = \Sigma_j <b_j|f_j|a_j>=\Sigma_j\, c_j \qquad \ldots\ldots. (2.66)$$

When, however, D_1 and D_2 differ in more than one ket in their compositions, the matrix elements of one-electron operators between D_1 and D_2 disappear.

b) Matrix elements of two-electron operators between dets.

Here one can expect four types of general results depending on the nature of the dets involved. Let us consider four determinants D_1, D_2, D_3, and D_4 constituting the wavefunction Ψ in their linear combination. The compositions of the dets in abbreviated notations (Eq. 2.51) are

$$\begin{aligned} D_1 &= (a_1a_2 \cdots a_ia_j \cdots a_n) \\ D_2 &= (a_1a_2 \cdots b_ia_j \cdots a_n) \\ D_3 &= (a_1a_2 \cdots b_ib_j \cdots a_n) \\ D_4 &= (a_1a_2 \cdots b_ib_jb_k \cdots a_n) \end{aligned} \qquad \ldots\ldots. (2.66)$$

The dets D_2, D_3 and D_4 differ each from D_1 by having one, two, and three different element electronic spin orbitals, i.e., microkets respectively. Diagonal and nondiagonal matrix elements taking D_1 in each case are

$$<D_1|\Sigma_{i<j}\ g_{ij}|D_1>=\Sigma_{i<j}\{<a_ia_j|g_{ij}|a_ia_j>-<a_ia_j|g_{ij}|a_ja_i>\delta_{msi,msj}\} \quad \ldots (2.67)$$

$$<D_1\left|\Sigma_{i<j}\, g_{ij}\right|D_2> = \Sigma_j\left\{<a_ia_j\left|g_{ij}\right|b_ia_j> - <a_ia_j\left|g_{ij}\right|a_jb_i> \delta_{msi,msj}\right\} \quad \text{.... (2.68)}$$

$$<D_1\left|\Sigma_{i<j}\, g_{ij}\right|D_3> = \left\{<a_ia_j\left|g_{ij}\right|b_ib_j> - <a_ia_j\left|g_{ij}\right|b_jb_i> \delta_{msi,msj}\right\} \quad \text{.... (2.69)}$$

$$<D_1\left|\Sigma_{i<j}\, g_{ij}\right|D_4> = 0 \quad \text{.... (2.70)}$$

The matrix elements of two-electron operator between any two dets differing in their compositions from each other by more than two electronic kets always vanish (cf. 2.70).

The initial forms of D_2, D_3, D_4 involved in the wavefunction Ψ may not be the same as in (2.66). If these be the case, then these initial forms should be subjected to exchange permutations of pairs of electronic coordinates to bring these to their respective standard forms D_2, D_3, D_4, of (2.66). The standard form is the one in which the diagonal term of the det concerned is in maximum possible conformity with the diagonal term of D_1 if the latter be initially chosen to be in its normal form. The standard forms of D_2, D_3 and D_4 will be related to their initial forms, employed in Ψ through a multiplicative constant of +1 or -1 depending on the even or odd number of permutational exchanges of pairs of electronic coordinates necessary to attain the standard forms from their respective initial forms.

2.8 Matrix Elements of The Hamiltonian Operator between Two Determinants

The general results of the matrix elements [Eqs. 2.63, 2.65, 2.67-2.69] are innocuous in appearance, but these assume quite a complex look, through in a straightforward manner, when applied to actual cases. As we shall very often engage ourselves with energy matrices, we try to concretize the general results as far as possible for the matrix elements of the Hamiltonian between dets. For an atom, the electronic Hamiltonian (Eq. 2.1) is

$$\mathscr{H} = -\left(\hbar^2/2m\right)\Sigma_i \nabla_i^2 + \Sigma_{i<j} e^2/r_{ij} - \Sigma Ze^2/r_i = \Sigma u_i + \Sigma_{i<j} e^2/r_{ij} \quad \text{.... (2.71)}$$

where the division of the operator into one-electron and two-electron components is apparent in (2.71).

Eqs. (2.63, 2.65,2.67-2.69) indicate that each individual integral involves one same ket or kets, two different kets, three different kets or at most four different kets and the summation, if any, extends over electronic tags *i, j*. Now exchanges permutation of electronic coordinates with the kets (spin orbitals) remaining fixed is equivalent to exchanging the particular quantum sets (n_i l_i m_{li} m_{si}) etc. of the kets keeping the electron tags fixed. In view of this, the summation over *i, j*, in the aforesaid general results may be replaced by summation over electronic kets with different quantum number labels, viz., (n_i l_i s_i m_{li} m_{si}) and (n_j l_j s_j m_{li} m_{si}) or viceversa.

We shall from now on use the symbols a, b, c, d for the four different kets required at any time in any individual integral with the quantum number tags being n_a, l_a, s_a, m_{la}, m_{sa} for the ket, $|a>$, and so on for the other kets.

The full expression for the ket is

$$|a> = R_{nala}(r) Y_{la\ mla}(\theta,\varphi).\{\alpha_a \ or \ \beta_a\} \quad \ldots\ldots. (2.72)$$

with similar expressions for $|b>, |c> and |d>$.

Turning now to the matrix elements of the one-electron part of the hamiltonian (2.71), Eqs. (2.63) and (2.65) may be arranged as

$$< D_1|\Sigma u_i|D_1> \ = \Sigma_a < a(1)|\ u_1|a(1)> \quad \ldots\ldots. (2.73)$$

$$< D_1|\Sigma u_i|D_2> \ = \Sigma_{a,b} < a(1)|\ u_1|b(1)> \quad \ldots\ldots. (2.74)$$

On using (2.72)

$$< D_1|\Sigma u_i|D_2> = \Sigma_{a.b} < R_{nala}\ Y_{lamla}|u_1|R_{nblb} Y_{lbmlb} >< \alpha_a / \beta_a|\alpha_b / \beta_b >$$

$$= \Sigma_{a,b} < R^{*}_{nala}\ u_1\ R_{nblb}\ r^2 > \delta(1_a,1_b)\delta(ml_a,ml_b)\delta(ms_a,ms_b)$$

$$= \Sigma_{a,b}\ 1(n_a,l_a,\ n_b l_b)\delta(1_a,1_b)\delta\,(ml_a,ml_b)\delta\,(ms_a,ms_b) \quad \ldots\ldots. (2.75)$$

The operator u_1, being a function of r and independent θ, φ, is included in the integral of the radial function part. The spherical harmonics and the spin parts are integrated to appear in terms of kronecker δ because of orthonormal properties of the harmonics and of the spin functions.

Of the different matrix elements of the two-electron operator part $\Sigma_{i<j}\ e^2 / r_{ij}$, of the hamiltonian (2.71), let us first consider the matrix element (2.69). For a two-electron system.

$$< D_1|\Sigma_{i<j}\ g_{ij}|D_3> = \{< a(1)b(2)|e^2 / r_{12}|c(1)d(2)> - < a(1)b(2)|e^2 / r_{12}|d(1)c(2)>\}$$

where in the first intregral $|a>$ and $|c>$ are functions of the coordinates ($r_1\ \theta_1\ \varphi_1\ \omega_1$) of electron 1, and the kets $|b>$ and $|d>$ are those of $(r_2\ \theta_2\ \varphi_2\ \omega_2)$ of electron 2. In the second integral of the RHS, the dependences of the kets $|c>$ and $|d>$ upon the coordinate sets are reversed.

Now, $< ab|e^2/r_{12}|cd> = < a\ b|e^2\ \Sigma_{k=0}^{\infty}\left(r_<{}^{k}/r_>{}^{k+1}\right)P_k(\cos\gamma)|cd>$

$= \Sigma_{k=0}^{\infty}\ \Sigma_{m=-k}^{k}\ e^2\{< ab|\left(r_<{}^{k}/r_>{}^{k+1}\right)(4\pi / 2k+1)Y_{km}^{*}(\theta_2\varphi_2)Y_{km}(\theta_1\ \varphi_1)|cd>\}$ from addition theorem, Eqs. (1.26 & 1.28) sec. 1.3.1

$$= e^2 \sum_{k=0}^{\infty}\Big[\int_0^{\infty}\int_0^{\infty} R_{nala}{}^{*} R_{nblb}{}^{*} R_{nclc}\, R_{ndld}\left(r_<^{k}/r_>^{k+1}\right) r_1^2 r_2^2 dr_1 dr_2 \times$$

$$(4\pi/2k+1)\sum_{m=-k}^{k}\Big\{\Big(\int_0^{\pi} \Theta_{la\,m\,la}(\theta_1)\Theta_{lc\,m\,lc}(\theta_1)\Theta_{km}(\theta_1)\sin\theta_1 d\,\theta_1\Big)\times$$

$$\int_0^{\pi}\ \Theta_{lb\,m\,lb}(\theta_2)\Theta_{ld\,m\,ld}(\theta_2)\Theta_{km}(\theta_2)\sin\theta_2 d\theta_2)\times$$

$$\Big(\int_0^{2\pi}\int_0^{2\pi}\Phi_{m\,la}{}^{*}(\varphi_1)\Phi_{m\,lc}(\varphi_1)\Phi_{m}(\varphi_1)\Phi_{m\,lb}{}^{*}(\varphi_2)\Phi_{m\,ld}(\varphi_2)\Phi_m^{*}(\varphi_2)d\varphi_1\, d\varphi_1\, d\varphi_2\Big)$$

$$\delta(ms_a, ms_c)\,\delta(ms_b, ms_d)\ \Big\}\ \Big] \qquad \ldots\ldots (2.76)$$

The above, apart from the radial integral part, consists of two integrals of triple product of Φ functions each and furthermore two integrals of triple product of Associated Legendre's functions each.

I. Evaluation of integral of the Φ function part:

Singlevalued nature of the involved Φ functions demands the satisfaction of the following for nonvanishing values of the integrals

$$m + m_{lc} - m_{la} = 0 = m_{ld} - m_{lb} - m$$

i.e., $m_{la} - m_{lc} = m = m_{ld} - m_{lb}$ for any single value of m

or, subject to the condition $m_{la} + m_{lb} = m_{lc} + m_{ld}$

The Φ integral part is thus equal to $1/(2\pi)\delta(m_{la} + m_{lb}, m_{lc} + m_{ld})$

II. The overall integral over the Θ and the Φ function parts reduces to

$$\Big\{\sqrt{(2/2k+1)}\int_0^{\pi}\Theta_{1a\,m\,1a}(\theta_1)\Theta_{lc\,m\,lc}(\theta_1)\Theta_{kmla-mlc}(\theta_1)\sin\theta_1 d\,\theta_1\Big\}\times$$

$$\Big\{\sqrt{(2/2k+1)}\int_0^{\pi}\Theta_{1b\,m\,1b}(\theta_2)\Theta_{ld\,m\,ld}(\theta_2)\,\Theta_{kmld-mlb}(\theta_2)\sin\theta_2 d\,\theta_2\Big\}\times$$

$$\delta(ms_a, ms_c)\,\delta(ms_b, ms_d)\,\delta(m_{la} + m_{lb}, m_{lc} + m_{ld})$$

$$= c^k(l_a m_{la}, l_c m_{lc})c^k(l_d m_{ld}, 1_b m_{lb})\delta(ms_a, ms_c)\,\delta(ms_b, ms_d)\,\delta(m_{la} + m_{lb}, m_{lc} + m_{ld})$$

The c^k's represent the values of the integrals of the triple product of the Associated Legendre's polynomicals multiplied by $\sqrt{(2/2k+1)}$ for any given integral value of k. The c^k's also obey the relation, viz.,

$$c^k(1m_l, 1' m_1') = (-1)^{ml-ml'} c^k(1' m_1'.1m_1)$$

$$\text{Thus } < ab|e^2/r_{12}|cd> = e^2 \sum_k \delta(ms_a, ms_c)\delta(ms_b, ms_d)\delta(m_{la}+m_{lb}, m_{lc}+m_{ld})\times$$

$$c^k(l_a m_{la}, l_c m_{lc})c^k(l_d m_{ld}, l_b m_{lb})R^k_{nala,nblb,nclc,ndld} \qquad \text{....... (2.77)}$$

where $R^k_{nala.....ndld} = e^2 \int_0^\infty \int_0^\infty R_{nala}{}^* R_{nblb}{}^* R_{ncle} R_{ndld} \left(r_<^k / r_>^{k+1}\right) r_1^2 r_2^2 dr_1 dr_2$

We shall shortly see that the c^k values are read off from tables already constructed for different electronic configurations. As to the R^k part of (2.77), these cannot be given numerical values theoretically unless the actual forms of the radial functions in the individual kets are specifically known. Eq. (2.77) thus displays the extent of concretizing the first individual integral occurring in the general result of Eq. (2.69) when the operator is the electronic repulsion one between a given pair of electrons. Obviously more such terms will accumulate when there are additional electrons setting up new pairs for electrostatic repulsion.

We shall now have a dig at Eq. (2.67) for the diagonal matrix elements and concretize it, as far as practicable, when the operator part of the hamiltonian (2.71) is the electrostatic repulsion. Using the kets $|a>, |b>$, etc. eq. (2.67) may be written as, remembering that only two electrons are involved,

$$< D_1|\sum_{i>j} e^2/r_{ij}|D_1> = |< ab|e^2/r_{12}|ab> - < ab|e^2/r_{12}|ba>| \qquad \text{....... (2.78)}$$

Consider each integral in it separately. In $< ab|e^2/r_{12}|ab>$, the roles of $|c>$ and $|d>$ of Eq. (2.77) are played by $|a>$ and $|b>$ respectively. Guided by (2.77), we can write down the result

$$< ab|e^2/r_{12}|ab> = \sum_{k=0}^{\infty}\{c^k(l_a m_{la}, l_a m_{la})c^k(l_b m_{lb}, l_b m_{lb})F^k(n_a l_a, n_b l_b)\} \qquad \text{....... (2.79a)}$$

$$= \sum_{k=0}^{\infty}\{a^k(l_a m_{la}, l_b m_{lb})F^k(n_a l_a, n_b l_b)\} \qquad \text{....... (2.79b)}$$

The distinction between a and a^k, despite the use of the same letter, is not to be missed. Here F^k, the radial integral is

$$F^k(n_a l_a, n_b l_b) = e^2 \int_0^\infty \int_0^\infty R_{nala}{}^* R_{nblb}{}^* R_{nala} R_{nblb} \left(r_<{}^k / r_>{}^{k+1}\right) r_1^2 r_2^2 d_{r1}\ d_{r2}$$

And $a^k(1_a m_{la}\ l_b m_{1b}) = c^k(l_a m_{la} m_{la})c^k(l_b m_{lb}, l_b m_{lb})$

This integral portion of the diagonal element is called the coulomb integral and is symbolized by J (a, b).

$$J(a,b) = < a\,b|e^2/r_{12}|ab> = \sum_{k=0}^{\infty}\{a^k(l_a m_{la}, l_b m_{lb})F^k(n_a l_a, n_b l_b)\} \qquad \text{....... (2.80)}$$

The other integral of (2.78), viz., $< ab|e^2/r_{12}|ba>$, called the exchange integral and symbolized by K (a, b), is

$$K(a,b) = < a\,b|e^2/r_{12}|ba> = \sum_{k=0}^{\infty}\{b^k(l_a m_{la} l_b m_{lb})G^k(n_a l_a, n_b l_b)\delta(ms_a, ms_b)\}$$

$$=\sum_{k=o}^{\infty}\{c^k(l_a m_{la} l_b m_{lb})c^k(l_a m_{la} l_b m_{lb})G^k(n_a l_{la}, n_b l_{lb})\delta(ms_a, ms_b)\}$$

where $Gk(n_a l_a, n_b l_b) = e^2 \int_0^\infty \int_0^\infty R_{nala} * R_{nblb} * R_{nblb} R_{nala} (r_< k / r_>{}^{k+1}) r_1^2\, r_2^2\, dr_1 dr_2$

and $b^k(l_n m_{la}, l_b m_{lb}) = \{c^k(l_n m_{la}, l_b m_{lb})\}^2$

(Note the different significances of b and b^k)

It is readily seen that for a pair of equivalent electrons $(n_a l_a = n_b l_b), F^k = G^k$.

Having shown the way for treating the individual integrals occurring in the diagonal and nondiagonal matrix elements [cf. Eqs. (2.69, 2.67)] of the hamiltonian operator, similar transformation of Eq. (2.68) is thus easily feasible. However, in many atomic problems we shall need, for evaluations of the energy levels, only the diagonal matrix elements of the electronic repulsion operator. We shall necessarily concentrate more on the diagonal term.

2.9 c^k Integrals and Condon Slater Parameters, F^k and G^k

Eqs. (2.79 a, b, 2.80 a, b) show clearly that the c^k integrals and hence a^k's and b^k's depend on the l and m_l values of the electrons. Each c^k integral is over a triple product of three Associated Legendre's functions of degrees l, l' and k. As the functions are arguments of cosθ, in integrand finally resolves itself into a product of trigonometric functions. Moreover, the final expressions for the matrix elements concerned demand a summation over k. Gaunt[16] has shown that $\sum_{k=0}^{\infty}$ is severely restricted to a summation over only a few terms due to limitation imposed on k by triangular condition. The latter requires k, l and l' equal the sides of a triangle of even perimeter. Analytically this triangular condition means

$$k + l + l' = 2g \text{ (g = an integer)}$$

$$|l - l'| \le k \le |l + l'| \qquad \text{....... (2.81)}$$

From this pair of equations it is at once evident that for two equivalent d-electrons the possible values of k to be considered for summation in c^k's are just 0, 2 and 4 only. A complete general result for such an integral was formulated by Gaunt and based on this, comprehensive tables containing numerical values of the c^k integrals as well as of a^k's and b^k's have been constructed for diverse categories of electrons with their l and m_l values. a^k and b^k tables are also drawn up for equivalent elements. These tables [16,17] will help the users to read off the values of the integrals directly with the allowed values of k [Eqs. (2.79a & 2.81)]

Table T2.1

C^k values for different two-electron configurations)

We write $c^k = \pm\sqrt{x/D_k}$, where D_k depends only on l and l'. In the table are listed only the sign preceding the radical and the value of x, D_x being given at the head of each column. Since $b^k = (c^k)^2, b^k = \pm x/D_x$. Note that $c^k(l'm_1', lm_1) = (-1)^{m_1-m_1'} c^k(lm1, l'm_1')$

$$c^k(lm_1, l'm_1')\,for\ l+l'odd$$

ll'	m_1	m_1'	k= 1	3	5
sp	0	±1	$-\sqrt{1/3}$		
	0	0	+1		
sf	0	±3		$-\sqrt{1/7}$	
	0	±2		+1	
	0	±1		-1	
	0	0		+1	
pd	±1	±2	$-\sqrt{6/15}$	$+\sqrt{3/245}$	
	±1	±1	+3	-9	
	±1	0	-1	+18	
	0	±2	0	+15	
	0	±1	-3	-24	
	0	0	+4	+27	
	±1	∓2	0	+45	
	±1	∓1	0	-30	
df	±2	±3	$-\sqrt{15/35}$	$+\sqrt{10/315}$	$-\sqrt{1/1524.6}$
	±2	±2	+5	-20	+5
	±2	±1	-1	+24	-15
	±2	0	0	-20	+35
	±1	±3	0	+25	-7
	±1	±2	-10	-15	+24
	±1	±1	+8	+2	-50
	±1	0	-3	+2	+80
	0	±3	0	+25	-28
	0	±2	0	0	+63
	0	±1	-6	-9	-90
	0	0	+9	+16	+100
	±2	∓3	0	0	-210
	±2	∓2	0	0	+126
	±2	∓1	0	+10	-70
	±1	∓3	0	0	-84
	±1	∓2	0	+25	+112
	±1	∓1	0	-15	-105

$c^k\left(lm_1, l'm_1'\right) for\ l + l'\ even$

ll'	m_1	m_1'	k= 0	2	4	6
ss	0	0	+1			
sd	0	±2		$+\sqrt{1/5}$		
	0	±1		-1		
	0	0		+1		
pp	± 1	±1	+1	$-\sqrt{1/25}$		
	± 1	0	0	+3		
	0	0	+1	+4		
	± 1	∓ 1	0	-6		
pf	± 1	±3		$+\sqrt{45/175}$	$-\sqrt{1/189}$	
	± 1	±2		-30	+3	
	± 1	±1		+18	-6	
	± 1	0		-9	+10	
	0	±3		0	-7	
	0	±2		+15	+12	
	0	±1		-24	-15	
	0	0		+27	+16	
	± 1	∓ 3		0	-28	
	± 1	∓ 2		0	+21	
	± 1	∓ 1		+3	-15	
dd	±2	±2	+1	$-\sqrt{4/49}$	$+\sqrt{1/441}$	
	±2	±1	0	+6	-5	
	±2	0	0	-4	+15	
	±1	±1	+1	+1	-16	
	±1	0	0	+1	+30	
	0	0	+1	+4	+36	
	± 2	∓ 2	0	0	+70	
	± 2	∓ 1	0	0	-35	
	± 1	∓ 1	0	-6	-40	
ff	± 3	±3	+1	$-\sqrt{25/225}$	$+\sqrt{9/1089}$	$-\sqrt{1/7361.64}$
	± 3	±2	0	+25	-30	+7
	± 3	±1	0	-10	+54	-28
	± 3	0	0	0	-63	+84
	± 2	±2	+1	0	-49	+36
	± 2	±1	0	+15	+32	-105
	± 2	0	0	-20	-3	+224
	± 1	±1	+1	+9	+1	-225
	± 1	0	0	+2	+15	+350
	0	0	+1	+16	+36	+400
	± 3	∓ 3	0	0	0	-924
	± 3	∓ 2	0	0	0	+462
	± 3	∓ 1	0	0	+42	-210
	± 2	∓ 2	0	0	+70	+504
	± 2	∓ 1	0	0	-14	-378
	± 1	∓ 1	0	-24	-40	-420

Although for our purpose, single-electron kets of only equivalent electrons/will mainly appear in the repulsion integrals, the c^k tables, presented here, include the results for non-equivalent electrons as well. This makes the tables more general and amenable to wider use for calculations in the cases of free atoms and ions.

Table T$_{2.2}$ (a_k, b_k Values)

The value of this coefficient is independent of the signs of m_1 and m_1'. As in the preceding table, we print the common denominator of several values but once at the beginning of each group. For l=0, $a^k(oo,l'm_l') = \delta(k,0)$ for all $l'm_l'$; $fork = 0, \alpha^o(lm_1, m_l') = 1$ for all values of the arguments; in the table we give values only for l, l' k>0. Note that

$$a^k(l'm_l', lm_l) = \alpha^k(m_l, l'm_l')$$

ll'	$\lvert m_1\rvert$	$\lvert m_1'\rvert$	$k=$ 2	4	ll'	$\lvert m_1\rvert$	$\lvert m_1'\rvert$	$k=$ 2	4	6
pp	1	1	1/25		*df*	2	3	10/105	3/693	
	1	0	- 2			2	2	0	-7	
	0	0	- 4			2	1	- 6	1	
						2	0	- 8	6	
pd	1	2	2/35			1	3	- 5	-12	
	1	1	- 1			1	2	0	28	
	1	0	- 2			1	1	3	-4	
	0	2	- 4			1	0	4	-24	
	0	1	2			0	3	- 10	18	
	0	0	4			0	2	0	-42	
						0	1	6	6	
pf	1	3	5/75			0	0	8	36	
	1	2	0							
	1	1	- 3		*ff*	3	3	25/225	9/1089	1/7361.64
	1	0	- 4			3	2	0	-21	-6
	0	3	- 10			3	1	- 15	3	15
	0	2	0			3	0	- 20	18	-20
	0	1	6			2	2	0	49	36
	0	0	8			2	1	0	-7	-90
						2	0	0	-42	120
dd	2	2	4/49	1/441		1	1	0	1	225
	2	1	- 2	- 4		1	0	12	6	-300
	2	0	- 4	6		0	0	16	36	400
	1	1	1	16						
	1	0	2	- 24						
	0	0	4	30						

We shall allow ourselves to indulge in exercises on c^k integrals in the last section of this chapter (Sec. 2.11).

The $F^k\left(lm_1, l'm_1'\right)$ and $G^k\left(lm_1, l'm_1'\right)$ integrals required in (2.79, 2.80) of the diagonal matrix elements are not directly evaluated. Since in our calculations pertaining to perturbed systems we shall need only the relative energy levels, numerical values of F^k and G^k will not be necessary. Hence F^k *'s* and G^k *'s* are retained in the final expressions of the relative energy levels in the forms of unevaluated parameters, viz., F_k *'s* and G^k *'s*. The relationship between F^k and F_k and between G^k and G_k hinge upon the product of the two c^k values associated with the F^k and the G^k respectively occurring in Eqs. (2.79 a, b & 2.80a, b). If D_k is the denominator occurring in the product of the two c^k values (obtainable from Table $T_{2.1}$) associated with the corresponding F^k or G^k then.

$$F_k = F^k/D_k \text{ and } G_k = G^k/D_k \qquad \text{.......} (2.82)$$

These F_k *'s* and G_k *'s* are called Condon Stater parameters[17]. Seldom, though in a few instances, would we need the actual values of the parameters. When one has to evaluate these, the latter can be obtained from suitable spectral data and ionization energies of the atoms[19,20]. In calculations, involving especially d – and f electrons, some authors[18,20,21] prefer to switch over from F_k *'s* to new parameters A, B and C called the Racah parameters[21]. The changeover is effected through the following proposed equalities:

$$A = F_0 - 49F_4\ ;\ B = F_2 - 5F_4;\ C = 35F_4 \qquad \text{.......} (2.83)$$

The usage of the Racah parameters makes relative energy expressions look simpler. In our treatment of the d-electron transition metal ions as well as of the atomic system in general, we shall use the parameters. A, B and C instead of the Condon Slater's.

2.10 Ancillary Aspects of Group Theory.

Group theoretical ideas and formulations are almost an inalienable part of problem dealing with perturbations in atoms and molecules. But while in certain cases of atomic perturbations, e.g., Russell Saunders' coupling, spin orbit interaction, Zeeman effect, we can skirt its direct use by adopting alternative easier techniques, in symmetry perturbations (Sec.1.1E) of the atomic or ionic environments of coordination complexes, concepts, theorems and formulations of symmetry transformation groups are inextricably mixed up in any sort of calculation.

What is essential for a follow up of the present text is a deep and comprehensive knowledge of the symmetry transformation groups. We shall assume that our readers are adequately equipped with the knowledge of this discipline. We shall, therefore, contend ourselves with just referring, without any elaboration, to only those aspects of the transformation groups that enter directly or indirectly in treating problems lying within the province of crystal field theory (Chapters 7-11). For detailed exposition, text on group theory and symmetry may be consulted [22, 23, 24, 25].

i) Finding the character set of a reducible representation using a basic set of functions. There exit short cut techniques, apart from the normal ones, for assessing the character set.

ii) To determine the number and types of IR's in a reducible representation, one uses the relation

$$n_\mu = 1/g[\Sigma_R \chi_\mu(R)\chi^{red}(R) \qquad \ldots\ldots (2.84)$$

iii) To construct IR symmetry adapted function by linear combination of basis functions :

This can be directly achieved in easier cases on a little reflection. In some cases, one needs to examine the effect of some selective transformation operators of the concerned group on the basis function. This gives an inkling of the composition of the symmetry adapted function. The general method, however, is to use the projection operator of appropriate symmetry on the basis functions. The expression for the projection operator is

$$P^\mu = (l_\mu/g)\Sigma_R \chi_\mu(R)R \qquad \ldots\ldots (2.85)$$

iv) The character $\chi^{(\mu x v)}$(R) of the product representation (dpr) $\Gamma^{\mu \times v}(R)$, is equal to the product of the characters of the component IR's.

$$\chi^{(\mu x v)}(R) = \chi_\mu(R).\chi_v(R) \qquad \ldots\ldots (2.86)$$

v) Resolution of a dpr of a transformation group as a direct sum of component IR's This is done with the help of (2.86) and (2.84). Results of many frequently occurring dpr's of some IR's of several important point groups are furnished in Table T 2.3

Ample instances of the application of the above principle appear in chapters 8, 9 and 10. Table T 2.3 trails section 2.11 of the present chapter.

vi) Molecular orbitals and molecular wavefunction transform each as bases of one of the IR's of the point group of the molecule. Energy degeneracy is always indicated by the dimensionality of the IR to which the orbital or the wavefunction belongs.

vii) Atomic orbitals of which the variables are measured in an axial system with the origin coincident with the fixed point of the point symmetry group of the species behave as bases of one or other IR's the point group.[cf. the atomic orbitals of the central ion in a coordination complex of different streochemical environments].

viii) Branching IR's due to descent in symmetry

Atoms belong to complete rotation inversion group SO (3). But when an atom is placed in an environment (eg., in a complex or a crystal) its point group changes to one of lower symmetry. Original degenerate IR's of the previous point group branch into IR's of lower dimensionalities.

ix) The vanishing or the nonvanishing of a transition moment integral, determining the nonfeasibility or the feasibility respectively of electronic transition, is controlled by the nature of the resolved symmetry type of the dpr of the integrand.

x) There exists a general form of the rotation matrix drawn up in a basis set of spherical harmonics of degree l (l = 0 or any positive integer). The matrix for a rotation of $(2\pi/n)$ about the quantising axis is diagonal and is of the order $(2l+1)x(2l+1)$.

This is the number of spherical harmonics serving as basis function for a given value of l and incidentally equal to the number of the different m_l values involved.

$$C_n = \begin{bmatrix} e^{-il2\pi/n} & & & & & & & \\ & e^{-i(l-1)2\pi/n} & & & & & & \\ & & \cdot & & & & & \\ & & & e^{-i(0)2\pi/n} & & & & \\ & & & & \cdot & & & \\ & & & & & \cdot & & \\ & & & & & & \cdot & \\ & & & & & & e^{i(l-1)2\pi/n} & \\ & & & & & & & e^{il2\pi/n} \end{bmatrix} \qquad \ldots\ldots (2.87)$$

For complete spherical groups, all rotational axes are equivalent.

xi) The character of this general rotation matrix about the quantizing axis is given by.

$$\chi(C_n) = \{sin(l+1/2)\beta\}/(sin\,\beta/2) \qquad \ldots\ldots (2.88)$$

where $\beta = 2\pi/n$. This relation is not valid for $\beta = 0$

xii) The transformation behaviour of spins under group operators and the group theoretical treatment of spin orbit interactions and of systems with half odd integral total angular momentum quantum number require the concept and use of the double groups and the resolution of the dpr's of the double groups into component IR's Most of the theorems pertaining to representations and characters of the simple point groups also apply to the double groups. (See chaps. 8 &11)

2.11 Exercises Involving c^k Integrals

As promised in sec. 2.9, we illustrate here how the electrostatic perturbation effect of $\Sigma e^2/r_{kl}$ can be computed using the $c^k/a^k, b^k$ table ($T_{2.1}$, $T_{2.2}$). It will turn out later that our main preoccupation will be the equivalent d^n electron systems and with this in view we prefer to confine ourselves to calculations involving d^2 systems.

Example 1:

Calculate the electrostatic perturbation effect on the energy levels of d^2 system expressed by the determinantal wavefunction $(d_{yz}^{+} d_{xy}^{+})$, the supersign (+) indicates that both the electrons are of $\alpha-$ spin.

The real forms of the orbitals d_{yz} and d_{xy} are given by (Appendix Chap 1)

$$d_{yz} = \frac{i}{\sqrt{2}}\{d_{-1} + d_1\} \text{ and } d_{xy} = \frac{i}{\sqrt{2}}\{d_{-2} - d_2\}$$

$$\in (d^2) = <(d_{yz}^{+}\, d_{xy}^{+})|e^2/r_{12}|(d_{yz}^{+} d_{xy}^{+})>$$

$= J(d_{yz}, d_{xy}) - K(d_{yz}, d_{xy})$ where Eq. (2.80 and 2.80a) are used remembering that both the electrons have $\alpha-$spins.

$J(d_{yz}, d_{xy}) =< d_{yz}d_{xy}|1/r_{12}|d_{yz}d_{xy} >$, where e^2 factor has been merged with the radial parts the d- orbitals involved.

$$= \quad 1/4 <\{d_{-1} + d_1\}\{d_{-2} - d_2\}|1/r_{12}|\{d_{-1} + d_1\}\{d_{-2} - d_2\}>$$

$$= \quad 1/4 <\{d_{-1}\, d_{-2}.|1/r_{12}|d_{-1} + d_{-2} > + < d_{-1}\, d_2|1/r_{12}|d_{-1}\, d_2 >$$

$$+ < d_1\, d_{-2}|1/r_{12}|d_1\, d_{-2} > + < d_1\, d_2|1/r_{12}|d_1\, d_2\}$$

where only the integrals as have $m_{la} + m_{lb} = m_{lc} + m_{ld}$, are nonvanishing and hence retained. The other four integrals, being of the vanishing type, have been dropped.

$$\therefore \quad J(d_{yz}, d_{xy}) \quad = \quad 1/4\{J(-1,-2) + J(-1, 2) + J(1,-2) + J(1, 2)\}$$

$= \quad 1/4\ [4A\ -8B+4C]$, on using Eq. (2.80), c^k/a^k, b^k table ($T_{2.1}$, $T_{2.2}$), and finally converting the Condon Stater parameters into Racah's via Eq. (2.80) {Note that in using Eq. (2.80) for (a,b)in the above equation, references to n and *l* values of the electrons are omitted as these are equivalent electrons. Only the m_l values are indicated. Thus in using the c^k/a^k, b^k tables, j(-1,-2) would mean

$$J(-1,-2) = \Sigma_{k=0}^{4}\, c^k(-1,-1)c^k(-2,-2)\, F^4 = \Sigma_{k=0}^{4}\, a^k(-1,-2)F^k$$

Thus the coulomb integral, $J(d_{yz}, d_{xy}) = A - 2B + C$

The exchange integral is

$K(d_{yz}, d_{xy}) =< d_{yz}\, d_{xy}|1/r_{12}|d_{xy}\, d_{yz} >$, with e^2 having been absorbed in the radial function $= 1/4 <\{d_{-1} + d_1\}\{d_{-2} - d_2\}|1/r_{12}|\{d_{-2} - d_2\}\{d_{-1} + d_1\}>$

$$= 1/4 < d_{-1}d_{-2} - d_{-1}d_2 + d_1d_{-2} - d_1d_2|1/r_{12}|d_{-2}d_{-1} - d_2d_{-1} + d_{-2}d_1 - d_2d_1 >$$

$$= 1/4[K(-1,-2)+K(-1,2)+K(1,-2)+K(1,2)]$$

= 3B+C, on using Tables T $_{2.1}$,T $_{2.2}$ and Eq. (2.83).

$$\therefore \;\; \in (d^2)=J(d_{yz},d_{xy})-K(d_{yz},d_{xy})=A-5B.$$

Example 2

A triplet state of d^2 system is given, in (LSM_LM_S) scheme, as $|3120>$. Find the perturbational coulomb repulsion energy in this state.

Since the energy is independent of Ms values, we stick to Ms = 0 and write down the expression for $|3120>$ in terms of two-electron determinants of single-electron spin orbitals. $^3F=|3120>=1\sqrt{2}\{(2^+0^-)+(2^-0^+)\}(cf.sec 4.2)$

$$\in(^3F) \;\; = \;\; 1/2<\{(2^+0^-)+(2^-0^+)\}\;|1/r_{12}|\;\{(2^+0^-)+(2^-0^+)\}>e^2$$

part having been included in the radial portion.

$$= \;\; 1/2\,[<(2^+0^-)|1/r_{12}|(2^+0^-)>+<(2^+0^-)|1/r_{12}|(2^-0^+)>$$
$$+<(2^-0^+)|1/r_{12}|(2^+0^-)>+<(2^-0^+)|1/r_{12}|(2^-0^+)>]$$

Because of the implied presence of the normalizing factor with each determinant, we can effectively choose the diagonal term only from each of the determinants for integration allowing simultaneously for exchange of the spin orbitals between the electron pair (cf. Eq. 2.78).

$$\therefore \in(^3F)=1/2[<20|1/r_{12}|20>-<20|1/r_{12}|02>+<20|1/r_{12}|20>-<20|1/r_{12}|02>]$$

the other integrals disappear due to $\delta(m_{sa},m_{sc})$ factor.

$=1/2\,[2J(2,0)-2K(2,0)]=A-8B$, on using the requisite tables.

It will be found very helpful later if we have direct tables of J and K integrals for equivalent d- electrons with varied m_l values expressed in Racah parameters. Such tables can be constructed directly with help of Eqs. (2.67, 2.78), c^k table (T $_{2.1}$) and the relationship between the Condon Slater parameters and Racah's.

Such computations have been done by a number of authors (cf. Griffith[25])

Table: $T_{2.4}$

(Coulomb and exchange integrals of equivalent d-electrons in Reach parameters)

m_l values	J (coulomb integral)	K (exchange integral)
* $\pm 2, \pm 2$	A+4B +2C	A+4B+2C
$\pm 2, \mp 2$	A+4B +2C	2C
$\pm 2, \pm 1$	A-2B +C	6B+C
$\pm 2, \mp 1$	A-2B +C	C
$\pm 2, 0$	A-4B +C	4B+C
* $\pm 1, \pm 1$	A+B +2C	A+B+2C
$\pm 1, \mp 1$	A+B +2C	6B+2C
$\pm 1, 0$	A+2B +C	B+C
* 0, 0	A+4B +C	A+4B+3C

Integrals marked with asterisk (*) before the m_l values give either the J integral or K integral, not both. The integral results remain unaffected even if the second set of m_l values interchange their positions with those of the first set.

Table: $T_{2.3}$

Resolution of dpr into component IR's of point groups valid for point groups O_h, O, T_d, D_{6h}, C_{6h}, D_3, D_{3d}, D_4, D_{4h}, C_{4v} C_4 and some others. For the last five point groups, E×E = $A_1+A_2+B_1+B_2$, an exception from the tabulated IR's.

IR's	A	A_1	A_2	B_1	B_2	E	E_1	E_2	T_1	T_2
A	A					E	E_1	E_2	T_1	T_2
A_1		A_1	A_2	B_1	B_2	E	E_1	E_2	T_1	T_2
A_2		A_2	A_1	B_2	B_1	E	E_1	E_2	T_2	T_1
B_1		B_1	B_2	A_1	A_2		E_2	E_1		
B_2		B_2	B_1	A_2	A_1		E_2	E_1		
E	E	E	E	E	E	A_1+A_2+E			T_1+T_2	T_1+T_2
E_1	E_1	E_1	E_1	E_2	E_2		$A_1+A_2+E_2$	$B_1+B_2+E_1$		
E_2	E_2	E_2	E_2	E_1	E_1		$B_1+B_2+E_1$	$A_1+A_2+E_2$		
T_1		T_1	T_2			T_1+T_2			$A_1+E+T_1+T_2$	$A_2+E+T_1+T_2$
T_2		T_2	T_1			T_1+T_2			$A_2+E+T_1+T_2$	$A_1+E+T_1+T_2$

Exercises

1. Write down Maxwell's field equations in c.g.s Gaussian forms. Utilise the equations to establish the electric and magnetic fields given by (2.9).
2. Show that the classical hamiltonian function for an electron moving in an electromagnetic field (E, H) is given by
   $H=(1/2m)\{(p_x+e/c\,A_x)^2+(p_y+e/c\,A_y)^2+(p_z+e/c\,A_z)^2\}-e\phi$ where the vector potential $A=\nabla\times H$ and ϕ is the scalar potential.
3. Write the possible dets (determinantal wavefunctions) of an excited He atom having the configuration ($1s^1\ 2p^1$). Use any one of the dets to estimate the Coulomb repulsion energy (perturbational part) using the c^k table T $_{2.1}$).
4. How and why is it that the hamiltonian of a molecule or of a complex ion ought to be totally symmetric in the group of a molecule or ion?
5. Of the following two repulsion integrals, which one will have the larger repulsion and by how much?
   $<(2^+1^+)|e^2/r_{12}|(2^+1^+)>$ and $<(2^+1^-)|e^2/r_{12}|(2^+1^-)>$ both the electrons being equivalent.
6. Utilise the principle of dpr to find the symmetry types in xyz and x^2yz when two functions are subjected to transformations under i) D_{4h} and ii) O_h respectively.
7. Show that $\Sigma_{i<j}\,e^2/r_{ij}$ commutes with the spin free hamiltonian of a multi electron atom.

Reference

1, 2, 7. H. F. Hameka: Advanced Quantum Chemistry (Addison Wesley)

3. H. Goldstein: Classical Mechanics (Addison Wesley)

4, 8. J. D. Jackson: Classical Electrodynamics (Wiley Eastern)

5, 9, 13, H. Eyring. J. Walter and G. E. Kimball: Quantum Chemistry (John Wiley and Song)

6. W. B. Boast: Vector Fields (Harper and Row)

10, 14, 17, E.U. Condon and G. H. Shortley: Theory of Atomic Spectra (Cambridge University Press)

11. F. L. Pilar: Elementary Quantum Chemistry (McGraw-Hill)

12, 23. S.C. Rakshit: Molecular Symmetry Groups and Chemistry (New Book Stall)

15. J. A. Pople and D. L. Beveridge: Approximate Molecular Orbital Theory (Mc

Graw-Hill)

16. J. A. Gaunt: Trans. Roy Soc., A 228, 194, 1929.

18. R. M. Golding: Applied Wave Mechanics (van Nostrand)

19. C.S. Moore: U. S. Nat. Bur. Std. Cire., 467, 1949.

20. 26. J. S. Griffith: The Theory of Transition Metal Ions (Cambridge Univ. Press)

21. A.P. B. Lever: Inorganic Electronic Spectroscopy (Elsevier)

22. D. S. Schonland: Molecular Symmetry (van Nostrand)

24. F. A. Cotton: Chemical Applicationing of Group Theory (Interscience)

25. D.M. Bishop: Group Theory and Chemistry (Clarendon Press)

Appendix 1

Spherical Harmonics in Rectangular Coordinates (Normalised) (Phases as presented by Condon and Shortley).

$$Y_0^0 = \sqrt{\frac{1}{4\pi}}$$

$$Y_1^{-1} = \sqrt{\frac{3}{8\pi}}\frac{x-iy}{r}$$

$$Y_1^0 = \sqrt{\frac{3}{4\pi}}\frac{z}{r}$$

$$Y_1^1 = -\sqrt{\frac{3}{8\pi}}\frac{x+iy}{r}$$

$$Y_2^{-2} = \sqrt{\frac{5}{4\pi}}\sqrt{\frac{3}{8}}\frac{(x-iy)^2}{r^2}$$

$$Y_2^{-1} = \sqrt{\frac{5}{4\pi}}\sqrt{\frac{3}{2}}\frac{z(x-iy)}{r^2}$$

$$Y_2^0 = \sqrt{\frac{5}{4\pi}}\sqrt{\frac{1}{4}}\frac{3z^2-r^2}{r^2}$$

$$Y_2^1 = -\sqrt{\frac{5}{4\pi}}\sqrt{\frac{3}{2}}\frac{z(x+iy)}{r^2}$$

$$Y_2^2 = \sqrt{\frac{5}{4\pi}}\sqrt{\frac{3}{8}}\frac{(x+iy)^2}{r^2}$$

$$Y_3^{-3} = \sqrt{\frac{7}{4\pi}}\sqrt{\frac{5}{16}}\frac{(x-iy)^3}{r^3}$$

$$Y_3^{-2} = \sqrt{\frac{7}{4\pi}}\sqrt{\frac{15}{8}}\frac{z(x-iy)^2}{r^3}$$

$$Y_3^{-1} = \sqrt{\frac{7}{4\pi}}\sqrt{\frac{3}{16}}\frac{(x-iy)(5z^2-r^2)}{r^3}$$

$$Y_3^0 = \sqrt{\frac{7}{4\pi}}\sqrt{\frac{1}{4}}\frac{z(5z^2-3r^2)}{r^3}$$

$$Y_3^1 = -\sqrt{\frac{7}{4\pi}}\sqrt{\frac{3}{16}}\frac{(x+iy)(5z^2-r^2)}{r^3}$$

$$Y_3^2 = \sqrt{\frac{7}{4\pi}}\sqrt{\frac{15}{8}}\frac{z(x+iy)^2}{r^3}$$

$$Y_3^3 = -\sqrt{\frac{7}{4\pi}}\sqrt{\frac{5}{16}}\frac{(x+iy)^3}{r^3}$$

$$Y_4^{-4} = \sqrt{\frac{9}{4\pi}}\sqrt{\frac{35}{128}}\frac{(x-iy)^4}{r^4}$$

$$Y_4^{-3} = \sqrt{\frac{9}{4\pi}}\sqrt{\frac{35}{16}}\frac{z(x-iy)^3}{r^4}$$

$$Y_4^{-2} = \sqrt{\frac{9}{4\pi}}\sqrt{\frac{5}{32}}\frac{z(x-iy)^2}{r^4}(7z^2-r^2)$$

$$Y_4^{-1} = \sqrt{\frac{9}{4\pi}}\sqrt{\frac{5}{16}}\frac{(x-iy)}{r^4}(7z^3-3zr^2)$$

$$Y_4^0 = \sqrt{\frac{9}{4\pi}}\sqrt{\frac{1}{64}}\frac{35z^4-30z^2r^2+3r^4}{r^4}$$

$$Y_4^1 = \sqrt{\frac{9}{4\pi}}\sqrt{\frac{5}{16}}\frac{(x+iy)}{r^4}(7z^3-3zr^2)$$

$$Y_4^2 = \sqrt{\frac{9}{4\pi}}\sqrt{\frac{5}{32}}\frac{(x+iy)^2}{r^4}(7z^2-r^2)$$

$$Y_4^3 = -\sqrt{\frac{9}{4\pi}}\sqrt{\frac{35}{16}}\frac{z(x+iy)^3}{r^4}$$

$$Y_4^4 = \sqrt{\frac{9}{4\pi}}\sqrt{\frac{35}{128}}\frac{(x+iy)^4}{r^4}$$

$$Y_5^{-5} = \sqrt{\frac{11}{4\pi}}\sqrt{\frac{63}{256}}\frac{(x-iy)^5}{r^5}$$

$$Y_5^{-4} = \sqrt{\frac{11}{4\pi}}\sqrt{\frac{315}{128}}\frac{z(x-iy)^4}{r^5}$$

$$Y_5^{-3} = \sqrt{\frac{11}{4\pi}}\sqrt{\frac{35}{256}}\frac{(x-iy)^3}{r^5}(9z^2-r^2)$$

$$Y_5^{-2} = \sqrt{\frac{11}{4\pi}}\sqrt{\frac{105}{32}}\frac{(x-iy)^2}{r^5}(3z^3-zr^2)$$

$$Y_5^{-1} = \sqrt{\frac{11}{4\pi}}\sqrt{\frac{15}{128}}\frac{(x-iy)}{r^5}(21z^4-14z^2r^2+r^4)$$

$$Y_5^0 = \sqrt{\frac{11}{4\pi}}\frac{1}{8}\frac{63z^5-70z^3r^2+15zr^4}{r^5}$$

$$Y_5^1 = -\sqrt{\frac{11}{4\pi}}\sqrt{\frac{15}{128}}\frac{x+iy}{r^5}(21z^4-14z^2r^2+r^4)$$

$$Y_5^2 = \sqrt{\frac{11}{4\pi}}\sqrt{\frac{105}{32}}\frac{(x+iy)^2}{r^4}(3z^2-2r^2)$$

$$Y_5^3 = -\sqrt{\frac{11}{4\pi}}\sqrt{\frac{35}{256}}\frac{(x+iy)^2}{r^4}(9z^2-r^2)$$

$$Y_5^4 = \sqrt{\frac{11}{4\pi}}\sqrt{\frac{315}{128}}\frac{z(x+iy)^4}{r^4}$$

$$Y_5^5 = -\sqrt{\frac{11}{4\pi}}\sqrt{\frac{63}{256}}\frac{(x+iy)^4}{r^4}$$

Appendix II

$\mathcal{D}_3$	E	$2C_3$	$3C_2'$		
A_1	1	1	1		$x^2+y^2;\ z^2$
A_2	1	1	−1	$T_z;\ R_z$	
E	2	−1	0	$(T_x, T_y);\ (R_x, R_y)$	$(x^2-y^2, xy);\ (xz, yz)$

$\mathcal{D}_4$	E	$2C_4$	C_2	$2C_2'$	$2C_2''$		
A_1	1	1	1	1	1		$x^2+y^2;\ z^2$
A_2	1	1	1	−1	−1	$T_z;\ R_z$	
B_1	1	−1	1	1	−1		x^2-y^2
B_2	1	−1	1	−1	1		xy
E	2	0	−2	0	0	$(T_x, T_y);\ (R_x, R_y)$	(xz, yz)

$\mathcal{D}_{3h}$	E	$2C_3$	$3C_2'$	σ_h	$2S_3$	$3\sigma_v$		
A_1'	1	1	1	1	1	1		$x^2+y^2;\ z^2$
A_2'	1	1	−1	1	1	−1	R_z	
E'	2	−1	0	2	−1	0	(T_x, T_y)	(x^2-y^2, xy)
A_1''	1	1	1	−1	−1	−1		
A_2''	1	1	−1	−1	−1	1	T_z	
E''	2	−1	0	−2	1	0	(R_x, R_y)	(xz, yz)

$\mathcal{D}_{4h}$	E	$2C_4$	C_2	$2C_2'$	$2C_2''$	i	$2S_4$	σ_h	$2\sigma_v$	$2\sigma_d$		
A_{1g}	1	1	1	1	1	1	1	1	1	1		$x^2+y^2;\ z^2$
A_{2g}	1	1	1	−1	−1	1	1	1	−1	−1	R_z	
B_{1g}	1	−1	1	1	−1	1	−1	1	1	−1		x^2-y^2
B_{2g}	1	−1	1	−1	1	1	−1	1	−1	1		xy
E_g	2	0	−2	0	0	2	0	−2	0	0	(R_x, R_y)	(xz, yz)
A_{1u}	1	1	1	1	1	−1	−1	−1	−1	−1		
A_{2u}	1	1	1	−1	−1	−1	−1	−1	1	1	T_z	
B_{1u}	1	−1	1	1	−1	−1	1	−1	−1	1		
B_{2u}	1	−1	1	−1	1	−1	1	−1	1	−1		
E_u	2	0	−2	0	0	−2	0	2	0	0	(T_x, T_y)	

$\mathscr{C}_{3v}$	E	$2C_3$	$3\sigma_v$		
A_1	1	1	1	T_z	x^2+y^2; z^2
A_2	1	1	−1	R_z	
E	2	−1	0	(T_x, T_y); (R_x, R_y)	(x^2-y^2, xy); (xz, yz)

$\mathscr{C}_{4v}$	E	$2C_4$	C_2	$3\sigma_v$	$2\sigma_d$		
A_1	1	1	1	1	1	T_z	x^2+y^2; z^2
A_2	1	1	1	−1	−1	R_z	
B_1	1	−1	1	1	−1		x^2-y^2
B_2	1	−1	1	−1	1		xy
E	2	0	−2	0	0	(T_x, T_y); (R_x, R_y)	(xz, yz)

$\mathscr{T}_d$	E	$8C_3$	$3C_2$	$6S_4$	$6\sigma_d$		
A_1	1	1	1	1	1		$x^2+y^2+z^2$
A_2	1	1	1	−1	−1		
E	2	−1	2	0	0		$(2z^2-x^2-y^2, x^2-y^2)$
T_1, F_1	3	0	−1	1	−1	(R_x, R_y, R_z)	
T_2, F_2	3	0	−1	−1	1	(T_x, T_y, T_z)	(xy, xz, yz)

O	E	$8C_3$	$3C_2$	$6C_4$	$6C_2'$		
A_1	1	1	1	1	1		$x^2+y^2+z^2$
A_2	1	1	1	−1	−1		
E	2	−1	2	0	0		$(2z^2-x^2-y^2, x^2-y^2)$
T_1, F_1	3	0	−1	1	−1	(R_x, R_y, R_z); (T_x, T_y, T_z)	
T_2, F_2	3	0	−1	−1	1		(xy, xz, yz)

O_h	E	$8C_3$	$3C_2$	$6C_4$	$6C_2'$	I	$8S_6$	$3\sigma_h$	$6S_4$	$6\sigma_d$		
A_{1g}	1	1	1	1	1	1	1	1	1	1		$x^2+y^2+z^2$
A_{2g}	1	1	1	−1	−1	1	1	1	−1	−1		
E_g	2	−1	2	0	0	2	−1	2	0	0		$(2z^2-x^2-y^2, x^2-y^2)$
T_{1g}, F_{1g}	3	0	−1	1	−1	3	0	−1	1	−1	(R_x, R_y, R_z)	
T_{2g}, F_{2g}	3	0	−1	−1	1	3	0	−1	−1	1		(xy, xz, yz)
A_{1u}	1	1	1	1	1	−1	−1	−1	−1	−1		
A_{2u}	1	1	1	−1	−1	−1	−1	−1	1	1		
E_u	2	−1	2	0	0	−2	1	−2	0	0		
T_{1u}, F_{1u}	3	0	−1	1	−1	−3	0	1	−1	1	(T_x, T_y, T_z)	
T_{2u}, F_{2u}	3	0	−1	−1	1	−3	0	1	1	−1		

Appendix III

Character Tables of Two-valued Representations of Point Groups

D_2'		E	R	$C_2(z)$ $RC_2(z)$	$C_2(y)$ $RC_2(y)$	$C_2(x)$ $RC_2(x)$
Γ'_1	A'	1	1	1	1	1
Γ'_2	B_1'	1	1	1	−1	1
Γ'_3	B_2'	1	1	−1	1	−1
Γ'_4	B'_3	1	1	−1	−1	1
Γ'_5	E'	2	−2	0	0	0

D'_3		E	R	C_3^2R C_3	C_3R C_3^2	$3C_2$	$3C_2R$
Γ'_1	A'_1	1	1	1	1	1	1
Γ'_2	A'_2	1	1	1	1	−1	−1
Γ'_3	E'_1	2	2	−1	−1	0	0
Γ'_4	E'_2	1	−1	−1	1	i	$-i$
		1	−1	−1	1	$-i$	i
Γ'_5	E'_3	2	−2	1	−1	0	0

D'_4		E	R	C_4^3R C_4	C_4R C_4^3	C_2R C_2	$2C_2'R$ $2C_2'$	$2C_2''R$ $2C_2''$
Γ'_1	A'_1	1	1	1	1	1	1	1
Γ'_2	A'_2	1	1	1	1	1	−1	−1
Γ'_3	B'_1	1	1	−1	−1	1	1	−1
Γ'_4	B'_2	1	1	−1	−1	1	−1	1
Γ'_5	E'_1	2	2	0	0	−2	0	0
Γ'_6	E'_2	2	−2	$\sqrt{2}$	$-\sqrt{2}$	0	0	0
Γ'_7	E'_3	2	−2	$-\sqrt{2}$	$\sqrt{2}$	0	0	0

O'		E	R	$4RC_3^2$ $4C_3$	$4RC_3$ $4C_3^2$	$3RC_2$ $3C_2$	$3RC_4^3$ $3C_4$	$3RC_4$ $3C_4^3$	$6RC_2'$ $6C_2'$
Γ'_1	A'_1	1	1	1	1	1	1	1	1
Γ'_2	A'_2	1	1	1	1	1	−1	−1	−1
Γ'_3	E'_1	2	2	−1	−1	2	0	0	0
Γ'_4	T'_1	3	3	0	0	−1	1	1	−1
Γ'_5	T'_2	3	3	0	0	−1	−1	−1	1
Γ'_6	E'_2	2	−2	1	−1	0	$\sqrt{2}$	$-\sqrt{2}$	0
Γ'_7	E'_3	2	−2	1	−1	0	$-\sqrt{2}$	$\sqrt{2}$	0
Γ'_8	G'	4	−4	−1	1	0	0	0	0

Chapter Three

One Valence /Optical Electron Systems.

Perturbational Eigenfunctions and Energy Values

(Spin Orbit Interactions and External Magnetic Field)

Atoms may be conceived as under perturbations ever from within and sometimes, under proper improvisations, from outside. The intraperturbations are in-built and the external perturbations man built.

The one valence/optical electron systems may be polyelectronic atom or ion, such as sodium atom or monoionized calcium or this may be a single electron system, viz., a hydrogen atom. The spectral energy levels and features of such one electron system are independent of the type of coupling of angular momenta undergone by all the electrons present in the cores (inner complete shells and subshells) or the polyelectronic atoms. The only type of intrapertubation acting on the central field (cf. Eq 2.25) unperturbed state of the one-valence/optical electron system is the spin – (same) orbit interaction, the hamiltonian for which is given by $\xi(r)\, l.s$ (Eq 2.27). This, however, assumes the form of $(\zeta_{nl}/\hbar^2) l.s$ in the matrix element of the operator after integration of the radial parts of the matrix element (cf. Eq. 2.28).

That such a perturbation of the type mentioned above is operative in sodium atom and the like ones was first inferred from the experimental observation of the multiplicity of spectral lines under high resolution. It was first learnt experimentally and later supported theoretically that the energy levels of all the one-optical electron systems are doublets. A one-optical electron system containing an s, p, d or f electron has the atomic state represented as ^2S, ^2P, ^2D or ^2F respectively, i.e., in general notation by ^{2s+1}L state as will be established in the present chapter.

3.1. Perturbational Energies and Eigenfunctions of Spin-Orbit Hamiltonian

We now try to find the perturbation effect in sodium-like atoms by considering two specific cases, viz. when the sole optical electron is i) a p – electron ii) a d electron and one more general case, viz., iii) when the electron concerned has the orbital momentum quantum number = *l*. The basis set of kets, which constitute the eigenfunctions of the central field hamiltonian $\mathscr{H}_{\text{cent}}$ of the unperturbed state, may be chosen either in the (L S M_L M_s) scheme or in (L, S, J, M_J) one. (sec 1.4, viii)

(i) For the ^2P state of a sodium – like atom containing an optical p-electron we choose the first scheme and denote the ket by indicating the m_l and m_s values, since *l* and *s* of the electron are evident. The m_s value is indicated by + or – signs depending on whether it is +1/2 or -1/2 respectively. The unperturbed kets are $|1^+>$, $|1^->$, $|0^+>$, $|0^->$, $|-1^+>$ and $|-1^->$. Since l_z or s_z operators do not commute with spin orbit interaction perturbation operator, $\xi(r)l.s$; $(\zeta/\hbar^2)l.s$ (Eq 2.28) these kets are not all diagonal under $(\zeta/\hbar^2)l.s$ or $\xi(r)l.s$. Let us find how these kets behave under $\xi(r)l.s$. Recalling Eq (2.30) and the Eqs (1.51, 1.52, 1.56, 1.57) we have

$$\xi(r)l.s|1^+> = \xi(r)l_z s_z\{|1^+\rangle\} + (\xi(r)/2)\{l_+\sqrt{1/2(1/2+1)-1/2(1/2-1)}|1^->+1\ .0\}\hbar$$

$$= (\xi(r)/2)\hbar^2|1^+> \qquad \text{......... (3.1)}$$

Again $\xi(r)l.s|-1^+> = \xi(r)\{l_z s_z + 1/2(1_+ s_- + l_- s_+)\ \}|-1^+>$

$$= \{-(\xi/2)|-1^+> + \sqrt{2}(\xi/2)|0^->\}\hbar^2 \qquad \text{......... (3.2)}$$

It is thus seen that the kets $|0^->$ and $|-1^+>$ mix under spin orbit perturbation.

Similarly $\xi(r)l.s|0^+> \sqrt{2}(\xi/2)\,|1^-> \hbar^2 \quad = (\sqrt{2}/2)\xi\,\hbar^2|1^->$ .... (3.3)

$$\xi(r)l.s|0^-> \ = (\sqrt{2}/2)\,\xi\hbar^2|-1^+> \qquad \text{......... (3.4)}$$

$$\xi(r)l.s|-1^+> = \{-(\xi/2)|-1^+> + (\sqrt{2}\,\xi/2)|0^->\}\hbar^2 \qquad \text{......... (3.5)}$$

$$\xi(r)\,l.s|-1^-> (\xi/2)\hbar^2|-1^+> \qquad \text{......... (3.6)}$$

Noting the mutual mixing of the kets under the spin orbit interaction operator as reflected in the relations (3.1-3.6) we sequentially number the kets.

$|1^+>|-1^->,|0^+>|1^->$, $|0^->$ and $|-1^+>$ as 1, 2, 3, 4, 5 and 6 forming the basis set of the perturbation matrix of the operator $\xi(r)l.s$

The mutual orthogonality of the kets and the results (3.1. to 3.6) enable us to write down the spin orbit perturbation matrix

$$\begin{bmatrix} \zeta/2 & \cdot & \cdot & \cdot & \cdot & \cdot \\ \cdot & \zeta/2 & \cdot & \cdot & \cdot & \cdot \\ \cdot & \cdot & \cdot & \sqrt{2}\,\zeta/2 & \cdot & \cdot \\ \cdot & \cdot & \sqrt{2}\,\zeta/2 & -\zeta/2 & \cdot & \cdot \\ \cdot & \cdot & \cdot & \cdot & \cdot & \sqrt{2}\,\zeta/2 \\ \cdot & \cdot & \cdot & \cdot & \sqrt{2}\,\zeta/2 & -\zeta/2 \end{bmatrix} \qquad \text{......... (3.7)}$$

where $\xi(r)$ changes, after integration of the radial parts of the kets in each matrix element, to $\zeta/\hbar^2$ (cf. Eq 2.28). The perturbation energy matrix thus gives two (1×1) matrices each with $\varepsilon = \zeta/2$ and two (2 ×2) matrices each with the same pair of energy values $\zeta/2$ and $-\zeta$. Thus the original six degenerate eigenfunctions of the unperturbed state now, on spin orbit interaction divide themselves into two sets of degenerate eigenfunctions, four with perturbation energy, $\zeta/2$ and two other eigenfunctions with energy $-\zeta$.

To find the eigenvector sets associated with energies $-\zeta$ and $\zeta/2$ of the (2×2) matrices, both of which are identical, let us consider the first (2×2) matrix. For $\varepsilon=-\zeta$. the following equation is obtainable

$$\begin{bmatrix} 0-(-\zeta) & \zeta/\sqrt{2} \\ \zeta/\sqrt{2} & -\zeta/2-(-\zeta) \end{bmatrix}\begin{bmatrix} c_1 \\ c_2 \end{bmatrix}=\begin{bmatrix} 0 \\ 0 \end{bmatrix} \qquad \text{......... (3.8)}$$

This equation, together with the orthonormality condition $c_1^2+c_2^2=1$ lead to

$c_1=-1/\sqrt{3}$ and

$c_2=\sqrt{2/3}$ or the pair of values with the sings reversed. .... (3.9)

We need to consider one pair only since the other one will yield linearly dependent solution. Similarly with eigenvalue, $\varepsilon=\zeta/2$. the first (2×2) matrix leads to

$$\begin{bmatrix} 0-(-\zeta/2) & \zeta/\sqrt{2} \\ \zeta/\sqrt{2} & -\zeta/2-\zeta/2 \end{bmatrix}\begin{bmatrix} c_1 \\ c_2 \end{bmatrix}=\begin{bmatrix} 0 \\ 0 \end{bmatrix} \qquad \text{......... (3.10)}$$

whence, again using the orthonoramality property,

$$\left.\begin{matrix} c_1=\sqrt{2/3} \\ c_2=1/\sqrt{3} \end{matrix}\right] \qquad \text{......... (3.11)}$$

or their negative pair, which is discarded for reasons of linear dependence. Thus the eigenfunctions (perturbational wavefunctions) from the first (2×2) matrix are

$$-1/\sqrt{3}\,|\,0^+>+\sqrt{2/3}\,|1^->\ \text{for}\ \varepsilon=-\zeta \qquad \text{......... (3.12a)}$$

$$\sqrt{2/3}\,|0^+>+\ 1/\sqrt{3}\,|1^->\text{for}\ \varepsilon=\zeta/2 \qquad \text{......... (3.13a)}$$

The eigenfunctions, recoupable from the second (2×2) matrix in a like manner, are

$$-1/\sqrt{3}\,|0^->+\ \sqrt{2/3}\,|-1^+>\text{for}\ \varepsilon=-\zeta \qquad \text{......... (3.12b)}$$

$$\sqrt{2/3}\,|0^->+1/\sqrt{3}\,|-1^+>\ \text{for}\ \varepsilon=\zeta/2 \qquad \text{......... (3.13b)}$$

The remaining two (1×1) matrices yield the eigenfunctions

$|1^+>$ for $\varepsilon=\zeta/2$ .... (3.13c)

and $|-1^->$ for $\varepsilon=\zeta/2$ .... (3.13d)

Obtaining the relations (3.12a, b) (3. 13a, b, c, d) is equivalent to having the spin orbit perturbation matrix diagonalized (strictly speaking block diagonalized in this case). The original unperturbed p-level thus splits into two fine structure levels with relative energies $\zeta/2$ and $-\zeta$ away from the baricenter. The atom is thus in a ^2P state.

i) Spin orbit interaction for a single d– optical electron of an atom :

Here to arrive at the final results, we choose the degenerate basis kets $|j\ m_j>$ as the unperturbed eigenfunctions of $\mathscr{H}_{cent}$. The (LSJM$_J$) scheme is chosen since the technique of treating the system expressed in (LSM$_L$M$_s$) scheme has already been illustrated in the previous example.

The tenfold degenerate unperturbed wavefunctions in the scheme are, for $j = l+s = 5/2$. (a d – electron)

$$|5/2, 5/2>, |5/2, 3/2>, |5/2, 1/2>, |5/2, -1/2>, |5/2, -3/2>$$

$$\text{and } |5/2, -5/2> \qquad \ldots\ldots\ldots\ldots (3.14)$$

and for $j = l - s = 3/2$

$$|3/2, 3/2>, |3/2, 1/2>, |3/2, -1/2>, |3/2, -3/2> \qquad \ldots\ldots\ldots (3.15)$$

These are all eigenfunctions of the operator l^2, s^2, j^2, j_z. Now the spin orbit hamiltonian operator $\xi(r)l.s$ i.e., $\xi(r)/2(j^2 - l^2 - s^2)$ commutes with each of the foregoing four operators. The five operators should thus have simultaneous set of eigenkets. The unperturbed kets (3.14 and 3.15), on testing are found to be eigenkets of $\xi(r)$ $l.s.$ as well. The matrix form of the perturbation operator should, therefore, be diagonal in the above basis set. The general form of the matrix elements of this perturbation energy matrix can be given as follows :

$$< j'm'_j|\xi(r)l.s|jm_j >= (\zeta/2\hbar^2)< j'm'_j|\ \{j(j+l) - l(l+1) - s(s+1)\ \}\hbar^2|j\ mj > \qquad \ldots\ldots(3.16)$$

$$= (\zeta/2)\{j(j+1) - l(l+1) - s(s+1)\}< j'm'_j|j\ \ mj > \delta_{j'j}\delta_{m'_j m_j} \qquad \ldots\ldots (3.17)$$

where $j = l+s$ or $l-s$, $l= 2$ and $s= 1/2$

Substituting $j = l+s = l+1/2$ in (3.17), we get the matrix element

$$< j'm'_j|\xi(r)l.s|j\ m_j > = (\zeta/2)\{\ (l+1/2)\,(l+1+1/2) - l(l+1) - 3/4\}\times$$

$$< j'm'_j|j\ mj > \delta_{j'j}\delta_{m'_j m_j}$$

$$= (\zeta/2)l < j'm'_j|j\ mj > \delta_{j'j}\delta_{m'_j m_j} \qquad \ldots\ldots\ldots (3.18)$$

Likewise, when $j = l\text{-}s = l\text{-}1/2$, Eq (3.17) yields

$$< j'm_j'|\xi(r)l.s|j\ mj> = -(\zeta/2)(l+1) < j'm_j'|j\ mj> \delta_{j'j}\delta_{m_j'm_j} \quad \ldots\ldots (3.19)$$

We can now write down the perturbation matrix with the aid of Eqs. (3.18 and 3.19). The ten elements of the diagonal matrix, taking $l = 2$ (d-electron) are written here in a row:

$$(\zeta, \zeta, \zeta, \zeta, \zeta, \zeta, -(3/2)\zeta, -(3/2)\zeta, -(3/2)\zeta - (3/2)\zeta) \quad \ldots\ldots\ldots (3.20)$$

It is seen from (3.20) that the original tenfold degeneracy is lifted only partially resulting into the following perturbational effect

i) Sixfold degenerate eigenstate of energy ζ with the six eigenkets of (3.14) and

ii) a fourfold degenerate eigenstate of energy $(-3/2)\zeta$ with the four eigenkets of (3.15)

Since, ultimately two fine structure levels $(\zeta\ and)(-3/2)\zeta)$ are obtained for a d-electron of an atom, the atomic state is ^{2}D.

iii) An atom with a single optical electron with $l = l$ general case:

The general treatment of the spin orbit interaction for such a case involves no new consideration. Acting within the (LSJM$_j$) scheme it follows that for such an electron, the perturbation energies in its possible different eigenstates are given by the expressions (cf. Eqs 3.18 and 3.19)

$$\varepsilon_{(l+1/2)} = (\zeta/2)l \text{ and } \varepsilon_{(l-1/2)} = -(\zeta/2)(l+1) \quad \ldots\ldots\ldots (3.21)$$

There are finally two sets of degenerate eigenfunctions with different appropriate m_l values characterized by energy values $(\zeta/2)l$ and $-(\zeta/2)(l+1)$ respectively for spin orbit perturbation of an atom with a single optical electron. Thus the atomic states are all fine-structure doublets, viz., ^{2}S, ^{2}P, ^{2}D, ^{2}F and so on, although ^{2}S is a single with j having only one value (0+1/2). i.e. 1/2.

The $\Delta\varepsilon$ value for the spin orbit fine structure doublets is (cf Eq 3.21)

$$\Delta\varepsilon = (\zeta/2)l - [-(\zeta/2)(l+1)] = (\zeta/2)(2l+1) \quad \ldots\ldots\ldots (3.22)$$

This hints at a possible experimental method for evaluating ζ the spin- orbit coupling constant from measurements of frequency differences of the structure lines.

It follows form Eq. (3.21) that the eigenstates of spin orbit energy $(\zeta/2)$ and $-\zeta$ are associated with j values 3/2 and 1/2 respectively of a p-electron. This can also be ascertained independently by subjecting the eigenfunction sets (3.13 a, b, c, d) and (3.12, a, b) to the operator action of j^2.

3.2. An Atom a Single Optical Electron in External Magnetic Field – Zeeman Levels.

Systems, like alkali metal atoms in magnetic field, have basically two perturbations playing upon them – one the intrinsic spin-orbit interaction and the other the externally imposed magnetic field. Such problems need be treated perturbationwise in two stages. If we assume that the spin-orbit interaction is large compared to that external magnetic field, we complete the spin-orbit perturbation study first. Now treating these results of perturbation energies and wavefunctions as our initial zero-order situations, there then ensures the treatment of magnetic field interaction.

Let us suppose that the external magnetic field H be applied parallel to the z direction. This will not affect the generality of the situation. Two perturbation hamiltonians are

$$\xi(r)l.s[Eq2.27] \text{ and } (eH_z/2mc)(l_z+2s_z)+(e^2H_z^2/2mc^2)(x^2+y^2)[\text{Eq } 2.37)$$

Of the two component parts of the external magnetic field perturbation we shall consider only the first one, called the paramagnetic perturbation hamiltonian as this one has a much larger effect than the other, called the diamagnetic hamiltonian.

Let us take the case of p-electron first. We have already obtained the eigenfunctions expressed in (LSM_LM_S) scheme, and energy eigenvalues of the spin orbit perturbation (sec. 3.1) Now to consider the additional external magnetic field perturbation we have to build up matrix form (i.e., the perturbation energy matrix) of the magnetic hamiltonian operator $(eH_z/2mc)(l_z+2s_z)$ in the basis set of the eigenfunction (3.12 a, b, 3.13 a, b, c, d) of sec 3.1.

The whole basis set comprises two degenerate series. With an eye to obtaining easily a blockdiagonal from of the energy matrix. We consider the mixed set instead of treating the degenerate sets separately. With this perspective, we sequentially number the six basis eigenfunctions 3.12a, 3.13a, 3.12b, 3.13b, 3.13c and 3.13d as 1, 2, 3, 4, 5 and 6 respectively. We can readily write down the elements of the perturbation matrix using the m_l, m_s values involved in the basis eigenfunctions and the following very apparent matrix element relation

$$< m_l' m_s' |\{(eH_z/2mc)(l_z+2s_z)\}| m_l m_s > = (e\hbar H_z/2mc)(m_l+2m_s)\delta_{m_l' m_l}\delta_{m_s' m_s} \quad .....(3.22)$$

$$= \beta H_z(m_l+2m_s)\delta_{m_l' m_l}\delta_{m_s' m_s} \quad .... (3.23)$$

where $\beta = (e\hbar/2mc)$ = Bohr magneton

The diagonal and off diagonal matrix elements are estimated using (3.23)

$$\left.\begin{array}{lll} H'_{11} = (1/3)\beta H_z, & H'_{12} = -(\sqrt{2}/3)\beta H_z, & H'_{1n} = 0 \; for \; n = 3,4,5,6 \\ H'_{21} = -(\sqrt{2}/3)\beta H_z, & H'_{22} = -(2/3)\beta H_z & H'_{2n} = 0 \; for \; n = 3,4,5,6 \\ H'_{33} = -(1/3)\beta H_z & H'_{34} = -(\sqrt{2}/3)\beta H_z, & H'_{3n} = 0 \; for \; n = 1,2,5,6 \\ H'_{44} = -(2/3)\beta H_z, & H'_{43} = -(\sqrt{2}/3)\beta H_z, & H'_{4n} = 0 \; for \; n = 1,2,5,6 \\ H'_{55} = 2\beta H_z, & H'_{5n} = 0 & for \; n = 1,2,3,4,6 \\ H'_{66} = -2\beta H_z, & H'_{6n} = 0 & for \; n = 1,2,3,4,5 \end{array}\right] \quad \ldots\ldots (3.24)$$

The secular determinant formed from (3.24) factor into two (2×2) dets and two (1×1) dets. The new energy eigenvalues, including $\zeta/2$ or $-\zeta$ of the unperturbed state (here the spin orbit interaction eigenstate), may now be worked out. We work out the solutions for one (2×2) secular determinant, viz,

$$\begin{vmatrix} -\zeta + (1/3)\beta H_z - \varepsilon & -(\sqrt{2}/3)\beta H_z \\ -(\sqrt{2}/3)\beta H_z & \zeta/2 + (2/3)\beta H_z - \varepsilon \end{vmatrix} = 0 \quad \ldots\ldots\ldots (3.25)$$

The quadratic equation can be solved directly or it can be done in the following indirect way. It is seen from this determinant that the unperturbed level $-\zeta$ of (3.12a) and (3.13a) get mixed under the influence of magnetic field interaction. Hence the perturbation theory covering non degenerate levels (sec 2.4) is here applicable to each separately. The first order corrections in energy appear in the corresponding diagonal terms and the second order ones in the off diagonal elements. The perturbation energy correction for the first level $(-\zeta)$ is

$$\varepsilon_l^{(1)} + \varepsilon_l^{(2)} = H'_{11} + (H'_{12} \cdot H'_{21})/(-\zeta - \zeta/2) = (1/3)\beta H_z - (4\beta^2 H_z^2)/27\zeta \quad \ldots (3.26)$$

and the energy correction to the second level $(\zeta/2)$ is

$$\varepsilon_2^{(1)} + \varepsilon_2^{(2)} = H_{22}{}' + (H'_{12} \cdot H'_{12})/\{+\zeta/2 - (-\zeta)\} = (2/3)\beta H_z + (4\beta^2 H_z^2)/27\zeta \; 3.27$$

Hence the suns of spin orbit and magnetic field interaction energies, using (3.26) and (3.27) separately and appropriately are

$$\varepsilon_1 = -\zeta + (1/3)\beta H_z - (4\beta^2 H_z^2)/27\zeta \quad \ldots\ldots\ldots (3.28)$$

$$\varepsilon_2 = -\zeta/2 + (2/3)\beta H_z + (4\beta^2 H_z^2)/27\zeta \quad \ldots\ldots\ldots (3.29)$$

The perturbed wavefunctions upto first order corrections would be $|cf \; sec \; Eq 2.41|$

$$\left.\begin{array}{l} \psi_1 = (3.12a) + \left| \{(\sqrt{2}/3)\beta H_z\}/(-\zeta - \zeta/2) \right| \times (3.13a) \\ \psi_2 = (3.13a) + \left| \{(\sqrt{2}/3)\beta H_z\}/\{\zeta/2 - (-\zeta)\} \right| \times (3.12a) \end{array}\right] \quad \ldots\ldots\ldots (3.30)$$

where (3.12a) and (3.13a) above represent the functions previously indicated by these numbers in sec. 3.1 and $\beta =$ Bohr magneton.

The couplete set of eigenvalues

$$\left.\begin{aligned}\varepsilon_5 &= \zeta/2+2\beta H_2\\ \varepsilon_2 &= \zeta/2+(2/3)\beta H_z+\left(4\beta^2H_z^2\right)/27\zeta\\ \varepsilon_4 &= \zeta/2-(2/3)\beta H_z+\left(4\beta^2H_z^2\right)/27\zeta\\ \varepsilon_6 &= \zeta/2-2\beta H_z\end{aligned}\right\} \quad \text{......... (3.31)}$$

and

$$\left.\begin{aligned}\varepsilon_1 &= -\zeta+(1/3)\beta H_z-\left(4\beta^2H_z^2\right)/27\zeta\\ \varepsilon_3 &= -\zeta-(1/3)\beta H_z-\left(4\beta^2H_z^2\right)/27\zeta\end{aligned}\right\} \quad \text{......... (3.32)}$$

These magnetically split levels are called Zeeman levels. The perturbed eigenfunctions corresponding to ε_3 and ε_4 may be worked out in a manner similar to the one that led to Eq (3.30) for the Zeeman levels ε_1 and ε_2. The wavefunctions corresponding to ε_5 and ε_6 are the unchanged functions $(3.13c)$ and $(3.13d)$ themselves.

Eqs. (3.31) and (3.32) indicate that the external magnetic field completely removes the fourfold and the twofold degeneracies that result after the spin orbit interaction in situ of the ^{2}P state of one-optical electron containing atom.

The following diagram shows the stagewise lifting of degeneracies with imposition of perturbations till Zeeman levels (first order) are obtained for a p-electron system.

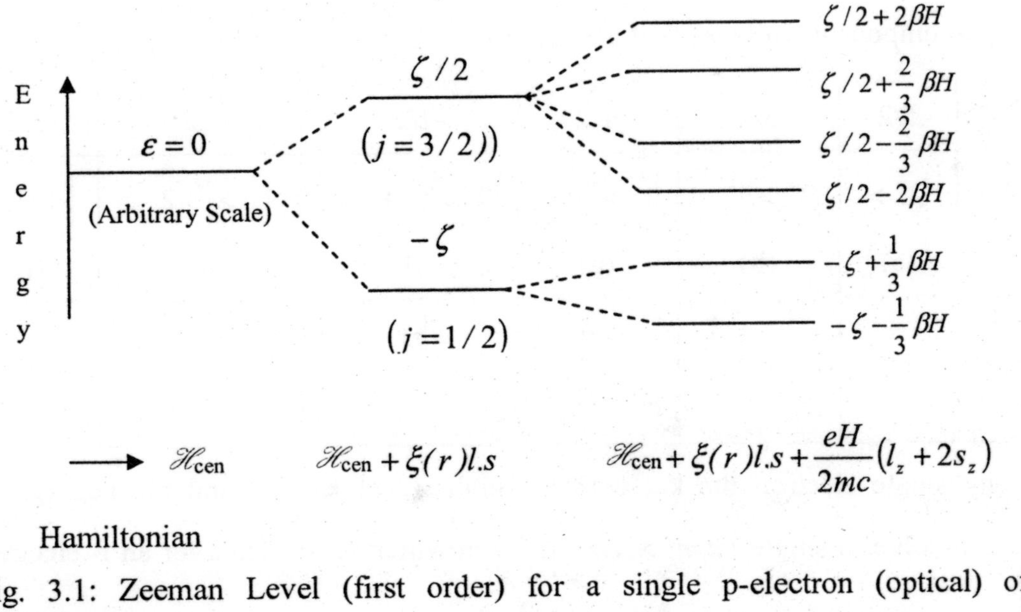

Fig. 3.1: Zeeman Level (first order) for a single p-electron (optical) of a multielectron atom

3.3. Zeeman Levels of a Multi-electron Atom with A Single Optical d-Electron

We have seen in the previous section that one makes a start with the eigenfunctions of $[\mathscr{H}_{cen+}\xi(r)l.s]$ operator in the (LSM_LM_S) scheme to work out the nature of Zeeman splitting. One may also start with initial eigenfunctions written in ($LSJM_J$) scheme as will be presently shown. The process of treating specific cases in such a scheme follows either of the two paths.

a) expressing the single electron kets, $|jm_j>$ in terms of $|m_l m_s>$ either by analysis or by using Clebsech Gordan coefficients and then applying the operator $\{eH/2mc\}(l_z+2s_z)$ to find the matrix elements of the perturbation matrix.

b) To find the general result of the matrix element $<LSJM_j|S_z|LSJM_j>$ and to utilize it in the specific cases under consideration to yield the elements of the perturbation matrix in the basis set of initial eigenfunctions in $|jm_j>$ forms of the operator $[\mathscr{H}_{cen+}\xi(r)l.s]$

Process(a)

The qualitative idea of the $|m_l m_s>$ components which are present in any particular $|jm_j>$ of a d electron can be obtained from the construction of the following type of tables. The construction is feasible with a little reflection remembering that for a single electron $j=l+s$ and $l-s$ and $m_j=m_l+m_s$.

Table $T_{3.1}$

$|m_l\ m_s>$ components in $|j\ m_j>$ of a d electron

J \ M_j	5/2	3/2	1/2	-1/2	-3/2	-5/2
5/2	$\|2^+>$	$\|2^->,\|1^+>$	$\|1^->,\|0^+>$	$\|0^->,\|-1^+>$	$\|-1^->,\|-2^+>$	$\|-2^->$
	(a)	(b)	(d)	(f)	(h)	(j)
3/2		(c)	(e)	(g)	(i)	
		$\|2^->,\|1^+>$	$\|1^->,\|0^+>$	$\|0^->,\|-1^+>$	$\|-1^->,\|-2^+>$	

For any single electron the ket with the highest values of j and m_j. i.e., $|j\ j>$ always involves a single $|m_l m_s>$.viz., $|l^+>$ in which $m_l = l$ Thus for an f- electron

$|7/2\ 7/2> \equiv |3^+>$ for a d-electron $|5/2\ 5/2> \equiv |2^+>$ for a p-electron $|3/2\ 3/2> \equiv |1^+>$ and so on. Using such quantitative forms of $|jj>$ expressed in $|1^+>$ we can find quantitatively the $|m_l m_s>$ components present in any $|j\ m_j>$ with the help of step-down operator j_- (Eq 1.4$'$). Starting with $|5/2\ 5/2>$ and allowing j_-, to operate on it one gets.

$$j_-|5/2\ 5/2> = (l_- + s_-)|2^+> \text{ since } j_- = (l_- + s_-)$$

or $$\sqrt{35/4 - 15/4}|5/2\ 3/2> = \sqrt{4}|1^+> + |2^->$$

i.e., $$|5/2\ 3/2> = \left(1/\sqrt{5}\ \right)|2^-> + \sqrt{4/5}|1^+> \quad \text{......... (3.33)}$$

which describes the $|j\ m_j>$ i.e., $|5/2\ 3/2>$ quantitatively as a linear combination of $|m_l m_s>'s$. In this way repeated operations by j_- can yield all the quantitative linear combinations of the kets in $|m_l m_s>$ form for each of the $|j\ m_j>$'s in upper half of the table T3.1. Each of the lower $|j\ m_j>'s$ of that table (corresponding to j = 3/2) can be obtained in $|m_l m_s>$ forms from the orthogonality relationship with respect to the $|j\ m_j>$ vertically above it. For example $|3/2\ 3/2>$, being similar to $|j\ mj>$ vertically above it. For example $|3/2\,3/2>$, being similar to $|5/2\ 3/2>$ in their qualitative compositions in $|m_l m_s>$ forms and additionally being orthogonal to the latter, will have the following quantitative linear combination of $|m_l m_s>'s$.

$$|3/2\ 3/2> = \sqrt{4/5}|2^-> - \left(1/\sqrt{5}\right)|1^+> \quad \text{......... (3.34)}$$

which is orthogonal to (3.33).

Applying the operator j_- to (3.34) or by applying the above principle of orthogonality, all the lower $|j\ m_j>'s$ can be obtained as linear combinations of $|m_l m_s>'s$.

We thus switch over (LSJM$_J$) scheme to (LSM$_L$M$_s$) scheme and adopt the procedure detailed in Sec. 3.2 to obtain the energyvalues and eigenfunctions of $\{[\mathscr{H}_{cen} + \xi(r)l.s] + (eH/2mc)(l_z + 2s_z)\}$operator. The perturbation matrix of $(eH/2mc)(l_z+2s_z)\}$in the basis set of the ten quantitative eigenfunctions, like those of (3.33), (3.34) of the operator $[\mathscr{H}_{cen} + \xi(r)l.s]$ will appear in block diagonal form provided we number the requisite eigenfunctions (a), (b), (c), (d) (e), (f), (g), (h), (i), (j) in a proper sequence. A little reflection will show that the pairs (bc), (de), (fg), (hi) will mix under (eH/2mc) (l_z+2s_z) [cf., Eqs 3.23 and 3.25] while a and j will remain separate. Thus the whole matrix will block diagonalize into two (1×1) and four (2×2) matrices yielding finally ten magnetically split levels (Zeeman levels) of the d-electron system. Details of the levels are worked out by process (b).

Process (b)

The process of finding the perturbational effect of the external magnetic field when the unperturbed eigenfunctions are given in the form of $|j, mj>'s$ depends on the value of the following special matrix element which can be rigorously established [2,3]

$$< l\ s\ j\ m_j | s_z | ls\ j\ m_j > = m_j \{j(j+1) - l(l+1) + s(s+1)\}/2j(j+1) \qquad \text{... (3.35)}$$

A more easily comprehensible treatment, avoiding the direct use of (3.35) is given in terms of "vector model" of an atom and precessional motions of the different vectors in external magnetic field and for this one may refer to any text on atomic spectra.[4,5]

We however, stick to the formal method, using (3.35), to find the general form of the matrix element of (*eH*/2*mc*) (l_z+2s_z) when the constant magnetic field is applied in the z direction.

$$< j' m_j' | (eH/2mc)(l_z + 2s_z) | j\ m_j >$$

$$= < j' m_j' | (eH/2mc)(j_z + s_z) | j\ m_j >$$

$$= \beta H \lfloor m_j + m_j \{j(j+1) - l(l+1) + s(s+1)\}/2j(j+1) \rfloor \delta_{j'j}\, \delta_{m_j' m_j}$$

$$= \beta H m_j \lfloor 1 + \{j(j+1) - l(l+1) + s(s+1)\}/2j(j+1) \rfloor \delta_{j'j}\, \delta_{m_j' m_j}$$

$$= g\beta m_j\ H \delta_{j'j}\, \delta_{m_j' m_j} \qquad \text{......... (3.36)}$$

wherein g, termed Lande *g* factor is given by

$$g = [1 + \{j(j+1) - l(l+1) + s(s+1)\}/2j(j+1)] \qquad \text{......... (3.37)}$$

Relation (3.36) indicates that the perturbation energy matrix is diagonal each value of j of the d-electron. The diagonal elements are written for economy of space in the form of row, for two possible values of j = 5/2 and 3/2

$$\left.\begin{array}{ll} \text{i) } g\beta H(5/2\ \ 3/2\ \ 1/2\ \ -1/2 -3/2 -5/2) & \text{for j= 5/2} \\ \text{ii) } g\beta H(3/2\ \ 1/2 -1/2 -3/2) & \text{for j = 3/2} \end{array}\right\} \qquad \text{......... (3.38)}$$

Therefore, the sixfold degenerate unperturbed level with relative energy ζ of the hamiltonian [$\mathscr{H}_{cen} + \xi(r) l.s$] of a-electron with j = 5/2 splits into six Zeeman levels and the fourfold degenerate eigenstate with relative energy $-(3/2)\zeta$ associated with j = 3/2 shakes off its degeneracy completely in external magnetic field. The diagram (Fig. 3.2) for Zeeman splitting appears below.

It is not necessary to know the actual initial energy value of a d^1 system regarded as an unperturbed eigenstate of $\mathscr{H}_{cen}$. Whatever it is, we set up an energy scale in which the zero level coincides with this initial state. This is referred to as an arbitrary scale in the diagram

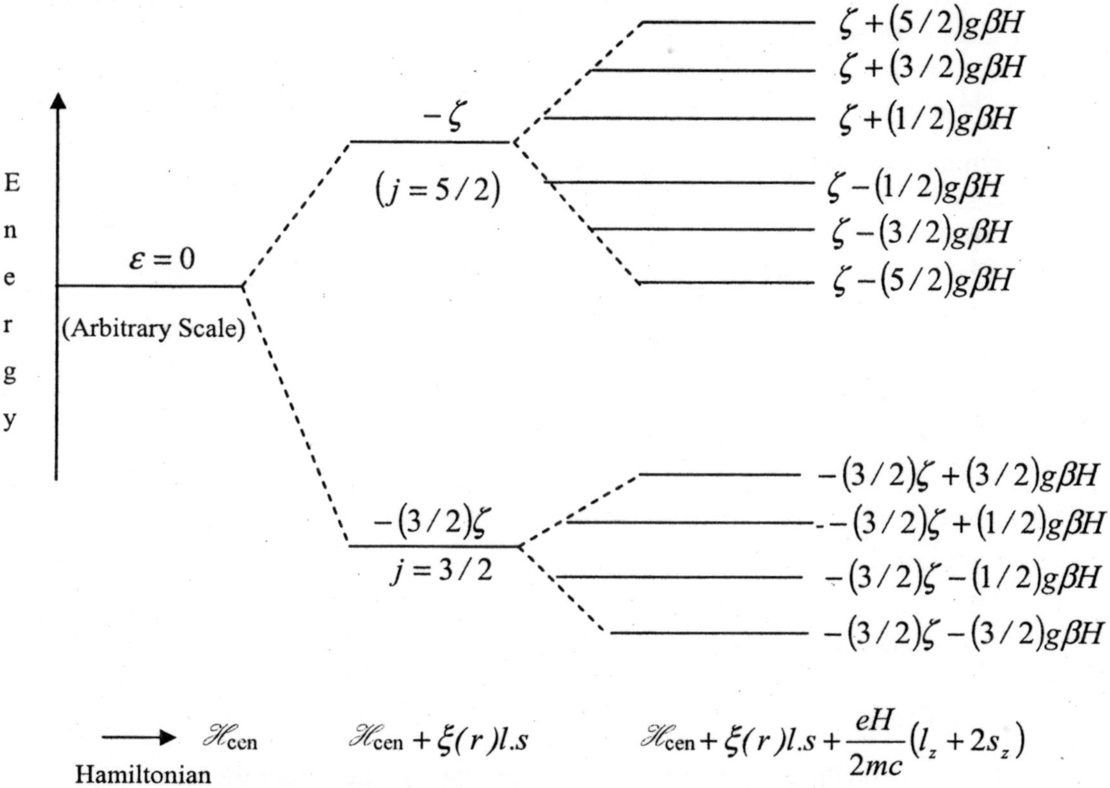

Fig. 3.2: Zeeman levels of single d-electron system

3.4. Energy Intervals of Split Levels (Alkali Metal Atoms and like Systems)

Spin orbit split levels

It is possible to lay down a general formulation for the energy gap between successive spin orbit levels of an atomic term (Eq. 3.22). The spin orbit interaction in one valence/optical electron atom (such as Li. Na) leads to a doublet term and the energy of each atomic doublet fine structure is dependent on the j value of the electron. The perturbation energy due to spin orbit interaction is given by

$$\varepsilon=(\zeta/2)[j(j+1)-l(l+1)-s(s+1)] \qquad \text{......... (3.17)}$$

For an atom with a single optical electron and belonging to a term ^{2s+1}L the electron has one of the two possible j values, viz., $(l+1/2)$ and $(l-1/2)$. From (3.17) it turns out

$$\Delta\varepsilon = \varepsilon_1 - \varepsilon_2 = (\zeta/2)l - [-(\zeta/2)(l+1)] = (\zeta/2)(2l+1) \qquad \text{........ (3.22)}$$

Zeeman level intervals:

Another type of splitting of each of the spin orbit levels due to interaction with the external magnetic fields, resulting in Zeeman fine structures, generates energy gaps between successive levels given by

$$\Delta E_{mag} = m_j gH\beta - m_j^{/} g\beta H = g\beta H, \text{ where } m_j = m_j^{/} + 1 \qquad \text{........ (3.38)}$$

The Zeeman levels, originating from single spin orbit level, are all equispaced and the successive energy gaps are proportional to g. the Lande' *g*-factor of the electron and the applied magnetic field *H* in the *z*-direction. Of course, the statement is valid if we neglect second order correction terms as exhibited by Eqs. 3.31 and 3.32.

Exercises

1. The two fine structure lines in the yellow doublet of sodium belong to the principal series and have the wavelengths $\lambda_1 = 5889$ Å and $\lambda_2 = 5863$ Å respectively, Calculate the spin orbitcoupling constant ζ of sodium. Will the coupling constant vary for different fine-structure levels of the same alkali metal?

2. Establish the matrix element relation

   $$< lsjm_j | s_z | lsjm_j > = m_j \{ j(j+1) - l(l+1) + s(s+1) \} / 2j(j+1)$$

3. Derive the energy expression of the magnetic fine structure levels (Zeeman levels) for one-optical electron system on the basis of "Vector model theory of an atom."

References

1. J. I. Powell and B. Crasemann : Quantum Mechanics (Addison Wesley)
2. E.U. Condon and G H. Shortley : Theory of Atomic Spectra (Cambridge University Press)
3. R. M. Golding ; Applied Wave Mechanics (Van Nostrand)
4. L. Pauling and S. Goudsmit : The Structure of Line-Spectra (Mc Graw-Hill)
5. H.E. While : Introduction to Atomic Spectra (Mc Graw-Hill)

Chapter Four

ATOMS WITH TWO OR MORE VALENCE/OPTICAL ELECTRONS. PERTURBATIONAL EIGENFUNCTIONS AND ENERGY EIGENVALUES

(Electrostatic and Magnetic Perturbations: Russell Saunder's Terms, Zeeman Levels)

4.1. General

Electrons, present outside filled up closed shells and sub-shells of atoms though normally called valence electrons, are also known as optical electrons since these are the ones that usually participate in electronic transitions. In a multi-electron atom, electrons in completely filled up shells and sub-shells give rise to a singlet S state (^{1}s). The net state of an atom is, therefore, decided by the nature of coupling and the number of optical electrons outside the fully filled up shells and sub-shells.

The spectral energy levels and terms of the alkali metals or of the singly ionised alkaline earth metals are all due to one optical electron. In view of Eqn. (2.25) of section 2.3, no Russell Saunders' coupling affects the one optical electron of these species or their spectral features irrespective of the type of coupling prevalent for the inner core electrons. Thus atoms or ions with two or more optical electrons (Ca, Al or transition series elements) come under our scrutiny here.

The electrostatic intra-perturbation, in the sense conveyed by Eqn. (2.25) and not by the total $\Sigma_{i<j} e^2 / r_{ij}$ of Eqn. (2.1), constitutes the real perturbation hamiltonian responsible for the splitting of the energy levels in the primary coupled states (prior to the secondary magnetic LS coupling) relative to that governed by $\mathscr{H}_{\text{cent}}$ (Eqn. 2.25) of the hypothetical unperturbed state of the atom, the major part of the original $\sum_{i<j} e^2 / r_{ij}$ having been absorbed in the latter. This discussion suggests that for calculational purposes we start with the matrix of the perturbation hamiltonian $\sum_{k<l} e^2 / r_{kl}$ (k&l optical electrons) in a basis set of suitable determinants describing the unperturbed atom. It means we just allow ourselves to be concerned with the effect of the primary coupling part of the Russell Saunders' scheme. The second stage of weak perturbation, due to magnetic LS coupling, of the primarily coupled states will be taken up in a later section of this chapter.

4.2 Electrostatic Perturbation in Multielectron Atom. Russell Saunders' Energy Levels and Eigenstates

To make our presentation clear and more concrete, we shall calculate the energy levels of some of the atomic terms yielded by Russell Saunders' coupling of the equivalent electronic configuration (p^2), (d^2), (d^3) and of the non equivalent configurations (np, n′p). At first two theorems on matrix elements for construction and solution of the energy matrices will be stated.

Theorem 1

If two hermitian operators α and β commute, then the matrix element of β between two different eigenstates of α with different eigenvalues vanishes i.e.

$<\phi_1|\beta|\phi_2 = 0$. if $[\alpha.\beta]=0$ and $\alpha\phi_1 = \lambda_1\phi_1 . \alpha\phi_2 = \lambda_2\phi_2$ such that $\lambda_1 \neq \lambda_2$

The proof runs as follows

$$<\phi_1|\alpha\beta|\phi_2> = <\phi_1|\beta\alpha|\phi_2> = \lambda_2 <\phi_1|\beta|\phi_2>$$

Again

$<\phi_1|\alpha\beta|\phi_2> = <\alpha\phi_1|\beta|\phi_2> = \lambda_1 <\phi_1|\beta|\phi_2>$ since α is hermitian and has, therefore real eigenvalues.

$\therefore \lambda_2 <\phi_1|\beta|\phi_2> = \lambda_1 <\phi_1|\beta|\phi_2>$ whence it follows that ......(4.1)

$<\phi_1|\beta|\phi_2> = 0$ since $\lambda_1 \neq \lambda_2$

Theorem 2

The trace (character) of a square matrix equals the sum of the eigenvalues of the matrix.

This theorem is called the diagonal sum rule and signifies

$$\chi(\mathbf{A}) = \Sigma_i a_{ii} = \Sigma_j \varepsilon_j \qquad ......(4.2)$$

where ε_j's are the eigenvalues of **A**.

This theorem is easy to prove remembering that a proper unitary (similarity) transformation generates the diagonal eigenvalue matrix and that the trace of a matrix remains unaltered under a unitary transformation.

i) (p^2) electronic configuration : Russell Saunders' energy levels and eigenstates:

We know from see 1.5.2 that a (p^2) configuration gives rise to ^3p, ^1D and a ^1S terms, one each. The overall ket forms of atomic terms in (L S M_l M_s) scheme are:

For $^1D \; |2020>, |2010>, |2000>, |20-10>, |20-20>$

For $^3P \; |1111>, \; |1110> . |111-1>$

$|1101>, \; |1100>, |110-1>$

$|11-11>, |11-10>, |11-1-1>$ *and*

for $^1S \qquad |0000>,$

The (LSM_LM_S) scheme is chosen because L and S serve more or less as good quantum numbers even after the primary coupling in the Russell Saunders' scheme and the states remain degenerate to within each specific value of J. In principle, these fifteen kets which are eigenkets of $\mathscr{H}_{cent.}$ should form the basis for the construction of the perturbation matrix of $\Sigma_{k<l} e^2/r_{kl}$. k, l etc being the optical electrons. But the task may be simplified by considering a smaller number of appropriate basis functions directed by theorem I (sec 4.2) instead of the total fifteen kets. It should be noted that since

$\left[\Sigma_{k<l} e^2/r_{kl}, S_z\right] = 0$ and also

$\left[\Sigma_{k<l} e^2/r_{kl}, L_z\right] = 0$. there can be no matrix element of $\Sigma_{k<l} e^2/r_{kl}$ between any two basis kets with different values of either M_L or M_s or both where M_L and M_S are the eigenvalues of L_z and S_z respectively (cf. Theorem 1).

It thus becomes apparent that, instead of dealing with a single (15×15) energy matrix, we may consider a number of small sized perturbation matrices of $\Sigma_{k<l} e^2/r_{kl}$ in the different basis sets of which no two or more kets of ^{1}D unperturbed atomic terms nor any two or more kets of ^{3}P can occur together in any single basis set. But specific D, P and S kets with identical values of both M_L and M_S occur in a basis set of an energy matrix.

A table is, therefore, given below showing the distribution of the foregoing fifteen two electron kets under different possible pairs of M_L, M_s values for (p^2) configuration alongwith the term symbols of the kets in each cell.

Table $T_{4.1}$

(Distribution of kets of (p^2) configuration under pairwise values of M_L and M_S)

M_L \ M_s	1	0	-1
2		$\|2020$ ^{1}D	
1	$\|1111>$ $\quad ^3P$	$\|2010>, \|1110>,$ $\quad ^1D, ^3P$	$\|111-1>$ $\quad ^3P$
0	$\|1101>$ $\quad ^3P$	$\|2000>, \|1100>, \|1000>$ $^1D, ^3P, ^1S$	$\|110-1>$ $\quad ^3P$
-1	$\|11-11>$ $\quad ^3P$	$\|20-10>, \|11-10>,$ $\quad ^1D, ^3P$	$\|11-1-1>$ $\quad ^3P$
-2		$\|20-20>,$ $\quad ^1D$	

In principle the ket contents of each individual cell of the above table ($T_{4.1}$) jointly can serve as separate basis sets for the different perturbation matrices. Thus there would be eight (1×1), two (2×2) and one (3×3) perturbation matrices. On little reflection, we can construct a table of the det forms of the contents (viz., the kets) of each cell of Table $T_{4.1}$. In constructing this table it is necessary to remember that the electrons are equivalent p-electrons, that Pauli's exclusion, principle is to be obeyed and the single- electron m_l m_s values are indicated in the dets.

Table $T_{4.2}$

(Microstate dets of (P^2) configuration under pairwise values of M_L, M_S

M_L \ M_s	1	0	-1
2		^1D ($1^+ 1^-$) A	
1	^3P ($1^+ 0^+$) B	^1D, ^3P ($1^+ 0^-$), ($1^- 0^+$) C	^3P ($1^- 0^-$) D
0	^3P ($1^+ -1^+$) E	^1D, ^3P, ^1S ($1^+ -1^-$), ($1^- -1^+$), (0^+0^-) F	^3P ($1^- -1^-$) G
-1	^3P ($0^+ -1^+$) H	^1D,.^3P ($0^+ -1^-$), ($0^- -1^+$) I	^3P ($0^- -1^-$) J
-2		^1D ($-1^+ -1^-$) K	

This changeover from table $T_{4.1}$ to table $T_{4.2}$ is necessary because the general result of the matrix element of the electrostatic repulsion operator $\sum_{k<l} e^2/r_{kl}$ are given (Eqns 2.77, 2.78) for the atomic states expressed in determinantal forms.

Energy values

To evaluate perturbational energies, there are two avenues. Firstly we can construct the perturbation matrices of the operator $\sum_{k<l} e^2/r_{kl}$ choosing the different basis sets from table $T_{4.2}$ and solve for the energy eigenvalues of these matrices to obtain the energies of ^1D, ^3P and ^1S Russell Saunders' terms relative to that of the unperturbed state. In so doing one has to use c^k/a^k, b^k table ($T_{2.1}$ and $T_{2.2}$).

The second process skirts the construction of actual matrices. Instead, it uses the 'diagonal sum rule' of theorem 2 to calculate energies. Noting the contents of the different cells of Table $T_{4.2}$ and applying the diagonal sum rule for the eigenvalue of the matrices (same as the solution of the dets) we may write,

considering cells A, B, F and Eqns. (2.78, 2.80, 2.80a),

$$\varepsilon(^1D) = \varepsilon(1^+1^-) = J(1,1) \qquad \ldots\ldots(4.3a')$$

$$\varepsilon(^3P) = \varepsilon(1^+0^+) = J(1,0) - K(1,0) \qquad \ldots\ldots(4.3b')$$

$$\varepsilon(^1D) + \varepsilon(^3P) + \varepsilon(^1S) = \varepsilon(1^+ -1^-) + \varepsilon(1^- -1^+) + \varepsilon(0^+0^-)$$

$$= J(1,-1) + J(1,-1) + J(0,0) \qquad \ldots\ldots(4.3c')$$

Subtracting (4.3a′) and (4.3b′) from (4.3c′)

$$\varepsilon(^1S) = J(1,-1) + J(1,-1) + J(0,0) - J(1,1) - J(1,0) + K(1,0) \qquad \text{......(4.3d′)}$$

Using c^k table (T $_{2.1}$) and the foregoing equations one gets

$$\varepsilon(^1D) = J(1,1) = \Sigma_k\, c^k(11,11)c^k(11,11)F^k$$

$$= F^0 + 1/25F^2 = F_0 + F_2 \qquad \text{......(4.3a)}$$

$$\varepsilon(^3P) = J(1,0) - K(1,0)$$

$$= \Sigma_k\, c^k(11,11)c^k(10,10)F^k - \Sigma_k c^k(11,10)c^k(11,10)F^k$$

$$= F_0 - 5F_2 \qquad \text{......(4.3b)}$$

$$\varepsilon(^1s) = J(1,-1) + J(1,-1) + J(0,0) - \varepsilon(^1D) - \varepsilon(^3P)$$

$$= \Sigma_k\, c^k(11,11)c^k(1-1,1-1)F^k + \Sigma_k\, c^k(11,11)c^k(1-1,1-1)F^k$$

$$+ \Sigma_k\, c^k(10,10)c^k(10,10)F^k - (F_0 + F_2) - (F_0 - 5F_2)$$

$$= 3F_0 + 6F_2 - F_0 - F_2 - F_0 + 5F_2 = F_o + 10F_2 \qquad \text{......(4.3c)}$$

The net energy levels relative to the main energy of the unperturbed level $(\varepsilon = 0)$ of the primary Russell Saunders' terms of (p^2) configuration are shown

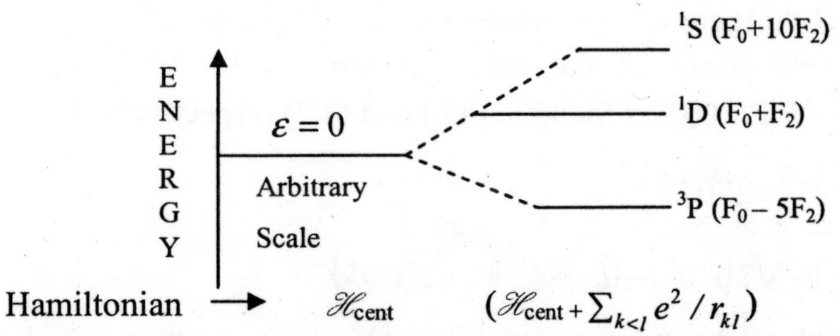

[Fig. 4.1 : Primary Russell Saunders' terms of (P^2) configuration]

Wavefunctions : (p^2 configuration)

The eigenstates of $(\mathscr{H}_{\text{cent}} + \sum_{k<l} e^2/r_{kl})$ i.e., the primary Russell Saunders' states, though obtainable from solutions of simultaneous equations originating from the energy matrices, are more easily derived by using the stepup and stepdown operators (1.4 Sec. 1.1) and the property of mutual orthogonality of wave functions.

It is evident from table $T_{4.2}$ that for each cell occupied by a single det, the latter is a wavefunction of that term (cells A, B, D, E, G, H, J, K), e.g.,

$$\psi(^1D) = |2020> \equiv (1^+1^-) \qquad \text{......4.4}$$

$$\psi(^3P)=|1111>\equiv(1^+0^+) \qquad \ldots\ldots 4.5$$

and so on for the others.

Now consider the cell C. Let us find the wavefunction corresponding to ^1D and ^{3}P that would emerge from this cell. Applying L_- to $\psi(^1D)$i.e., on the $|2020>$ of (4.4), we get, via Eqn. (1.53), after elimination $\hbar$ from both sides,

$$L_-\psi(^1D) = L_-|2020> = (l_{1-}+l_{2-})\{(1^+1^-)\}$$

$$\text{or}\sqrt{4}|2010> = \sqrt{2}(0^+1^-)+\sqrt{2}(1^+0^-)$$

$$\therefore \psi\{^1D,|2010>\} = 1/\sqrt{2}\{(1^+0^-)-(1^-0^+)\} \qquad \ldots\ldots(4.6)$$

The ^{3}P wavefunction of this cell (cell c) can be readily written down as it ought to be orthogonal to $\psi(^1D)$ of (4.6) above

$$\therefore \psi\{^3P,|1110>\} = 1/\sqrt{2}\{(1^+0^-)+(1^-0^+)\} \qquad \ldots\ldots(4.7)$$

Alternatively, the latter wave function can also be derived by allowing spin step down operator S_- to act on $\psi\{^3P,|1111>\}$ of cell B of table $T_{4.2}$.

To continue with further illustrations of calculation of wavefunctions, let us focus our attention on cell F. There are 3 wavefunctions, $\psi(^1D)$, $\psi(^3P)$, $\psi(^1S)$ built up with linear combinations of the dets appearing in F. Apply L_- to $\psi\{^1D,|2010>\}$ and to $\psi\{^3P|1110>\}$ as found in (4.6) and (4.7) respectively. Thus

$$L_-|2010> = L_- 1/\sqrt{2}\{(1^+0^-)-(1^-0^+)\}$$

$$\text{i.e., } \sqrt{6}|2000> = (l_{1-}+l_{2-})1/\sqrt{2}(1^+0^-)-(l_{1-}+l_{2-})1/\sqrt{2}(1^-0^+)$$

$$\text{or } \sqrt{6}|2000> = 1/\sqrt{2}\{\sqrt{2}(0^+0^-)+\sqrt{2}(1^+-1^-)-\sqrt{2}(0^-0^+)-\sqrt{2}(1^--1^+)\}$$

$$\psi\{^1D|2000>\} = 1/\sqrt{6}\{2(0^+0^-)+(1^+-1^-)-(1^--1^+)\} \qquad \ldots\ldots(4.8)$$

Similarly from (4.7)

$$L_-|1110> = (l_{1-}+l_{2-})1/\sqrt{2}\ \{(1^+0^-)+(1^-0^+)\}$$

$$\sqrt{2}|1100> = 1/\sqrt{2}\{\sqrt{2}(0^+0^-)+\sqrt{2}(1^+-1^-)+\sqrt{2}(0^-0^+)+\sqrt{2}(1^--1^+)\}$$

$$\psi\{^3P|1100>\} = 1/\sqrt{2}\{(1^+-1^-)+(1^--1^+)\} \qquad \ldots\ldots(4.9)$$

The third wavefunction $\psi\{^1S|0000>\}$, which ought to be orthogonal both to (4.8) and (4.9) can be directly written down on a little reflection even if one avoids the formal mathematical route. This function is

$$\psi\left\{{}^1S,|0000>\right\}=1/\sqrt{3}\left\{\left(0^+0^-\right)-\left(1^+-1^-\right)+\left(1^--1^+\right)\right\} \quad ......4.10$$

Before passing onto examples of other electronic configurations two general statements may be made.

1. For calculating the energy levels we need to construct the M_L M_S – microstate det tables (cf table $T_{4.2}$) only upto the row. $M_L = 0$. The lower half of the table can be inferred from the upper moiety.
2. For arriving at wavefunction of a specific term, one should start from the known det form of the wavefunction made evident by a single det containing cell of a suitable atomic term. One should then make either

a) vertical sojourns up or down the table with L_+ or L_- operators respectively or

b) horizontal sojourns rightwards or leftwards along the table with S- or S_+ operators

respectively till one reaches the target cell. Sometimes both vertical and horizontal sojourns may be necessary.

ii) (d^2) electronic configuration

This equivalent electronic configuration gives rise to ^3F, ^3P, ^1G, ^1D and ^1S terms (sec 1.5.2). To illustrate the calculation of energy let us compute the energy gap between ^1S and ^1D terms and for wavefunctions, we shall choose one of the degenerate ground level terms, $|3120>$. of ^3F. We start from the following half compiled M_L M_S- microstate det table $T_{4.3}$.

Table $T_{4.3}$ M_L M_s pair values and microstate dets of (d^2) configuration

M_L \ M_S	1	0	-1
4		^1G $(2^+ 2^-)$ A	
3	^3F $(2^+ 1^+)$ B	^1G, ^3F $(2^+ 1^-)$, $(2^- 1^+)$ C	^3F $(2^- 1^-)$ D
2	^3F $(2^+ 0^+)$ E	^1G, ^3F, ^1D $(2^+ 0^-)$, $(2^- 0^+)$, (1^+1^-) F	^3F $(2^- 0^-)$ G
1	^3F ^3P $(2^+ -1^+)$ (1^+0^+) H	^1G, ^3F ^3P $^1D$$(2^+ -1^-)$, $(2^- -1^+)$ (1^+0^-) (1^-0^+) I	^3F ^3P $(2^- -1^-)$ (1^-0^-) J
0	^3F ^3P $(2^+ -2^+)$ $(1^+ -1^+)$ K	^1G, ^3F ^3P $^1D$$^1S$ $(2^+ -2^-)$, $(2^- -2^+)$ $(1^+ -1^-)$ $(1^- -1^+)$ (0^+0^-) L	^3F ^3P $(2^- -2^-)$ $(1^- -1^-)$ M

From table $T_{4.3}$ we can write the following energy relations, using the diagonal sum rule (Theoem 2 sec 4.2)

Cell A $\varepsilon({}^1G)=\varepsilon(2^+2^-)$

Cell B $\varepsilon({}^3F)=\varepsilon(2^+1^+)$

Cell F $\varepsilon({}^1G)=\varepsilon({}^3F)+\varepsilon({}^1D)=\varepsilon(2^+0^-)+\varepsilon(2^-0^+)+\varepsilon(1^+1^-)$

$$\therefore \varepsilon({}^1D)=\varepsilon(2^+0^-)+\varepsilon(2^-0^+)+\varepsilon(1^+1^-)-\varepsilon(2^+2^-)-\varepsilon(2^+1^+) \qquad ......(4.11)$$

Using Eqns. (2.78,2.80 and 2.81)

$$\varepsilon({}^1D)=J(2{,}0)+J(2{,}0)+J(1{,}1)-J(2{,}2)-J(2{,}1)+K(2{,}1)$$

We now write the c^k equivalents using only the m_l's in the parentheses. Thus

$$\varepsilon({}^1D)=\Sigma_k\, c^k(22)\,c^k(00)F^k+\Sigma_k\, c^k(22)\,c^k(00)F^k+\Sigma_k\, c^k(11)c^k(11)F^k$$
$$-\Sigma_k\, c^k(22)\,c^k(22)F^k-\Sigma_k\, c^k(22)\,c^k(11)F^k+\Sigma_k\, c^k(21)c^k(21)F^k$$
$$=F_0-3F_2+36F_4 \qquad ......(4.12)$$

to arrive at which table $T_{2.1}$ Eqns (2.81) and (2.82) have been used.

To find $\varepsilon({}^1S)$ we work upon the cells 1 and L to find energy as

$$\varepsilon({}^1S)=F_0+14F_2+126F_4 \qquad ......(4.13)$$

whence, from (4.12) and (4.13)

$$\varepsilon({}^1S)-\varepsilon({}^1D)=17F_2+90F_4=17B+5C \qquad ......(4.14)$$

The final part of the present assignment is to find the wavefunction of the ^{3}F term having two-electron ket symbol $|3120>$. Starting from the apparently evident wavefunction of ^{3}F, having the ket viz., $|3121>$ of cell E and allowing S_- to operate we get

$$S_-\psi\{{}^3F,|3131>\}=(s_{1-}-s_{2-})\{(2^+0^+)\}$$
$$\sqrt{2}|3120>=\sqrt{3/4+1/4}(2^-0^+)+\sqrt{3/4+1/4}(2^+0^-)$$
$$\text{Thus } \psi\{{}^3F,|3120>\}=1/\sqrt{2}\,\{(2^+0^-)+(2^-0^+)\} \qquad ......(4.15)$$

iii) (d^3) electronic configuration

It is seen from the foregoing discussion that the electrostatic perturbation calculations in Russell Saunders' coupling depend upon prior construction of a suitable M_L M_S- microstate det' table. This job becomes a lengthy and laborious process as the number of optical electrons increases. All the same the process remains unchanged. For (d^3) configuration, there occur both doublets and quartets Russell Saunders' terms (sec 1.5.2) Amongst the doublets is a ^{2}H term and there is a ^{4}F, the ground state, among the quartets. Let us find the wavefunctions and energies of two atomic kets, viz., $|3\ 3/2\ 3\ 3/2>$ of ^{4}F and $|5\ 1/2\ 4\ 1/2>$ belonging to ^{2}H of d^3 configuration.

Since 3 electrons are involved, the microstate det form of ${}^4F$ corresponding to $M_s = 3/2$, i.e., of the given ket $|3\ 3/2\ 3\ 3/2>$ is

$$\psi\{{}^4F,|3\ 3/2\ 3\ 3/2>\} = (2^+1^+0^+)$$

Employing Eqn. (2.78) where summation extends over pairs of spin orbitals (or equivalently over pair of electrons) and using c^k table ($T_{2.1}$) one gets

$$\varepsilon\{{}^4F.|3\ 3/2\ 3\ 3/2\rangle\} = \langle(21)|1/r_{12}|(21)\rangle + \langle(20)|1/r_{13}|(20)\rangle + \langle(10)|1/r_{23}|(10)\rangle$$

$$= <21|1/r_{12}|21> + <20|1/r_{13}|20> + <10|1/r_{23}|10>$$

$$= <21|1/r_{12}|12> - <20|1/r_{13}|02> - <10|1/r_{23}|01>$$

$$= J(2.1) - K(2.1) + J(2.0) - K(2.0) + J(1.0) - K(1.0)$$

$$= \Sigma_k c^k(2.2)c^k(1.1)F^k - \Sigma_k c^k(2.1)c^k(2.1)F^k + \Sigma_k c^k(2.2)c^k(0.0)F^k$$

$$- \Sigma_k c^k(2.0)c^k(2.0)F^k + \Sigma_k c^k(1.1)c^k(0.0)F^k - \Sigma_k c^k(1.0)c^k(1.0)F^k$$

$$= 3F_0 - 15F_2 - 72F_4 = 3A - 15B \qquad ......(4.16)$$

The wavefunction of this state is evidently $(2^+\ 1^+0^+)$. Turning to the other atomic ket. viz., $|5\ {}^1\!/_2\ 4\ {}^1\!/_2>$ of ${}^2H$ state, we note that it is the immediate lower rung member of the ket $\{{}^2H,|5\ {}^1\!/_2\ 5\ {}^1\!/_2>\}$ The latter has only following microdet

$$\psi\{{}^2H,|5\ {}^1\!/_2\, 5\ {}^1\!/_2> = (2^+2^-1^+) \qquad ......4.17$$

$$L_-|5\ {}^1\!/_2\, 5\ {}^1\!/_2> \ = (l_{1-} + l_{2-} + l_{3-})\{2^+2^-,1^+)\}$$

i.e., $\sqrt{10}\,|5\ {}^1\!/_2\, 4\ {}^1\!/_2> = \sqrt{4}(1^+2^-1^+) + \sqrt{4}(2^+1^-1^+) + \sqrt{6}(2^+2^-0^+)$

The first det on RHS is inadmissible (Pauli's exclusion principle).

$$\therefore \psi\{{}^2H,|5\ {}^1\!/_2\, 4\ {}^1\!/_2>\} = \sqrt{2/5}(2^+1^-1^+) + \sqrt{3/5}(2^+2^-0^+) \qquad ......4.18$$

Having found the form of the required wavefunction (4.18) we may evaluate the energy of the ${}^2H$ state either by using (4.17) or (4.18), since the state is degenerate within the electrostatic perturbation and the energy is independent of M_L and M_s values. Choosing the simple *r* mode (namely det 4.17) using Eqn. (2.78), remembering three electrons are involved and applying c^k table ($T_{2.1}$) we have

$$\varepsilon\{{}^2H,|5\ {}^1\!/_2\, 5\ {}^1\!/_2>\} = <2^+2^-|1/r_{12}|2^+2^-> + <2^+1^+|1/r_{13}|2^+1^+> + <2^-1^+|1/r_{23}|2^-1^+>$$

$$- <2^+1^+|1/r_{13}|1^+2^+>$$

$$=J(2.2)+J(2.1)+J(2.1)-K(2.1)$$

$$=\Sigma_k c^k(2,2)\,c^k(2,2)F^k+\Sigma_k c^k(2,2)\,c^k(1,1)F^k+\Sigma_k c^k(2,2)\,c^k(1,1)F^k$$

$$-\Sigma_k c^k(2,1)\,c^k(2,1)F^k$$

$$=3F_0-6F_2-12F_4 \quad = \quad 3A-6B+3C \qquad \ldots\ldots(4.19)$$

iv) (np, n′p) nonequivalent electronic configuration

The Russell Saunders' terms possible for such electronic configuration are ^3D, ^3P, ^3S and ^1D, ^1P and ^1S (cf. Sec 1.5.l). Since the electrons have different n's, all possible pairs of m_l, m_s, values for the two electrons, such as (1^+1^+), (1^+1^-) (1^-1^+) etc., are permissible. Taking this into consideration, the 'M_LM_S- microstate det' table is constructed which appears below

Table $T_{4.4}$

$M_L M_S$ - Microstate dets of (np, n′p) electronic configuration

M_L \ M_S	1	0	-1
2	(1^+1^+) A	(1^+1^-), (1^-1^+) B	(1^-1^-) C
1	(1^+0^+), (0^+1^+) D	(1^+0^-), (1^-0^+), (0^+1^-), (0^-1^+) E	(1^-0^-),(0^-1^-) F
0	(0^+0^+), (1^+-1^+), (-1^+1^+) G	(0^+0^-), (0^-0^+), (1^+-1^-), (1^--1^+) (-1^+1^-), (-1^-1^+) H	(0^-0^-),(1^--1^-) (-1^-1^-) I
-1	(-1^+0^+) (0^+-1^+) J	(-1^+0^-), (-1^-0^+) (0^+-1^-), (0^--1^+) K	$(-1^-0^-)(0^--1^-)$ L
-2	(-1^+-1^+) M	(-1^+-1^-) (-1^--1^+) N	(-1^--1^-) O

Let us find the wavefunctions and energies of the states relevant to cell F of table ($T_{4.4}$). The atomic terms that are involved are ^3D and ^3P having the kets $|211-1>$ and $|111-1>$ respectively.

Det form ^3D viz. $|211-1>$

$$\psi\{^3D,|212-1>\}=(1^-1^-)\text{ cell C}$$

$$L_-|212-1> \;=(l_{1-}+l_{2-})\{(l^-1^-)\}$$

i.e., $$\sqrt{4}|211-1> \;=\sqrt{2}(0^-1^-)+\sqrt{2}(1^-0^-)$$

$\therefore$ $$\psi\{^3D,|211-1>=1/\sqrt{2}\{(1^-0^-)+(0^-1^-)\}\} \qquad \ldots\ldots(4.20)$$

Since the ^{3}P component of cell F must be orthogonal to ^{3}D component i.e., (4.20)

$$\psi\{^3p,|111-1>=1/\sqrt{2}\{(1^-0^-)-(0^-1^-)\}\qquad\ldots\ldots(4.21)$$

The wavefunctions having thus been found, we can now apply the diagonal sum rule for energy evaluation. Thus

$$\varepsilon(^3D) = \varepsilon(1^-1^-) \qquad \text{cell C}$$

$$\varepsilon(^3D)+\varepsilon(^3P) = \varepsilon(1^-0^-)+\varepsilon(0^-1^-)$$

$$\therefore\ \varepsilon(^3P) = \varepsilon(1^-0^-)+\varepsilon(0^-1^-)-\varepsilon(1^-1^-)$$

Thus the diagonal matrix elements of (1$^-$ 1$^-$), (1$^-$ 0$^-$), (0$^-$ 1$^-$), when the perturbation operator is e^2/r_{kl} can all be evaluated from the c^k table thus furnishing the energy data for ^{3}D and ^{3}P states.

An alternative way of calculating the $\varepsilon\left(^3P\right)$is to find the diagonal matrix element of the wavefunction (4.21) with respect to the perturbation operator e^2/r_{kl}

4.3. The Spin Orbit Intraperturbation: The Second Stage in Russell Saunders' Coupling – General

The two – or more than two optical electron systems (atoms) undergoing Russell Saunders' coupling are inherently subject to two types of intraperturbations. The main one, especially for lighter elements and first transition series elements, is the electrostatic perturbation. $\Sigma_{k>l}e^2/r_{kl}$ and the relatively weak one is the spin orbit magnetic perturbation, $\Sigma_i\,\xi(r_i)l_i, s_i\ \{Eqn.2.31]or(\lambda/\hbar^2)$L.S [Eqn. 2.32]. In the previous section we have already explained in detail as to how the electrostatic perturbation energies and wavefunctions of the Russell Saunders' state at the primarily coupled stage can be obtained. The attainment of these perturbed energy values along with their corresponding wavefunction is equivalent to stating that the coulomb perturbation matrix has been diagonalised or block diagonalised.

The previously obtained atomic terms, where L and S serve as good quantum numbers, are all degenerate within the J values. This degeneracy is lifted to give rise to fine structures of the atomic terms in the second stage of weak coupling due to spin orbit (LS) interaction. We shall consider each primarily coupled Russell Saunders' term [eg, ^{3}P or ^{1}D of (p^2) configuration] as an initial state to which LS perturbation is applied.

Before proceeding with the formal treatment of this second perturbation, we must clarify a number of points.

i) Taking the case of ^{1}D term, it is orbitally fivefold degenerate. Hence the perturbation matrix is of (5×5) order. If we continue to work within the (LSM$_L$M$_s$)

scheme of wavefunctions used in the primarily coupled stage of Russell Saunders' terms, the perturbation hamiltonian, $\Sigma_i \xi(r_i) l_i, s_i$ does not commute with $\Sigma_{k<l} e^2 / r_{kl}$ nor with L_z of which the initial wavefunctions are eigenfunctions. Therefore, there cannot be any splitting of the (5×5) dimensional energy matrix into a number of smaller dimensional ones to facilitate solutions on the basis of the previously stated theorem (Theorem 1 sec. 4.2). But, as we shall (see sec 4.4), there is a way to circumvent it.

ii) Although it is theoretically possible for a P state to mingle under perturbation with D state, we shall neglect such contribution coming from P basis sets. Note that electronstatic perturbation allows such mixing (cf. Table $T_{4.2}$ cell F).

iii) A higher (n' = high) quantum level D state can mix with a low quantum level (n = low) D term under $\Sigma_i \xi(r_i) l_i.s_i$. But the net contribution to perturbation energy emanating form such mixing is so low as to warrant its neglect in a simple treatment of the spin orbit interaction.

As we shall primary be interested in equivalent electronic configurations, the wavefunctions of the Russell Saunders' atomic terms caused by electrostatic perturbation of (p^2) configuration are assembled in table ($T_{4.5}$) and a few of those of (d^2) configuration in table ($T_{4.5}$) for our immediate and future use. The results are compiled by following the procedure enunciated in sec 4.2

Table$_{4.5}$ Wavefunctions and Energies of Primary Russell Saunders' terms (p^2) Configuration (Unperturbed Central field energy equal to zero)

Atomic Term		Atomic Ket $\lvert LSM_L M_S >$	Wavefunctions in det forms	Energy (Relative to unperturbed level)
^1S		$\lvert 0000>$	$1/\sqrt{3}\{\lvert 0^+0^-\rvert - \lvert 1^+ -1^-\rvert + \lvert 1^- -1^+\rvert\}$	$F_0 + 10F_2$
	a	$\lvert 1111>$	(1^+0^+)	
	b	$\lvert 1110>$	$1/\sqrt{2}\{\lvert 1^+0^-\rvert + \lvert 1^-0^+\rvert\}$	
	c	$\lvert 111-1>$	(1^-0^-)	
	d	$\lvert 1101>$	(1^+-1^+)	
^3P				F_0-5F_2 (ninefold degenerate)
	e	$\lvert 1100>$	$1/\sqrt{2}\{\lvert 1^+ -1^-\rvert + \lvert 1^-1^+\rvert\}$	
	f	$\lvert 110-1>$	$(1^- -1^-)$	
	g	$\lvert 11-11>$	$(0^+ -1^+)$	
	h	$\lvert 11-10>$	$1/\sqrt{2}\{\lvert 0^+ -1^-\rvert + \lvert 0^-1^+\rvert\}$	
	i	$\lvert 11-1-1>$	$(0^- -1^-)$	

(contd.)

^1D	a	$\|2020>$	$(1^+ 1^-)$	F_0+F_2 (Fivefold degenerate)
	b	$\|2010>$	$1/\sqrt{2}\{(1^+0^-)-(1^-0^+)\}$	
	c	$\|2000>$	$1/\sqrt{6}\{(1^+ -1^-)-(1^- -1^+)+2(0^+0_-)\}$	
	d	$\|20-10>$	$1/\sqrt{2}\{(0^+ -1^-)-(0^- -1^+)\}$	
	e	$\|20-20>$	$(-1^+ -1^-)$	

Table $T_{4.6}$

A few Wavefunctions and energies of primary Russell Saunders' terms (d^2) Configuration (Energy relative to unperturbed state energy equal to zero)

$A= F_0-49F_4$ $B = F_2-5F_4$. $C = 35F_4$

Atomic Term	Some specific atomic kets	Wavefunctions in det forms	Energy (Relative to unperturbed level)
^1G	$\|4040>$	(2^+2^-)	$F_0+ 4F_2+F_4$ $= A +4B+2C$
	$\|4030>$	$1/\sqrt{2}\{(2^+1^-)-(2^-1^+)\}(-2^+ -2^-)$	
	$\|40-40>$	(-2^+-2^-)	
^3F	$\|3131>$	(2^+1^+)	$F_0- 8F_2-9F_4$ $= A -8B$
	$\|3121>$	(2^+0^+)	
	$\|3130>$	$1/\sqrt{2}\{(2^+1^-)+(2^-1^+)\}$	
	$\|3120>$	$1/\sqrt{2}\{(2^+0^-)+(2^-0^+)\}$	
	$\|3111>$	$\sqrt{3/5}\{(2^+ -1^+)+\sqrt{2/5}(1^+0^+)\}$	
^3P	$\|1111>$	$\sqrt{2/5}\{(2^+ -1^+)-\sqrt{3/5}(1^+0^+)$	$F_0+ 7F_2-84F_4$ $= A +7B$
^1D	$\|2020>$	$\sqrt{2/7}\{(2^+0^-)-\sqrt{2/7}(2^-0^+)$ $-\sqrt{3/7}(1^+1^-)$	$F_0 - 3F_2+36F_4$ $= A -3B+2C$
	$\|2010>$	$\sqrt{12/28}(2^+ -1^-)-\sqrt{12/28}(2^- -1^+)$ $-1.417/\sqrt{28}(1^+0^-)+1.417/\sqrt{28}(1^-0^+)$	
^1S	$\|0000>$		$F_0+ 14F_2+126F_4$ $= A +14B+7C$

4.4. Spin Orbit Perturbation of Russell Saunders' Terms – Calculation:

We shall illustrate the application of this second perturbation by considering the eigenfunctions (Table $_{4.5}$ and $T_{4.6}$) of the electrostatic repulsion matrix of (p^2) configuration as the initial unperturbed states subjected to spin orbit interaction. $\sum_i \xi(r_i) l_i, s_i$. These initial functions are eigenfunctions of $\sum_{k<l} e^2 / r_{kl}$ as well as of $\mathscr{H}_{\text{cent}}$

i) Of the two spin orbital Hamiltonians (sec. 2.11 Eqns. (2.4) and (2.5) we shall use the relatively more important and the large one, viz., Eqn. (2.4) in its equivalent form $\sum_i \xi(r_i) l_i, s_i$ of Eqn. (2.27)

ii) Since we are not considering any mixing of terms of different L values (sec 4.3). the eigenfunction sets of ^{3}P and ^{1}D are to be considered separately. And since each set is degenerate perturbation theory for degenerate state (sec. 2.4 Eqn. 2.45) is to be applied both for ^{3}P and ^{1}D states. ^{1}S is, however, a non-degenerate singlet state.

iii) The apparent difficulty of reduction the perturbation matrix of $\sum_i \xi(r_i) l_i, s_i$ in the basis set written (LSM_LM_S) scheme, as mentioned in sec. 4.3 (i) can be tided over in the following manner. The operator $\sum_i \xi(r_i) l_i, s_i$ commutes with J^2 and J_z Although the nine atomic kets of ^{3}P are written in (LSM_LM_S) scheme, each such ket will have an unwritten M_j value. Hence, those degenerate states, though represented by kets in (LSM_LM_S) scheme, having the same M_j value will mix under $\sum_i \xi(r_i) l_i, s_i$ according to theorem 1 of sec. 4.2. But those kets having different M_j's will remain as separate basis sets for the perturbation matrix.

iv) We, therefore, collect the kets ^{3}P of (p^2) configuration appearing in table ($T_{4.5}$) in groups of different M_j values. Remembering J= L+S and $M_j = M_L+M_s$, the foregoing kets fall under the following groups :

M_j	2	1	0	-1	-2
	(a)	(b) & (d)	(c), (e) & (g)	(f) & (h)	(I)
Kets	$\vert 1111>$	$\vert 1110>$, $\vert 1101>$	$\vert 111-1>$ $\vert 1100>$ $\vert 11-11>$	$\vert 110-1>$, $\vert 11-10>$	$\vert 11-1-1>$

It is, therefore, predictable that the (9×9) order perturbation matrix will block diagonalize into two (1×1), two (2×2) and one (3×3) matrices the basis sets for each of which are evident from the above groupings.

v) We, then, construct and solve these matrices for the ^{3}P state. We show, in detail, the solutions for one (2×2) matrix. The results for rest are similarly worked out and recorded in table $T_{4.7}$.

Two dimensional matrix with basis set kets,

$|1110>$ and $|1101>$: Solution

From table $T_{4..5}$

$$|1110>=1/\sqrt{2}\{(1^+0^-)+(1^-0^+)\} \qquad \text{......(4.22)}$$

and

$$|1101>=(1^+-1^+) \qquad \text{......(4.23)}$$

The perturbation hamiltonian, in terms of Eqns. (2.31) assumes the form (in the matrix element integral after integration of the radial parts of the equivalent electrons)

$$\Sigma_i(\zeta_i/\hbar^2)l_i\cdot s_i=(\zeta/\hbar^2)\Sigma_i l_i\cdot s_i=(\zeta/\hbar^2)\Sigma_i\{(l_{zi}\ s_{zi})+(1/2)(l_{i+}s_{i-}+l_{i-}s_{i+})\} \qquad \text{...(4.24)}$$

We now find the diagonal and the off diagonal matrix elements of the operator (4.24) with respect to the kets (4.22). Let us first calculate

$$(\zeta/\hbar^2)\Sigma_i l_i\cdot s_i(1^+-1^+)=(\zeta/\hbar^2)\ [(l_{z1}\ s_{z1})+(1/2)(l_{1+}\ s_{1-}+l_{1-}s_{1+})$$

$$+(l_{z2}\ s_{z2})+(1/2)(l_{2+}s_{2-}+l_{2-}s_{2+})](1^+-1^+)$$

$$=(\zeta/2\hbar^2)\{\hbar\, l_{1+}(1^--1^+)+\hbar\, l_{2+}(1^+-1^-)\}$$

$$=(\zeta/2\hbar^2)\{0+\sqrt{2}\,\hbar^2(1^+0^-)\}$$

$$=(\zeta/2)\sqrt{2}(1^+0^-) \qquad \text{......(4.25)}$$

Using Eq (4.25) the diagonal matrix element

$$<(1^+-1^+)|\Sigma_i(\zeta_i/\hbar^2)l_i.s_i|(1^+-1^+)>=(\zeta/2)\sqrt{2}<(1^+-1^+)|(1^+0^-)>$$

$$= 0\text{, due to orthogonality} \qquad \text{......(4.26)}$$

The off diagonal element

$$<1/\sqrt{2}\{(1^+0^-)+(1^-0^+)\}|(\zeta_i/\hbar^2)l_i.s_i|(1^+-1^-)>$$

$$1/\sqrt{2}\{(1^+0^-)+(1^-0^+)\}|(\zeta/2)\sqrt{2}(1^+0^-)>=\zeta/2 \qquad \text{......(4.27)}$$

where Eqn. (4.25) as well as the orthonormal property of the dets have been utilized. The remaining diagonal matrix element, when similarly treated yields

$$<1/\sqrt{2}\{(1^+0^-)+(1^-0^+)\}|\Sigma_i(\zeta_i/\hbar^2)l_i.s_i|1/\sqrt{2}\{(1^+0^-)+(1^-0^+)\}>=0 \qquad \text{......(4.28)}$$

From (4.26), (4.27) and (4.28), there emerges the following secular det

$$\begin{vmatrix} -\varepsilon & \zeta/2 \\ \zeta/2 & -\varepsilon \end{vmatrix} = 0$$

whence $\varepsilon_1 = \zeta/2$ and $\varepsilon_2 = -\zeta/2$ ......(4.29)

and the eigenvector matrices are

$$\begin{pmatrix} c_1 \\ c_2 \end{pmatrix} = \begin{pmatrix} 1/\sqrt{2} \\ 1/\sqrt{2} \end{pmatrix} \qquad \text{for } \varepsilon = \zeta/2$$

$$\begin{pmatrix} c_1 \\ c_2 \end{pmatrix} = \begin{pmatrix} 1/\sqrt{2} \\ -1/\sqrt{2} \end{pmatrix} \qquad \text{for } \varepsilon = -\zeta/2$$

Hence the eigenfunctions (perturbational wavefunctions of final LS states) are

$$1/\sqrt{2}\left|1/\sqrt{2}\left\{\left(1^+0^-\right)+\left(1^-0^+\right)\right\}+\left(1^+-1^+\right)\right|$$

i.e., $1/2\left|\left\{\left(1^+0^-\right)+\left(1^-0^+\right)\right\}+\sqrt{2}\left(1^+-1^+\right)\right\}$ for $\varepsilon = \zeta/2$ ......(4.30)

and $1/2\left|\left\{\left(1^+0^-\right)+\left(1^-0^+\right)\right\}-\sqrt{2}\left(1^+-1^+\right)\right]$ for $\varepsilon = -\zeta/2$ ......(4.31)

It is to be borne in mind that this second perturbation effect is calculated by taking the eigenfunction of ($\mathscr{H}_{\text{cent}} + \sum_{k<l} e^2/r_{kl}$) as the initial unperturbed state. Hence, the energy value $\varepsilon = \zeta/2$, $-\zeta/2$ (Eqn 4.29) are relative to the primary Russell Saunders' energy level of the degenerate ^{3}P term, viz., $F_o - 5F_2$ (Table T4.5) Hence, the final energy levels of LS states, relative to the central field energy $\varepsilon_0 = 0$ are

$F_0 - 5F_2 + \zeta/2$ for the eigenfunction (4.30)

and $F_0 - 5F_2 - \zeta/2$ for the eigenfunction (4.31)

These eigenfunctions, being linear combinations of the perturbational eigenfunctions of the electrostatic hamiltonian, $\sum_{k<l} e^2/r_{kl}$, are simultaneous eigenfunctions of the latter operator as well and hence ultimately of the full operator ($\mathscr{H}_{\text{cent}} + \sum_{k<l} e^2/r_{kl} + \sum_i \xi(r_i) l_i, s_i$)

The other spin orbit matrices of ^{3}P state, when solved for energy values and eigenfunctions show three different energy values viz.

$-\zeta/2$ and $\zeta/2$ for the second (2×2) matrix

$\zeta/2$, $\zeta/2$ for the two (1×1) matrices

and $\zeta/2$, $-\zeta/2$, $-\zeta$ for the one (3×3) matrix

Table 4.7 shows the distributions of the eigenfunctions over the degenerate energy values of the spin orbit perturbation of the primary Russell Saunders' ^3P state of (p^2) configuration.

It should be borne in mind, however, that the foregoing spin-orbit interactions are allowed to operate within each primary coupled state and are unable to intermix any two latter states.

Table $T_{4.7}$

Gross Atomic Term	Final L.S term	Energy (Relative to unperturbed level $\varepsilon_0 = 0$)	Eigenfunctions
^3P	$^3P_2$ (J=2)	$F_0 - 5F_2 + \zeta/2$ (Five fold degenerate)	a) $1/2\left[(1^+0^-)+(1^-0^+)+\sqrt{2}(1^+-1^+)\right]$ b) $1/2\left[(0^+-1^-)+(0^--1^+)+\sqrt{2}(1^--1^-)\right]$ c) (1^+0^+) d) $(0^- -1^-)$ e) $1/\sqrt{6}\left[\sqrt{2}\{(1^+-1^-)+(1^--1^+)\}\right.$ $\left.+\{(1^-0^-)+(0^+-1^+)\}\right]$
^3P	$^3P_1$ (J=1)	$F_0 - 5F_2 - \zeta/2$ (Three fold degenerate)	f) $1/2\left[\{(1^+0^-)+(l^-0^+)\}-\sqrt{2}(1^+-1^+)\right]$ g) $1/\sqrt{2}\{(1^-0^-)-(0^+-1^+)\}$ h) $1/2\left[\sqrt{2}\{(1^--1^-)-\{(0^--1^+)+(0^+-1^-)\}\right]$
^3P	$^3P_0$ (J=0)	$F_0 - 5F_2 - \zeta$ (nondegenerate)	i) $1/\sqrt{6}\left[\sqrt{2}\{(1^-0^-)+(0^+-1^+)\}\right.$ $\left.-\{(1^+-1^-)+(1^--1^+)\}\right]$

The values of J recorded here for the set of eigenfunctions are the ones obtainable from operation of J^2 on the listed eigenfunctions. It is evident from table ($T_{4.5}$) that the M_J values of the five atomic kets of the primary Russell Sanders state ^1D are all different. Hence each atomic ket (in its microstate det form) will form a single base function. The spin orbit interaction energy works out to be zero in each case. This is also true for the ^1S state. Table $T_{4.8}$ is a summary of these results:

Table $T_{4.8}$

Gross Atomic Term	Final L.S term	Energy (Relative to unperturbed level $\varepsilon_0 = 0$)	Eigenfunctions (after spin orbit perturbation)
^1D	$^1D_2$ (J=2)	Same as in Table $T_{4.5}$	Same as in Table $T_{4.5}$
^1S	$^1S_0$ (J=0)	Same as in Table T $_{4.5}$	Same as in Table $T_{4.5}$

4.5 External Magnetic Field Pertrubation Zeeman Levels of Russell Saunders' Fine structures:

In the previous sections we have considered two successive applications of intra-perturbations, viz., electrostatic and spin orbit, on polyelectron eigenkets of $\mathscr{H}_{\text{cent}}$ and have finally obtained the eigenfunctions and eigenvalues of the fine structures (spin orbit interaction structures) of Russell Saunders' primary states of ^3P, ^1D and ^1S. We now study the possible effect on such spin orbit fine structures due to a third perturbation arranged by setting up a constant magnetic field acting parallel to the Z. direction, H= H_z, H_x = H_y =0. The magnetic field perturbation is represented by the hamiltonian (cf Eqn. 2.37 $eH/2mc\sum_i(l_{zi}+2sz_i)$ of which the matrix elements between pairs of basis eigenfunctions of ($\mathscr{H}_{\text{cent}}$ $+\sum_{k<l} e^2/r_{kl}+\sum_i \xi(r_i)l_i \cdot s_i$) appearing in tables ($T_{4.7}$ and $T_{4.8}$) have to be found for construction of the perturbation matrix.

Sticking to (p^2) configuration, it is evident that the matrix elements of magnetic field operator between two identical eigenfunctions or between two eigenfunctions having at least one common det component may have non-zero value. This is due to the fact that each such two-electron det is an eigendet of $eH/2mc\sum_i(l_{zi})+2s_{zi}$. Nonsurvival of matrix elements between pairs of basis eigenfunctions with no common two-electorn det components results from the orthogonality of two different dets.

As an aid to calculations of the relevant matrix elements involving the eigenfuncitons, it is advantageous if we have ready on hand the matrix elements of the operator between the individual two-electron dets forming the various wavefunction of tables $T_{4.2}$ $T_{4.5}$ $T_{4.7}$ and $7_{4.8}$. Thus selecting the det of cell B (Table $T_{4.2}$) we have

$$<(1^+0^+)|eH/2mc\sum_i(l_{zi}+2sz_i)|(1^+0^+)>$$

$$=eH/2mc<(1^+0^+)|(l_{z1}+2sz_1+l_{z2}+2sz_2)|(1^+0^+)>$$

$$=(e\hbar/2mc)H(1+2.1/2+0+2.1/2)=3\beta H \qquad \text{......(4.32)}$$

Similarly, choosing another two-electron ket, say, (1^+-1^-) of cell F (Table$T_{4.2}$), the matrix element will be

$$< \left(1^+ - 1^+\right)\left| eH/2mc\,\Sigma_i\left(l_{zi} + 2sz_i\right)\right| \left(1^+ - 1^-\right)>$$
$$= \left(e\hbar/2mc\right)H\left[1 + 2.1/2 - 1 + 2.(-1/2\right] < \left(1^+ - 1^-\right)\left| \left(1^+ - 1^-\right)\right.>$$
$$= \beta H \times 0 = 0 \qquad \ldots\ldots(4.33)$$

Values of other such matrix elements can be written by just mentally ascertaining the eigen values of $\Sigma_i\left(l_{zi} + 2sz_i\right), i.e., \left(m_{11} + 2m_{s1} + m_{12} + 2m_{s2}\right)$ when the former acts on the det $\left(m_{11}^{ms1}\ m_{12}^{ms2}\right)$.

Looking for common det components in the eigenfuncions of table ($T_{4.7}$) we serially number the function a, f, b, h, e, g, i, c and d from 1to 9. This is so done since (a and f), (b and h), (e, g and i) having one or more det components, will mix under eH/2mc $\Sigma_i\left(l_{zi} + 2s_{zi}\right)$ operation whereas c and d will remain separate. Thus the (9×9)perturbation matrix will block-diagonalise into two (2×2) matrices [basis sets (af) and (bh)], one (3×3) matrix [basis set (egi)] and two (1×1) matrices with independent basis sets c and d respectively, Considering the basis sets involved, it is clear that in the first two cases, non-degenerate level mix, in the third case two degenerate eigenfunctions emerge as non degenerate ones under the operation of the magnetic hamiltonian.

We shall work out the first (2×2) matrix in detail and list the results of the other matrices with other basis sets.

$$\left.\begin{aligned}
H'_{22} = H'_{ff} &= (1/4)\beta\left[1 + 2.1/2 + 2.(-1/2)\ \right] + (1/4)\beta H\left[1 + 2.(-1/2) + 2..1/2\right] \\
&\qquad + (2/4)\beta H\left[2.1/2 + 2.1/2\right] \\
&= (3/2)\beta H = H'_{11} = H'_{aa} \qquad \ldots\ldots(4.34) \\
H'_{12} = H'_{21} = H'_{fa} &= (1/4)\beta H\left[1 + 2.1/2 + 2.(-1/2)\ \right] + (1/4)\beta H\left[1 + 2.(-1/2) + 2.1/2\right] \\
&\qquad - (2/4)\beta H\left[1 + 2.1/2 - 1 + 2.1/2\right] = -1/2\,\beta H
\end{aligned}\right\}$$

The (2×2) matrix is $\beta H \begin{pmatrix} 3/2 & -1/2 \\ -1/2 & 3/2 \end{pmatrix}$ ......(4.35)

The secular determinant, taking into account the initial state energies $F_0 - 5F_2 + \zeta/2$ of (a) and $F_0 - 5F_2 - \zeta/2$ of (f) (table $T_{4.7}$), is

$$\begin{vmatrix} F_0 - 5F_2 + \zeta/2 + (3/2)\beta H - \varepsilon & (-1/2)\beta H \\ (-1/2)\beta H & F_0 - 5F_2 - \zeta/2 + (3/2)\beta H - \varepsilon \end{vmatrix} = 0 \qquad \ldots\ldots(4.36)$$

The new energies ε containing first order corrections come from the diagonal terms and corrections of second order from the off-diagonal ones. Since the original energy levels involved are nondegenerate

$$\varepsilon_a = \varepsilon_a^{(1)} + \varepsilon_a^{(2)} = H'_{11} + \left(H'_{12}H'_{21}\right)/(-\zeta/2 - \zeta/2) \qquad \ldots\ldots(4.37)$$

The energies of the two Zeeman levels (referred to ε_0 of $\mathscr{H}_{\text{cent}} = 0$)

$$\varepsilon_a = F_0 - 5F_2 + \zeta/2 + (3/2)\beta H - (1/4)\beta^2\ H/\zeta \qquad \ldots\ldots(4.38)$$

$$\varepsilon_f = F_0 - 5F_2 + \zeta/2 + (3/2)\beta H + (1/4)\beta^2\ H/\zeta \qquad \ldots\ldots(4.39)$$

The eigenfunctions can also be written down, but these will scarcely be needed for our discussion. The last two terms in (4.38) and (4.39) are second order corrections. The complete list of Zeeman levels upto first order corrections calculated in like manner, are

$$\left.\begin{aligned}
\varepsilon_a &= F_0 - 5F_2 + \zeta/2 + (3/2)\beta H &&= \varepsilon\left({}^3P_1\right) + (3/2)\beta H \\
\varepsilon_f &= F_0 - 5F_2 - \zeta/2 + (3/2)\beta H &&= \varepsilon\left({}^3P_2\right) + (3/2)\beta H \\
\varepsilon_h &= F_0 - 5F_2 + \zeta/2 - (3/2)\beta H &&= \varepsilon\left({}^3P_1\right) - (3/2)\beta H \\
\varepsilon_b &= F_0 - 5F_2 + \zeta/2 - (3/2)\beta H &&= \varepsilon\left({}^3P_2\right) - (3/2)\beta H \\
\varepsilon_g &= F_0 - 5F_2 - \zeta/2 &&= \varepsilon\left({}^3P_1\right) \\
\varepsilon_e &= F_0 - 5F_2 + \zeta/2 &&= \varepsilon\left({}^3P_2\right) \\
\varepsilon_i &= F_0 - 5F_2 - \zeta &&= \varepsilon\left({}^3P_0\right) \\
\varepsilon_c &= F_0 - 5F_2 + \zeta/2 + 3\beta H &&= \varepsilon\left({}^3P_2\right) + 3\beta H \\
\varepsilon_d &= F_0 - 5F_2 + \zeta/2 - 3\beta H &&= \varepsilon\left({}^3P_2\right) - 3\beta H
\end{aligned}\right\} \qquad \ldots\ldots(4.40)$$

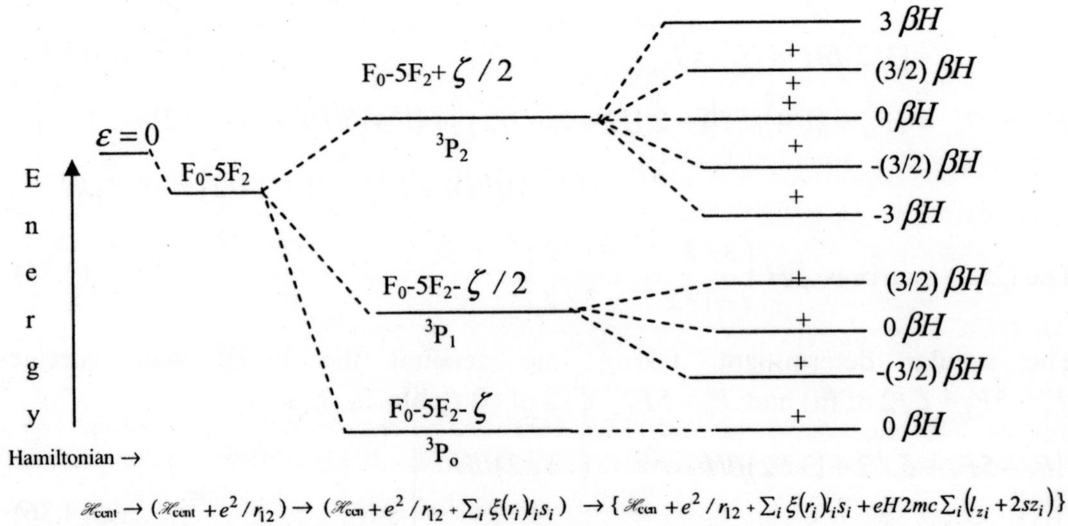

Fig. 4.2: Zeeman Levels of Russell Saunders' ${}^3P$ States)

Let us now find the Zeeman levels of $^1D_2$ and $^1S_0$ levels of (p^2) configuration. Scrutiny of table $T_{4.8}$, providing the eigenfunctions of the initial states characterized by the hamiltonian $(\mathscr{H}_{cen} + e^2/r_{12} + \Sigma_i \xi(r_i) l_i \cdot s_i)$ shows no common det component in the $^1D_2$ functions. Although ^1S and ^1D functions, viz., function of the table $(T_{4.5})$, have common det components the electrostatic perturbation widely separates to two levels. Hence one may easily rule out their mixing under an external magnetic field, especially when the latter is not too strong. We, therefore, need to consider only the diagonal matrix elements of the operator $eH/2mc \; \Sigma_i (l_{zi} + 2sz_i)$ with respect to the eigenfunctions of the $^1D_2$ and $^1S_0$ sets of table $(T_{4.8})$. The diagonal elements can immediately be written down as

$$H'_{aa} = 2\beta H, \quad H'_{bb} = H'_{cc} = H'_{dd} = 0. \; H'_{ee} - 2\beta H$$

The diagonal matrix element involving the ^1S_o function is also zero. The magnetic splitting of $^1D_2$ and $^1S_0$ of (p^2) is shown in the following diagram

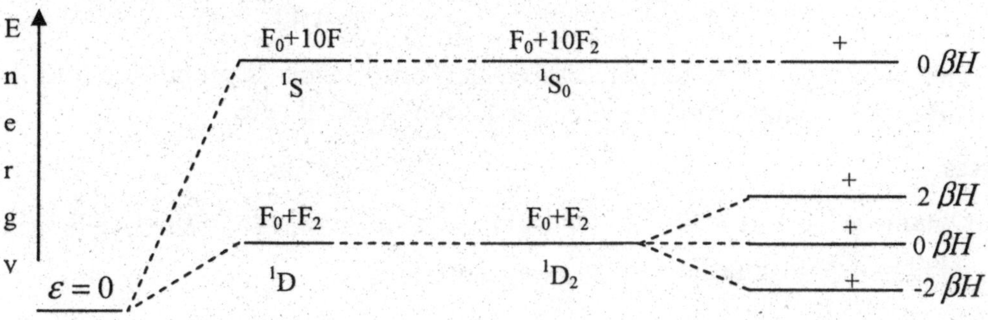

Hamiltonian →

$\mathscr{H}_{cent}$ → $(\mathscr{H}_{cent} + e^2/r_{12})$ $(\mathscr{H}_{cen} + e^2/r_{12} + \Sigma_i \xi(r_i) l_i s_i)$ $\{\mathscr{H}_{cen} + e^2/r_{12} + \Sigma_i \xi(r_i) l_i s_i + eH/2mc \, \Sigma_i (l_{zi} + 2sz_i)\}$

Fig. 4.3: Zeeman $^1D_2$ and $^1S_0$ of (p^2) configuration

It is thus seen that while the ninefold degeneracy of the primary ^3P Russell Saunders' state is completely lifted in an external magnetic field, the fivefold degenerate ^1D state's response is exhibited in partial removal of degeneracy. The Zeeman levels are two nondegenerate and one triply degenerate.

4.6 Energy Intervals between Russell Saunders' Fine Structures

The spin orbit interaction effect (eg, $^3p_2$, $^3p_1$, $^3p_0$ etc.) ^3p may be deemed as the fine structures of the primarily (electrostically) coupled Russell Sunders' term. The optical state of an atom belonging to the overall term $^{(2s+1)}L_j$ carries with it three quantum numbers J, L and S of multielectron atom. The radial part, containing the principal quantum numbers of individual electrons do not affect the optical state, except in its overall energy which can be taken care of through parameters.

The spin orbit perturbation energy can be expressed by (sec. 2.3)

$$\varepsilon = \lambda / 2[J(J+1) - L(L+1) - S(S+1)] \qquad \text{......(4.41)}$$

which results from one of the two equivalent alternative (2.31) and (2.32). For a term $^{(2s+1)}L_j$, J has the possible (L+S), (L+S-1). ....$|(L-S)|$ Hence for two values of J=L+S and J/= L+S-1, we have from (4.41),

$$\varepsilon_1 = \lambda LS \qquad \text{for } J = L+S$$

$$\varepsilon_2 = \lambda LS - \lambda(L+S) \qquad \text{for } J' = L+S-1$$

$$\Delta\varepsilon = \varepsilon_1\left({}^{2s+1}L_J\right) - \varepsilon_2\left({}^{2s+1}L_J\right) = \lambda(L+S) = \lambda J$$

......4.42

The energy gap between successive spin orbit levels is proportional to the higher of the two J values. This is called the Lande' interval rule.

Exercises

1. Obtain $\psi(^3P)$ of cell C (Table$_{4.2}$) starting from the atomic ket, viz., $|1111>$ of ^3P (Table $T_{4.1}$) of (p^2) configuration.

   [Hint : Write the det form of the atomic ket and locate the corresponding cell of table $T_{4.2}$ from which to make a start]

2. Establish the form of $\psi\{^1S\ |1000>\}$ of Eq. (4.10) and find the electrostatic perturbation energy in this state.

3. Find the electrostatic energy interval between the two triplet terms of (d^2) configuration. Express your results in terms of Recah parameters.

4. Find the spin orbit perturbation effect on the electrostatically perturbed degenerate states represented, in (LSM_LM_S) scheme of (p^2) configuration, by the atomic kets viz., $|111-1>, |1100>$ and $|11-11>$.

Chapter Five

j-j Coupled States, Energies and Eigenfunctions

5.1 General:

The two-step vector model treatment of j-j coupling (sec. 1.5, 1.5.1) corresponds to only a qualitative viewing of results achievable through quantitative treatment of two successive perturbations acting on a model central field atom. The main intraperturbation here is the spin orbit interaction, $\Sigma\xi(r_i)l_i.s_i$ and the weaker one is the coulombic repulsion, $\Sigma_{k<l}e^2/r_{kl}$ amongst the valence/optical electrons. We have seen previously, in chaps. 3 and 4, that the spin orbit term is applied in actual treatment in two forms, viz.

$(\zeta_i/\hbar^2)[\Sigma_i\{(l_{zi}.s_{zi})+(1/2)(l_{i+}s_{i-}+l_{i-}s_{i+})\}]$ in the matrix elements when the unperturbed basis set eigenfunctions are given in (LSM_LM_S) scheme and secondly as $(\zeta/2\hbar^2)\Sigma_i(j_i^2-l_i^2-s_i^2)$ when the basis kets are provided in $LSJM_J$) scheme. Both these alternative processes will be illustrated here with (p^2) electronic configuration. Furthermore, (d^2) case will also be surveyed for evaluation of energy.

A great deal of this chapter will have much in common, both in substance and description, with materials of chapter 3 and 4. However, in order to make the presentation clear and independent by itself, it is not encumbered by frequent back references to the foregoing chapters to prepare grounds for skipping even though it will invite some repetitive monotony.

5.2 Spin Orbit Perturbation. Energies and Eigenfunctions of Primarily Perturbed States. (LSM_LM_s) scheme

We know (p^2) configuration yields fifteen microstate dets (cf see 1.5.2, Table T1.2) that constitute the degenerate unperturbed wavefunctions of the hypothetical central field Hamiltonian, $\mathscr{H}_{cent}$ of an atom containing two equivalent p-electrons as valence/optical electrons. The two-electron dets of which the elements consists of single electron microkets, $|m_1m_s>$ i.e., spinorbitals, are cast into groups (Table T5.1) of different possible M_J values. This enables us to predetermine which of the dets are likely to mix under $\Sigma\xi(r_i)l_i\cdot s_i$ thus helping block diagonalization of the perturbation matrix (for reasons of Sec 4.4) iii) drawn up in the basis set of these two-electron dets.

Table $T_{5.1}$

Distribution of two-electron dets of (p^2) under possible M_J values

M_j	2	1	0	-1	-2
	(1^+1^-)	(1^+0^-)	$(1^+\text{-}1^-)$	$(1^-\text{-}1^-)$	$(\text{-}1^+\text{-}1^-)$
Two	(1^+0^+)	$(1^+\text{-}1^+)$	(1^-0^-)	$(0^+\text{-}1^-)$	$(0^-\text{-}1^-)$
electron		(1^-0^+)	$(1^-\text{-}1^+)$	$(0^-\text{-}1^+)$	
dets			(0^+0^-)		
			$(0^+\text{-}1^+)$		
	(a)	(b)	(c)	(d)	(e)

We number the functions serially as follow:

Column (a)	1,2	Serially	downward
Column (b)	3,4,5	,,	,,
Column (c)	6,7,8,9,10	,,	,,
Column (d)	11,12,13	,,	,,
Column (e)	14,15,	,,	,,

It is thus comprehensible that the (15×15) perturbation matrix is resolvable into two (2×2), two (3×3) and one (5×5) matrices with basis dets shown in columns a , e, b, d and e respectively. We shall work out separately the two matrices that would emerge from the basis dets of column (a) and of (b). To find the requisite matrix elements, it will be convenient if we have ready on hand the results of operations upon these dets by the perturbation hamiltonians $\Sigma\xi(r_i)l_i \cdot s_i$ or simply $\Sigma_i\, l_i \cdot s_i$ Thus

$$\Sigma_i\, l_i \cdot s_i\left(1^+1^-\right) = \left\{l_{zl}s_{zl} + (1/2)\left(l_{1+}s_{1-} + l_{1-}s_{1+}\right)\right\}\left(1^+1^-\right) + \left\{l_{z2}s_{z2} + (1/2)\left(l_{2+}s_{2-} + l_{2-}s_{2+}\right)\right\}\left(1^+1^-\right)$$

$$= \left\{(1/2)\left(1^+1^-\right)\hbar^2 + 0 + 0 - (1/2)\left(1^+1^-\right)\hbar^2 + \right.$$

$$\left. (1/2)l_{2-}\sqrt{s_2(s_2+1) + (1/2)(-1/2+1)}\left(1^+1^+\right)\hbar\right\}$$

$$= \left(\sqrt{2}/2\right)\left(1^+0^+\right)\hbar^2 \qquad \text{.........(5.1)}$$

Similarly

$$\Sigma_i l_i \cdot s_i(1^+0^+) = (1/2)(1^+0^+)\hbar^2 + 0 + 0 + 0 + (1/2) l_{2+}(1^+0^-)\hbar + 0$$

$$= (1/2)(1^+0^+)\hbar^2 + (\sqrt{2}/2)(1^+1^-)\hbar^2 \qquad \text{..........(5.2)}$$

Also

$$\Sigma_i l_i \cdot s_i(1^+0^-) = (1/2)(1^+0^-)\hbar^2 + (1/2) l_{2-}(1^+0^+)\hbar$$

$$= (1/2)(1^+0^-)\hbar^2 + (\sqrt{2}/2)(1^+ -1^+)\hbar^2 \qquad \text{..........(5.3)}$$

$$\Sigma_i l_i \cdot s_i(1^+ -1^+) = (1/2)(1^+ -1^+)\hbar^2 + 0 + 0 - (1/2)\hbar^2(1^+ -1^+) + (1/2)\hbar\, l_{2+}(1^+ -1^-) + 0$$

$$= (\sqrt{2}/2)(1^+1^-)\hbar^2 \qquad \text{..........(5.4)}$$

And finally

$$\Sigma_i l_i \cdot s_i(1^-0^+) = (1/2)(1^-0^+)\hbar^2 + 0 + (\sqrt{2}/2)(0^+0^+)\hbar^2 + 0 + (\sqrt{2}/2)(1^-1^-)\hbar^2 + 0$$

$$= -(1/2)(1^-0^+)\hbar^2 \qquad \text{..........(5.5)}$$

since the dets (0^+0^+) and (1^-1^-) are ruled out by Pauli's exclusion principle. However the det $l_{2-}(1^+0^+)$ leading to Eq (5.1) is not eliminated because the operator l_{2-} is still to act.

Utilising relations (5.1 to 5.5), the perturbation matrices of $\Sigma \xi(r_i) l_i \cdot s_i$ pertaining to columns a) and b) of Table $T_{5.1}$ are

$$\begin{bmatrix} 0 & \sqrt{2}/2\zeta \\ \sqrt{2}/2\zeta & (1/2)\zeta \end{bmatrix} \qquad \text{..........(5.6)}$$

$$\text{with eigenvalues} \quad \left.\begin{aligned} \varepsilon_1 &= \zeta \\ \varepsilon_2 &= -\zeta/2 \end{aligned}\right] \qquad \text{..........(5.7)}$$

and

$$\begin{bmatrix} \zeta/2 & \sqrt{2}/2\zeta & 0 \\ \sqrt{2}/2\zeta & 0 & 0 \\ 0 & 0 & -\zeta/2 \end{bmatrix} \qquad \text{..........(5.8)}$$

with eigenvalues

$$\left.\begin{aligned} \varepsilon_3 &= \zeta \\ \varepsilon_4 &= -\zeta/2 \\ \varepsilon_5 &= -\zeta/2 \end{aligned}\right] \qquad \text{..........(5.9)}$$

The eigenvector sets corresponding to all such energy values are found by solutions of the requisite simulataneous equations and the eigenfunctions of $\Sigma\xi(r_i)\,l_i.s_i$ [and hence of $\mathscr{H}_{cent} + \Sigma\xi(r_i)\,l_i.s_i$] ascertained. This exercise, when carried through all the component perturbation matrices yielded by the basis sets of Table $T_{5,1}$ renders only three distinct energy eigenvalues, viz., $-2\zeta, \zeta$ and $-\zeta/2$ with their respective degenerate eigenfunctions for the latter two energy states. These are enlisted in Table $T_{5.2}$ and represent the energies and wavefunctions of the primary coupled stage of the spin orbit interaction with the weak (electrostatic) interaction yet to occur to complete the j-j coupled state of the atom.

Table $T_{5.2}$

Spin-Orbit Coupling in (p^2) Configuration

Energy	Eigenfunctions
-2ζ	$1/3[\sqrt{2}\{(1^- 0^-) + (0^+ {-1}^+)\} - 2(1^- {-1}^+) - (0^+ 0^-)]$
-2ζ (Eight–fold degenerate)	$1/\sqrt{3}[\sqrt{2}\{(1^+ 1^-) - (1^+ 0^+)\}]$; $1/\sqrt{3}[\sqrt{2}\{(1^- {-1}^-) - (0^+ {-1}^-)\}]$; $1/\sqrt{3}[\sqrt{2}\{(1^+ {-1}^+) - (1^+ 0^-)\}]$; $1/\sqrt{3}[\sqrt{2}({-1}^+ {-1}^-) - (0^- {-1}^-)]$; $(1^- 0^+)$; $(0^- {-1}^+)$; $(1/\sqrt{2})[(1^- 0^-) - (0^+ {-1}^+)]$; $(2/3)[(\sqrt{2}/4)\{(1^- 0^-) + (0^+ {-1}^+)\} + (1^- {-1}^+) - (0^+ 0^-)]$
ζ (six–fold degenerate)	(a) $1/\sqrt{3}[(1^+ 1^-) + \sqrt{2}\,(1^+ 0^+)]$; (b) $1/\sqrt{3}[\sqrt{2}(1^+ 0^-) + (1^+ {-1}^+)]$; (c) $1/3[\sqrt{2}\{(1^- 0^-) + (0^+ {-1}^+)\} + (1^- {-1}^+) + 2(0^+ 0^-)]$; (d) $(1^+ {-1}^-)$; (e) $1/\sqrt{3}[\sqrt{2}(0^+ {-1}^-) + (1^- {-1}^-)]$; (f) $1/\sqrt{3}[\sqrt{2}(0^- {-1}^-) + ({-1}^+ {-1}^-)]$

5.3 Spin Orbit Coupling in (j-j) process

Basis set ($LSJM_J$) Scheme Wavefunction

The primary coupling stage in (j-j) coupling process dealt with in the previous section involves the use of dets built up with single-electron kets (i.e., spin orbitals) in the (LSM_LM_S) scheme. It is also possible to use multielectron determinantal wavefunctions comprising single-electron kets in the ($LSJM_J$) scheme as elements of the determinants.

The perturbation hamiltonian $\Sigma\xi(r_i)\,l_i.s_i \equiv \Sigma\xi/2(r_i)\left(j_i^2 - l_i^2 - s_i^2\right)$commutes with J^2 and J_z. The unperturbed eigenfunctions of $\mathscr{H}_{cen}$ in ($LSJM_J$) scheme constitute the basis functions for constructing the perturbation matrix and are themselves eigenfunctions of $\Sigma\xi/2\,(r_i)\left(j_i^2 - l_i^2 - s_i^2\right)$. For example, if.

$\psi = (a_1\ a_2\ a_3\ \ldots\ldots\ldots\ a_n)$ where a_1, a_2 etc are all single-electron kets $|l_l s_l j_l m_{j_l}>$ etc.

then

$$\Sigma \xi / 2(r_i)\left(j_i^2 - l_i^2 - s_i^2\right)\psi = \hbar^2 \Sigma \xi / 2(r_i)\left\{j_i(j_i+1) - l_i(l_i+1) - s_i(s_i+1)\right\}\psi \quad \ldots\ldots\ldots(5.10)$$

Therefore, the only types of matrix elements that will survive are the diagonal elements rendering the perturbation matrix a diagonal one. As in the pervious section, we shall illustrate with an equivalent (p^2) configuration undergoing j-j coupling. Such a configuration yields the following terms (Sec.1.5.2) in the usual notational scheme $(j_1\ j_2)_j$. Each such term, however, is degenerate within the range of M_J values permissible for a given J.

Terms	Terms in detail	Dets	Terms	Terms in detail	Dets
$(j_1, j_2)_J$	$(j_1, j_2)_J^{M_J}$	$(j_1^{m_{j1}}, j_2^{m_{j2}})_J$	$(j_1, j_2)_J$	$(j_1, j_2)_J^{M_J}$	$(j_1^{m_{j1}}, j_2^{m_{j2}})_J$
	$(\frac{3}{2},\frac{3}{2})_2^2$	$(\frac{3}{2}^{3/2},\frac{3}{2}^{1/2})_2$		$(\frac{3}{2},\frac{1}{2})_2^2$	$(\frac{3}{2}^{3/2},\frac{1}{2}^{1/2})_2$
	$(\frac{3}{2},\frac{3}{2})_2^1$	$(\frac{3}{2}^{3/2},\frac{3}{2}^{-1/2})_2$		$(\frac{3}{2},\frac{1}{2})_2^1$	$(\frac{3}{2}^{3/2},\frac{1}{2}^{-1/2})_2$
	$(\frac{3}{2},\frac{3}{2})_2^0$	$(\frac{3}{2}^{3/2},\frac{3}{2}^{-3/2})_2$	$(\frac{3}{2},\frac{1}{2})_2$	$(\frac{3}{2},\frac{1}{2})_2^0$	$(\frac{3}{2}^{1/2},\frac{3}{2}^{-1/2})_2$
$(\frac{3}{2},\frac{3}{2})_2$		$(\frac{3}{2}^{1/2},\frac{3}{2}^{-1/2})_2$		$(\frac{3}{2},\frac{1}{2})_2^{-1}$	$(\frac{3}{2}^{-1/2},\frac{1}{2}^{-1/2})_2$
	$(\frac{3}{2},\frac{3}{2})_2^{-1}$	$(\frac{3}{2}^{1/2},\frac{3}{2}^{-3/2})_2$			$(\frac{3}{2}^{-3/2},\frac{1}{2}^{1/2})_2$
	$(\frac{3}{2},\frac{3}{2})_2^{-2}$	$(\frac{3}{2}^{-1/2},\frac{3}{2}^{-3/2})_2$		$(\frac{3}{2},\frac{1}{2})_2^{-2}$	$(\frac{3}{2}^{-3/2},\frac{1}{2}^{-1/2})_2$
$(\frac{3}{2},\frac{3}{2})_0$	$(\frac{3}{2},\frac{3}{2})_0^0$	$(\frac{3}{2}^{1/2},\frac{3}{2}^{-1/2})_0$	$(\frac{3}{2},\frac{1}{2})_1$	$(\frac{3}{2},\frac{1}{2})_1^1$	$(\frac{3}{2}^{3/2},\frac{1}{2}^{-1/2})_1$
		$(\frac{3}{2}^{3/2},\frac{3}{2}^{-3/2})_0$			$(\frac{3}{2}^{1/2},\frac{1}{2}^{1/2})_1$
(1/2,1/2)	$(\frac{1}{2},\frac{1}{2})_0^0$	$(\frac{1}{2}^{1/2},\frac{1}{2}^{-1/2})_0$	$(\frac{3}{2},\frac{1}{2})_1$	$(\frac{3}{2},\frac{1}{2})_1^0$	$(\frac{3}{2}^{1/2},\frac{1}{2}^{-1/2})_1$
				$(\frac{3}{2},\frac{1}{2})_1^{-1}$	$(\frac{3}{2}^{-1/2},\frac{1}{2}^{-1/2})_1$
					$(\frac{3}{2}^{-3/2},\frac{1}{2}^{1/2})_1$

..........(5.11)

Energy Eigenvalues: The energy values are readily obtained from the diagonal perturbation matrix consequent upon Eq (5.10). The general expression for energy is $\zeta/2\Sigma\{j_i(j_i+1)-l_i(l_i+1)-s_i(s_i+1)\}$. Since for the two p-electrons $l_i = 1\ and\ s_i = 1/2$ each, it is evident that the energies are just dependent on the J_i values and the radial parameter ζ. From an examination of (p^2) configuration expressed in the form of (j-j) coupled terms of (5.11), it is clear that energy values are three in number having (j_1j_2)values (3/2, 3/2), (3/2, 1/2) and (1/2, 1/2) respectively. Using these values of j_1, j_2 in the energy expressions for two p electrons, one gets

$$\left.\begin{aligned} \varepsilon_1 &= \zeta && \text{(six fold degenerate)} \\ \varepsilon_2 &= -\zeta/2 && \text{(eightfold degenerate)} \\ \varepsilon_3 &= -2\zeta && \text{(non degenerate)} \end{aligned}\right\} \qquad \text{..........(5.12)}$$

These values and degeneracies agree with those obtained in the previous section using the unperturbed basis dets in (LSM_LM_s) scheme.

Eigenfunctions :

All the two-electron dets with (j_1, j_2) as (3/2, 3/2) and having different M_J values are eigendets of the perturbation operator $\Sigma\xi/2\,(r_i)\left(j_i^2-l_i^2-s_i^2\right)$ .Of these, however, the two dets, corresponding to $(3/2,\ 3/2)_2^0$ and another two dets corresponding to $(3/2,\ 3/2)_0^0$ form eigenfunctions in their suitable linear combinations. Such considerations also apply to dets emanating from tems $(3/2,\ 1/2)_2$, $(3/2,\ 1/2)_1$. Whenever more than one det occurs with similar values of j_i's and M_J, a linear combination will constitute the eigenfunction. Where no such plurality occurs, the individual dets is the eigenfunction. The det corresponding to the term $(1/2, 1/2)_0^0$ of (5.11) is itself a non degenerate eigenfunction.

It is possible to convert these eigenfunctions of ($LSJM_J$) scheme into those (LSM_LM_S) by using either step-up, step-down operators or using Clebsch Gordan Coeffcients [1,2] and Wigner's table. When so done, these will be the same as in Table $T_{5.2}$ or in a few cases some linear combinations of the degenerate eigenfunctions of that table.

Turning to (d^2) equivalent electronic configuration, it is easy to determine the perturbational energies due to spin orbit coupling part of the (j-j) mode. The following (j-j) coupled terms arise from (d^2) configuration (sec 1.5.2)

i) $(5/2,\ 5/2)_4$, $(5/2,\ 5/2)_2$, $(5/2,\ 5/2)_0$

ii) $(5/2,\ 3/2)_4$, $(5/2,\ 3/2)_3$, $(5/2,\ 3/2)_2$, $(5/2,\ 3/2)_1$

iii) $(3/2,\ 3/2)_2$, $(3/2,\ 3/2)_0$

These give us three distinct values of energies due to spin-orbit perturbation for the three different pairs of (j_1, j_2) values in the set of terms obtainable from (d^2) configuration (cf Eq 5.11) The energies are

$$\varepsilon_1 = 2\zeta \text{ for } (5/2, 5/2) \text{ set}$$

$$\varepsilon_2 = -3\zeta \text{ for } (5/2, 3/2) \text{ set} \quad \text{..........(5.13)}$$

$$\varepsilon_3 = -\zeta / 2 \text{ for } (3/2, 3/2) \text{ set}$$

5.4 Electrostatic Perturbations in second Stage of (j-j) Coupling Energyeigenvalues and Eigenfunctions

The spin orbit interaction is the main intrapertubation of a polyelectronic atom undergoing (j-j) coupling. We have already obtained spin orbit interaction matrix, energy eigenvalues and eigenfunctions in the previous section (cf Table $T_{5.2}$).

A fuller Hamiltonian contains the electrostatic repulsion $\sum k < l\, e^2 / r_{kl}$ as the second weak perturbation to be reckoned with in the full (j-j) scheme. Since this perturbation is weak compared to the one due to spin orbit, we shall assume that the electrostatic perturbation will not be able to intermix the eigenfunctions belonging to the different eigenvalues, viz., $\zeta, -\zeta/2$ and -2ζ of the primary perturbed states. We have, therefore, to consider this secondary perturbation by separately selecting the basis sets of the primary eigenfunctions belonging to different eigenvalues (cf Table $T_{5.2}$) of (p^2) configuration. We shall however treat one case fully, viz., the perturbative effect on the levels belonging to ζ and leave out the rest.

i) The eigenfunctions belonging to ζ (Table $T_{5.2}$) are sixfold degenerate. Since $\sum_{k<l} e^2 / r_{kl}$ commutes with L^2 and L_z of a central field atom, the eigenfunctions (Table $T_{5.2}$) will also the eigenfunctions of $\sum_{k<l} e^2 / r_{kl}$. Only if the degeneracy of ζ be not lifted completely, do we have the need to consider linear combinations of two or more functions of the set to serve as the final eigenfunctions in the perturbed state.

ii) To evaluate preturbational energies, we shall have to fall back on the method employed in Chapter-4, finding the diagonal and the off diagonal matrix elements J (ab), K (ab) and $< ab \left| e^2 / r_{kl} \right| cd >$ in terms of Condon Slater parameters with the two-electron eigenfunctions of ζ serving as basis set.

iii) Since the electrostatic matrix elements survive or disappear depending on a number of kronecker δ terms, viz., $\delta_{msa}, \delta_{msb}$ in K (ab) also on $\delta_{(mla+mlb),(mlc+mld)}, \delta_{msa,msc}, \delta_{msb,msd}$ etc. (sec 2.8 Eq 2.77, 2.80a), this will help us in weeding out the nonsurviving matrix elements from our calculations.

iv) Diagonal Matrix Elements (Table $T_{5.2}$)

a) $$<1/\sqrt{3}\{(1^+1^-)+\sqrt{2}(1^+0^+)\}|e^2/r_{12}|1/\sqrt{3}\{(1^+1^-)+\sqrt{2}(1^+0^+)\}>$$

$$= 1/3J(1,1)+2/3J(1,0)-2/3K(1,0)+\sqrt{2}/3<(1^+1^-)\}|e^2/r_{12}|(1^+0^+)>$$

$$+\sqrt{2}/3<(1^+0^+)|e^2/r_{12}|(1^+1^-)>$$

$$= 1/3\sum_k c^k(1,1)c^k(1,1)F^k+2/3\sum c^k(1,1)c^k(0,0)F^k-2/3\sum_k c^k(1,0)c^k(1,0)F^k \qquad \text{..........(5.14)}$$

= F_0-$3F_2$, on consulting the c^k table (Table $T_{2.1}$)

b) $$<1/\sqrt{3}\{\sqrt{2}(1^+0^-)+(1^+-1^+)\}|e^2/r_{12}|1/\sqrt{3}\{\sqrt{2}(1^+0^-)+(1^+-1^+)\}>$$

$$= \quad 2/3\,J(1,0)+1/3\,J(1,-1)-1/3K(1,-1)$$

$$= \quad 2/3F^0+2/3\,(-2/25)F^2+1/3F^0+1/3\,(1/25)F^2-1/3\,(6/25)F^2$$

$$= \quad 2/3F_0\quad -4/3F_2+1/3F_0+1/3F_2-2F_2$$

$$= \quad F_0-3F_2 \qquad \text{..........(5.15)}$$

c) $$<1/3\left[\sqrt{2}\{(1^-0^-)+(0^+-1^+)\}+(1^--1^+)+2(0^+0^-)\right]|e^2/r_{12}|$$

$$1/3\left[\sqrt{2}\{(1^-0^-)+(0^+-1^+)\}+(1^--1^+)+2(0^+0^-)\right]>$$

$$= \quad 2/9<(1^-0^-)|e^2/r_{12}|(1^-0^-)>+2/9<(0^+-1^+)|e^2/r_{12}|(0^+-1^+)>$$

$$+1/9<(1^--1^+)|e^2/r_{12}|(1^--1^+)>+4/9<(0^+\,0^-)|e^2/r_{12}|(0^+\,0^-)>$$

$$-2/9<(-1^+1^-)|e^2/r_{12}|(0^+0^-)>-2/9<(0^+\,0^-)|e^2/r_{12}|(-1^+1^-)>$$

$$= \quad 2/9J\,(1,0)-2/9k(1,0)+2/9J(0,-1)-2/9K(0,-1)+1/9J(1,-1)$$

$$+4/9J(0,0)-2.2/9\sum_k c^k(-1,0)c^k(0,1)F^k$$

$$= 2/9\sum_k c^k(1,1)c^k(0,0)F^k-2/9\sum_k c^k(1,0)c^k(1,0)F^k+2/9\sum_k c^k(0,0)c^k(-1,-1)F^k$$

$$-2/9\sum_k c^k(0,-1)c^k(0,-1)F^k+1/9\sum_k c^k(1,1)c^k(-1,-1)F^k$$

$$+4/9\sum_k c^k(0,0)c^k(0,0)F^k-4/9\sum_k c^k\,(-1,0)c^k(0,1)F^k$$

$$= \quad F_0+F_2 \qquad \text{..........(5.16)}$$

on substitution of values from C^k tables and remembering the relation between F^k and F_k

d) $<\left(1^+ -1^-\right)\left|e2/r_{12}\right|\left(1^+ -1^-\right)> = J(1,-1) = F^0 + (1/25)F^2 = F_0 + F_2$ ....(5.17)

e) $<1/\sqrt{3}\left\{\sqrt{2}\left(0^+ -1^-\right)+\left(1^- -1^-\right)\right\}\left|e^2/r_{12}\right|1/\sqrt{3}\left\{\sqrt{2}\left(0^+ -1^-\right)+\left(1^- -1^-\right)\right\}>$

$= 2/3\, J(0,-1) + 1/3\, J(1,-1) - 1/3 K(1,-1)$

$= 2/3 \Sigma_k c^k(0,0) c^k(-1,-1) F^k + 1/3 \Sigma_k c^k(1,1) c^k(-1,-1) F^k - 1/3 \Sigma_k c^k(1,-1)$
$c^k(1,-1) F^k$

$= 2/3\ F^0 + 2/3\ (-\ 2/25)\ F^2 + 1/3\ F^0 + 1/3\ (1/25)\ F^2 - 1/3\ (6/25)\ F^2$

$= F_0 - 3\ F_2$ ....(5.18)

f) $<1/\sqrt{3}\left\{\sqrt{2}\left(0^- -1^-\right)+\left(-1^+ -1^-\right)\right\}\left|e^2/r_{12}\right|1/\sqrt{3}\left\{\sqrt{2}\left(0^- -1^-\right)+\left(-1^+ -1^-\right)\right\}>$

$= 2/3\, J(0,-1) - 2/3\, K(0,-1) + 1/3 J(-1,-1)$

$= F_0 - 3F_2$ ....(5.19)

Non diagonal matrix elements:

All such nondiagonal matrix elements are zeros except the one between the eigenfunctions(c) and (d) of Table $T_{5.2}$. Thus

$<1/3\left[\sqrt{2}\left\{\left(1^- 0^-\right)+\left(0^+ -1^+\right)\right\}+\left(1^- -1^+\right)+2\left(0^+ 0^-\right)\right]\left|e^2/r_{12}\right|\left(1^+ -1^-\right)>$

$= 2/3 < \left(0^+ 0^-\right)\left|e^2/r_{12}\right|\left(1^+ -1^+\right)> - 1/3 < \left(-1^+ 1^-\right)\left|e^2/r_{12}\right|\left(1^+ -1^-\right)>$

$= 2/3 \Sigma_k c^k(0,1) c^k(-1,0) F^k - 1/3 \Sigma_k c^k(-1,1)\, c^k(-1,1) F^k$

$= -\ 4F_2$ ....(5.20)

The foregoing results (3.14-5.20) show that the coulomb repulsion matrix is made up of four (1×1) matrices each with eigenvalue F_0-$3F_2$ and one (2×2) matrix, the elements of which are given by (5.16), (5.17), (5.20) and (5.20), the last two furnishing the offdiagonal elements. The eigenvalues of this (2×2) matrix can be obtained from

$$\begin{bmatrix} F_0 + F_2 - \epsilon & -4F_2 \\ -4F_2 & F_0 + F_2 - \epsilon \end{bmatrix} = 0 \qquad \text{..........(5.21)}$$

which yields $\epsilon_1 = F_o - 3F_2$ and $\epsilon_2 = F_o - 5F_2$ with the eigenfunctions

$\left(1/3\sqrt{2}\right)\left[\sqrt{2}\left\{\left(1^- 0^-\right)+\left(0^+ -1^+\right)\right\}+\left(1^- -1^+\right)+2\left(0^+ 0^-\right)+3\left(1^+ -1^-\right)\right]$

and

$$\left(1/3\sqrt{2}\right)\left[\sqrt{2}\left\{\left(1^-0^-\right)+\left(0^+-1^+\right)\right\}+\left(1^--1^+\right)+2\left(0^+0^-\right)-3\left(1^+-1^-\right)\right] \quad \text{..........(5.22)}$$

respectively.

The final set of eigenvalues and eigenfunctions incorporating coulomb interactions within the term given by the spin orbit value, ζ of the (j-j) coupled (p^2) configuration are shown in Table $T_{5.3}$. The remaining set of (j-j) coupled (p^2) configuration eigenfunctions and eigenvalues can similarly be worked out. A schematic diagram for splitting of energy levels in (j-j) coupling for (p^2) is given (5.1).

Table $T_{5.3}$

Energy eigenvalues and eigenfunctions of (j-j) coupled terms of (p^2) after electrostatic splitting within the term with the spin orbit energy ζ (Energies are relative to the original unperturbed central field energy, $\in = 0$)

Energy eigenvalue	Eigenfunctions
$\zeta + F_0 - 3F_2$	i) $1/\sqrt{3}\left\{\left(1^+1^-\right)+\sqrt{2}\left(1^+0^+\right)\right\}$
	ii) $1/\sqrt{3}\left\{\left(1^+-1^-\right)+\sqrt{2}\left(1^+0^-\right)\right\}$
	iii) $\left(1/3\sqrt{2}\right)\left[\sqrt{2}\left\{\left(1^-0^-\right)+\left(0^+-1^+\right)\right\}+\left(1^--1^+\right)+2\left(0^+0^-\right)+3\left(1^+-1^-\right)\right]$
	iv) $1/\sqrt{3}\left\{\left(1^--1^-\right)+\sqrt{2}\left(0^+-1^-\right)\right\}$
	v) $1/\sqrt{3}\left\{\left(-1^+-1^-\right)+\sqrt{2}\left(0^--1^-\right)\right\}$
$\zeta + F_0 + 5F_2$	vi) $\left(1/3\sqrt{2}\right)\left[\sqrt{2}\left\{\left(1^-0^-\right)+\left(0^+-1^+\right)\right\}+\left(1^--1^+\right)+2\left(0^+0^-\right)-3\left(1^+-1^-\right)\right]$

It should be noted that the remarks made immediately before table $T_{4.7}$ apply here in the reverse sense, viz. the weak coulombic interaction cannot cause any mixing of any two district spin-orbit states.

After weak electrostatic perturbation of the spin-orbit eigenstates, fifteen wavefunctions distribute themselves over five different energy levels, two of which are non-degenerate, one triply degenerate, and the remaining two are each five-fold degenerate, as shown in Fig. 5.1 and partially in the foregoing table.

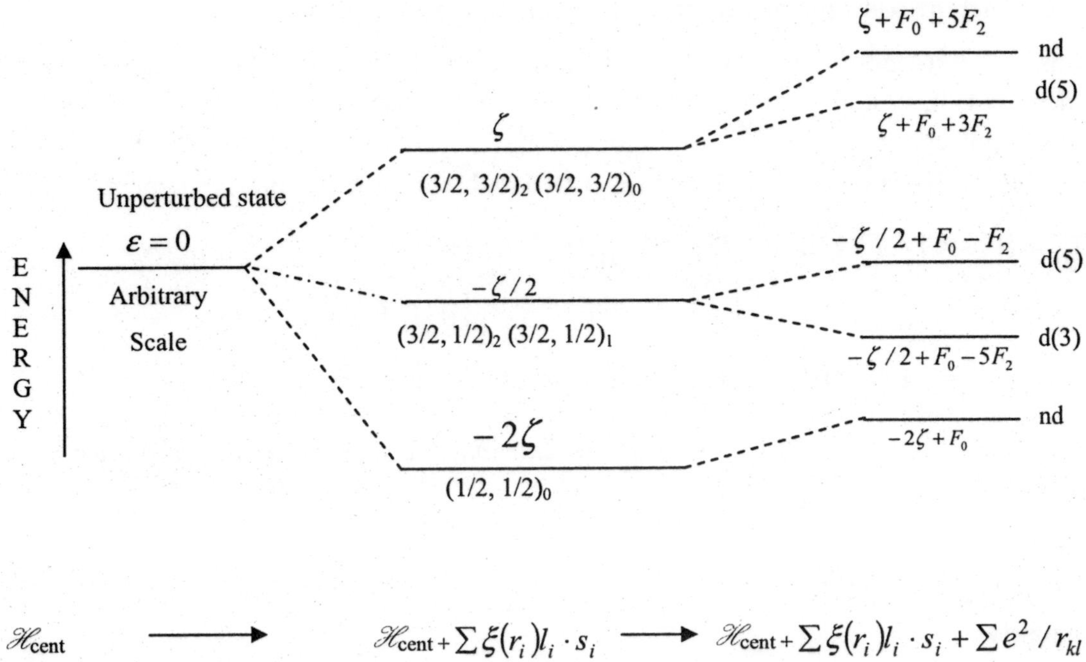

nd= nondegenerate d(n) = n-fold degenerate

Fig. 5.1: Splitting of levels in jj-coupled (p^2) configuration

5. External Magnetic Field Perturbation of (j-j) coupled Terms, Zeeman levels

We have already discussed in the previous three sections the effect of application of two successive perturbations, viz., $\sum \xi(r_i) l_i \cdot s_i$ and $\sum e^2 / r_{kl}$ leading to different (j-j) coupled terms of (p^2) configuration. We now impose a third perturbation, viz., an external constant magnetic field, H, applied parallel to z-direction and see how this affects the terms already obtained in the (j-j) coupling of (p^2) configuration. This third perturbation has the Hamiltonian, $eH / 2mc \sum (l_{zi} + 2s_{zi})$ and the initial basis functions are the eigenfunctions of both the spin orbit interaction and coulomb interaction Hamiltonians of Table $T_{5.3}$. From the chosen basis set, it is evident that we are going to examine the nature of Zeeman splitting of the (j-j) levels stemming from only one of the initial spin orbit states with energy ζ of the (p^2) configuration.

It is pointed out for the benefit of the readers that a large chunk of the general discussion in the initial part of sec 4.5 applies in the present case also including observation following relations (4.32) and (4.33) there. With this background, if we now look out for the two-electron det components, present in the eigenfunctions of Table $T_{5.3}$ that will mix under the magnetic perturbation giving non vanishing matrix elements, we are led to the following conclusion :

a) that the four basis functions, i), ii), iv) and v) of Table T5.3 will not mix amongst themselves or with any other.

b) that the basis functions, iii) and vi) mix under the magnetic perturbation.

Thus the (6×6) perturbation matrix will resolve itself into four (1×1) matrices (basis functions i, ii, iv and v, and one (2×2) matrix in the basis set of functions iii) and vi) with initial energies $\zeta + F_0 - 3F_2$ and $\zeta + F_0 + 5F_2$ respectively. Let us serialize the functions i), ii), iv), v), iii) and vi) from 1 to 6 and work out the matrix elements

$$H'_{11} = H'_{ii} = <1/\sqrt{3}\,\{(1^+1^-) + \sqrt{2}(1^+0^+)\}|eH/2mc\,(l_{zl} + 2s_{zl} + l_{z2} + 2s_{z2})|1/\sqrt{3}\,\{(1^+1^-) + \sqrt{2}(1^+0^+)\}>$$

$$= 1/3 < (1^+1^-)\,|eH/2mc\,(m_{11} + 2m_{sl} + m_{12} + 2m_{s2})|\,(1^+1^-)>$$

$$+ 2/3 < (1^+0^+)|eH/2mc\,(m'_{11} + 2m'_{sl} + m'_{12} + 2m'_{s2})|\,(1^+0^+)>$$

$$= 1/3\,(1 + 2.1/2 + 1 - 2.1/2)\,\beta H + 2/3\,(1 + 2.1/2 + 0 + 2.1/2)\,\beta H$$

$$= 8/3\beta H \qquad \text{.........(5.23)}$$

Similarly

$$\left.\begin{aligned} H'_{22} &= H'_{ii,ii} = 4/3\beta H \\ H'_{33} &= H'_{iv,iv} = -4/3\beta H \\ H'_{44} &= H'_{v,v} = -8/3\beta H \end{aligned}\right\} \qquad \text{.........(5.24)}$$

Again all the matrix elements of the (2×2) matrix with the basis function 5 and 6 (i.e., iii and vi of table T5.3) are all zeros, the matrix being a null matrix.

$$H'_{55} = H'_{iii,iii} = H'_{66} = H'_{vi,vi} = 0 \text{ and } \quad H'_{56} = H'_{iii,vi} = H'_{65} = 0$$

The external magnetic field interaction matrix in thus diagonal, the diagonal elements of which are written in a row [from (5.23), (5.24) and (5.25)]

$$(8/3\,\beta H, 4/3\,\beta H, -4/3\,\beta H, -8/3\beta H, 0, 0) \qquad \text{.........(5.26)}$$

The degeneracy of the original levels is thus completely lifted leading to a final set of six energy levels, i.e., Zeeman levels referred to energy scale $\in = 0$ for the original unperturbed state.

$$\in'_1 = \zeta + F_0 - 3F_2 + 8/3\beta H; \qquad \in'_2 = \zeta + F_0 - 3F_2 + 4/3\beta H$$

$$\in'_3 = \zeta + F_0 - 3F_2 - 4/3\beta H; \qquad \in'_4 = \zeta + F_0 - 3F_2 - 8/3\beta H$$

$$\in'_5 = \zeta + F_0 - 3F_2; \qquad \in'_6 = \zeta + F_0 - 5F_2$$

A diagramatic sketch of Zeeman splittings of one of the primary spin orbit states in the (j-j) coupling of (p^2) configuration appears in fig. 5.2

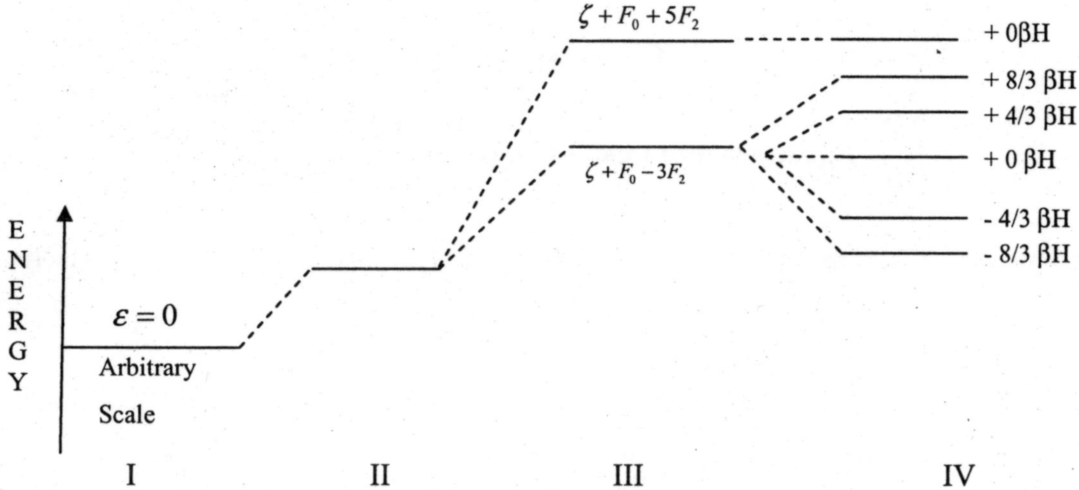

$$\mathscr{H}_{\text{cent}} \rightarrow \{1 + \Sigma\xi(r_i)l_i s_i\} \rightarrow \{11 + \Sigma e^2 / r_{kl}\} \rightarrow \left\{111 + \frac{eH}{2mc}\Sigma(l_{zi} + 2s_{zi})\right\}$$

Fig. 5.2: Zeeman levels of the uppermost jj coupled terms of (p^2) configuration

Exercises

1. A (d^2) system consists of 45 dets (cf $T_{4.3}$) in (LSMLMS) scheme. What principle would you use to simplify the solution of (45×45) matrix form of the operator $\Sigma\xi(r_i)l_i \cdot s_i$ in the basis set of these dets by tackling a number of small-sized matrices?

2. Cerium ion in an appropriate valence state has the configuration (f^2). If the coupling mode be of the (j-j) type, can you find any of the energies of its primary multiple spin-orbit levels relative to its unperturbed state energy level taken as zero?

   How is this particular spin-orbit level going to be affected in the second stage of the (j-j) coupling scheme?

References

1. J.I. Powell and B. Crasemann : *Quantum Mechanics* (Addison Wesley)
2. M. Tinkham : Group *Theory and Quantum Mechanics* (Tata Mc- Graw Hill)

   A.W. Joshi : *Elements of Group Theory for Physicists* (Wiley Eastern, New Delhi)
3. R. M. Golding : Applied Wave Mechanics (Addison Wesley)
4. H. Eyring, J. Walter and G. E. Kimball : *Quantum Chemistry* (John Wiley and Sons)

Chapter Six

Energy Levels of (p^n) and (d^n) Configurations, Energy Intervals, Fitting of Equations and Evaluation of Racah Parameters

6.1 Qualitative Implications of Hole-Particle Formalism for l^n and l^{4l+2-n} Electron Systems

In the above caption 1^n means $P^n\ (l=1), d^n\ (l=2), f^n\ (l=3)$ equivalent electron systems. It is known that the multiplet term types in l^n are identical with those emerging from l^{4l+2-n} electron systems (cf Sec. 1.5.2 Chap 1) The reason for this as well as the qualitative idea about the respective energies of the terms of the two configurations, l^n and, l^{4l+2-n} can be gathered from the hole-particle formalism.

According to Dirac[1], the absence of an electron is equivalent to the presence of a "hole" in something. This hole plays the role of a positron. Let us now explain the idea by confining ourselves to d^n electron systems.

The systems (d^1 and d^9), (d^2 and d^8), (d^3 and d^7) and (d^4 and d^6) display the same type of multiplets for each configuration of any such given pair within a parenthesis. Speaking of d^1 and d^9 the latter may be deemed as a one-electron lacking d^{10} system or equivalently a d^{10} system with one hole furrowed in it. i.e., one positron imposed on an original d^{10} system. Hence a d^9 system may be thought as a perturbed state of d^{10} system (^1S state), the perturbation being caused by the positron imposed on it. Similarly a d^1 system can be viewed as a perturbed from of d^0 system or ^1S state of an initial field, the sole d-electron imposed on the central field causing the perturbation.

Viewed in the same light, a d^2 system is an electrostatically perturbed state of a d^0 central field system giving rise to multiplets ^3F, ^3P, ^1D, ^1G and ^1S. Analogously a d^8 system is a two-electron deficient, i.e., two-positron imposed d^{10} ($\equiv$ ^1S) electron system. The electrostatic perturbations caused by these two positrons on d^{10} state are quantitatively similar, both in nature and magnitude, to those caused by two electrons of d^2 system on an initial d^0 central field state of ^1S. The total energies of the terms of d^2 systems are the ones when the respective repulsion energies to the two electrons in F, P, D, G, S states added to the newly emerged hypothetical central field 1s state of the two-hole containing d^{10} (i.e. a d^8) born from original ^1S state of d^{10}, central field ^1S state of the two-hole containing d^{10} (i.e. a d^8) born from original ^1S state of d^{10}, yield the total energy values of the multiples. Since there is no qualitative or quantitative difference in the repulsion energies between either the two protons or the two electrons the relative energies of the multiplets of d^2 and d^8 will be identical. The actual energies will, however, differ by a constant magnitude, viz. the energy difference of the two hypopthetical central field systems of d^2 and d^8. This argument when extended to (d^3 and d^7) and (d^4 and d^6) enables us to conclude that the relative energies of the multiplets, referred to some suitably defined zero level energy, are the same for such configuration pairs. Such reasonings will also apply to the pairs of (p^n, p^{6-n}) and (f^n, f^{14-n}) of equivalent p and f electron systems. As to d^5, it will have a highest spin state S of d^{10}. Besides d^5 will have additional multiplets as against none more of d^{10}.

6.2 Energy Levels and Energy Intervals of p^n Multiplets

We are already acquainted with the technique of calculating electrostatic energies of the multiplet terms of (p^n) systems (chap 4). There we have already obtained the energy expressions in terms of Slater Condon parameters. F_k's. These for (p^2) and hence for (P^4) electron system are [Eqs. 4.3a, b, c].

$$\left.\begin{aligned} \varepsilon({}^3p) &= F_0 - 5F_2 && \text{..........(a)} \\ \varepsilon({}^1D) &= F_0 + F_2 && \text{..........(b)} \\ \varepsilon({}^1S) &= F_0 + 10F_2 && \text{..........(c)} \end{aligned}\right\} \qquad \text{........(6.1)}$$

Van Vleck[2] developed an operator technique by which the energy expressions of (p^3) multiplets can all be derived in terms of components of the (p^2) multiplets. The wavefunctions of ^3P, ^1D and ^1S states of (p^2), being the sum of two-electron dets (cf Chap 4) are eigenfunctions of L^2, S^2, $l_1^2, l_2^2, s_1^2, s_2^2$ and hence also of $l_1.\ l_2$ and $s_1.\ s_2$ operators. This is because

$$\left.\begin{aligned} l_1.l_2 &= (1/2)\left(L^2 - l_1^2 - l_2^2\right) \\ &\text{and} \\ s_1.s_2 &= (1/2)\left(S^2 - s_1^2 - s_2^2\right) \end{aligned}\right\} \qquad \text{........(6.2)}$$

It is now easy to establish the following matrix elements (in units of $\hbar^2$)

$$<^3P|l_1.l_2|^3P> = -1 \text{ and } <^3P|s_1.s_2|^3P> = 1/4 \qquad \text{........(6.3)}$$

$$<^1D|l_1.l_2|^1D> = 1 \text{ and } <^1D|s_1.s_2|^1D> = -3/4 \qquad \text{........(6.4)}$$

$$<^1S|l_1.l_2|^1S> = -2 \text{ and } <^1S|s_1.s_2|^1S> = -3/4 \qquad \text{........(6.5)}$$

Van Vleck showed that it is possible to write down an operator ω_{12} for a pair of electrons [as are present in (p^2) system] viz,

$$\omega_{12} = \{F_0 - (5 + 3l_1.l_2 + 12s_1.s_2)F_2\} \qquad \text{........(6.6)}$$

such that the diagonal matrix elements of this operator between the multiplets of (p^2)

are the same as the repulsion operator e^2/r_{12} of the electron pair. Thus applying (6.6) along with (6.3), (6.4) and (6.5) we find (cf Chap 4)

$$<^3P|\omega_{12}|^3P> = F_0 - 5F_2 = \varepsilon(^3P) = <^3P|e^2/r_{12}|^3P>$$

and similarly $<^1D|\omega_{12}|^1D> = F_0 + F_2 = \varepsilon(^1D) = <^1D|e^2/r_{12}|^1D>$ ....(6.7)

$$<^1S|\omega_{12}|^1S> = F_0 + 10F_2 = \varepsilon(^1S) = <^1S|e^2/r_{12}|^1S>$$

The general form of (6.7) is

$$<^{2s+1}L|\omega_{12}|^{2s+1}L> = \varepsilon(^{2s+1}L) \qquad \text{........(6.8)}$$

Van Vleck further showed that with more electron pairs (as in p^3) the diagonal matrix elements of $\Sigma_{k<l} e^2/r_{kl}$ of the different multiplets can be obtained by using the operator W = $\Sigma_{k<l}\,\omega_{kl}$.

$$W = \Sigma_{k<l}\,\omega_{kl} = \sum_{k<l}^{pairs}\{F_0 - (5 + 3l_k.1_l + 12s_k.s_l)F_2\}$$

$$= (1/2)n(n-1)F_0 - \{(5/2)n(n-1) + (3/2)L^2 - (3/2)\Sigma_{k=l}^{n} l_k^2 + 6S^2 - 6\Sigma_k s_k{}^2\}F_2 \quad \text{...(6.9)}$$

where it should be noted that

$\sum^{pairs} 3(L^2 - l^2{}_k - l^2{}_j) = (3/2)\left[L^2 - \sum_{k=1}^{n} l^2{}_k\right]$ since L is not affected by the number of pairs. The same reasoning applies to $\sum^{pairs} 12(S^2 - s_k^2 - s_1^2)$.Thus for equivalent p-electron system, The diagonal matrix elements are obtainable from (6.8) and (6.9).

$$<^{2s+1}L|W|^{2s+1}L> = (1/2)n(n-1)F_0 - (1/2)\{5n^2 - 5n + 3L(L+1) - 6n + 12S(S+1) - 9n\}F_2$$

$$= (1/2)n(n-1)F_0 - (1/2)\{5n^2 - 20n + 3L(L+1) + 12S(S+1)\}F_2 \qquad \text{....(6.10)}$$

For the (p^3)electronic configuration, the Russell Saunders' multiplets are ^{4}S, ^{2}D and ^{2}P. We can thus evaluate the diagonal matrix elements of the electrostatic perturbation operator $\Sigma_{k<l} e^2/r_{kl}$ through the agency of W operator, i.e., via (6.10)

Thus for ^{4}S with L = 0, S= 3/2 and n=3

$$<^4S|W|^4S> = 3F_0 - 15F_2 = \varepsilon(^4S)$$

and similarly

$$\left.\begin{aligned} &<^2D|W|^2D> = 3F_0 - 6F_2 = \varepsilon(^2D) \\ &<^2P|W|^2P> = 3F_0 \qquad = \varepsilon(^2P) \end{aligned}\right\} \quad ....(6.11)$$

Energy Intervals

The equations (6.2), (6.11) for (p^2), (p^4) and (p^3) configurations allow us to derive the rations of the energy gaps of the different Russell Saunders' terms of the configurations. If r_2, r_4, and r_3 represent the interval ratios of the successive term energies in descending order of (p^2), (p^4) and (p^3) respectively, then

$$r_2 = r_4 = \{\varepsilon(^1S) - \varepsilon(^1D)\}/\{\varepsilon(^1D) - \varepsilon(^3P)\} = 3/2$$

and $$r_3 = \{\varepsilon(^2P) - \varepsilon(^2D)\}/\{\varepsilon(^2D) - \varepsilon(^4S)\} = 2/3$$

The values are somewhat larger than the experimental values for the elements of the first short period and are in fair agreement with those of the second period of the periodic table. However, further down the table, disagreement pops up again. The main reason is that in computing these ratios, the spin orbit perturbations have not taken into consideration. These constitute an inbuilt feature of the experimental values alongwith the electrostatic effect. Some authors have shown that the interval gap ratios can be made to agree fairly well the experimental values if one introduces a two-electron dependent correction factor vaguely termed as polarization correction, in the treatment.

Readers are aware of Hund's rule[4] for determining the ground term of a configuration. It states that the ground of a configuration is the one with the highest possible value of spin and with as high a value of orbital angular momentum as is compatible with the first requirement. Eq (6.10) gives a quantitative indication as to why this should be so for (p^n) though it does not tell why more premium should be conferred on the spin value first than on orbital moment.

The overall physical reason as to why the highest spin states to be associated with terms of the lower repulsion energy is the presence of the largest number of exchange integral terms (K integrals) with negative sign in the theoretical expression for repulsion energies for any given electronic configuration. Strictly speaking Hund's ground state rule is valid if the configuration contains a ground configuration. A more relaxed form of Hund's rule states that for a given set of terms arising from a configuration, the terms may be arranged in groups on the basis of spin multiplicity first and the group with higher spin will be lower in energy. Now within any particular spin group, the term with highest L (orbital angular momentum quantum number) will lie deepest in that group. As an example, d^2 gives the term in descending order of energies as $^1S > ^1G > ^1D > ^3P > ^3F$. This shows Hund's rule (relaxed form) is obeyed mostly except for the deviation created by ^1G and ^1D between themselves.

6.3 Energy Levels of d^n Systems

To compute the electrostatic energies of any d^n system we may proceed in a manner followed for d^2 (sec 4.2). But this process gets complicated as n acquires higher and higher values ultimately terminating at ten. We may, instead, follow a method, like that of p^n sec 6.2, for the d^n case. But the technique will be applicable only to the highest spin states of d^n and not any of the lower spin state terms.

Let us first write down the electrostatic repulsion energies of the highest spin states of (d^2) viz. (Table T4.6)

$$\left.\begin{aligned} {}^3F &= A-8B \\ {}^3P &= A+7B \end{aligned}\right] \qquad \text{....(6.12)}$$

These are the repulsion energies of a single electron pair each with (say) α - spin and hence these constitute a measure of $\{J(12)-K(12)\}$ where J and K are coulomb and exchange integrals respectively. Incidentally these amounts of electrostatic repulsion energies are obtainable from the diagonal matrix elements of an operator $\omega_{12} = A-(8+3l_1 \cdot l_2)B$)cf Eq (6.6) of p^2 system) between the ^{3}F states and between the p^3 states respectively. The operator ω_{12} thus deputises for the actual operator e^2/r^{12}. If we have n electrons, all of α spins, d^n will change into highest spin multiplet terms of (d^n) configuration, viz. $({}^{n+1}L)$terms.

Now W which is a sum of n(n-1)/2 pair of operators of type ω_{12} is

$$W = (1/2)n(n-1)[A-8B]-(3/2)\sum^{pairs}\left(L^2-l_k^2-l_1^2\right)B$$

$$= (1/2)n(n-1)[A-8B]-(3/2)\left[L^2-\Sigma_k^n l_k^2\right]B \qquad \text{....(6.13)}$$

$$\therefore \langle {}^{n+1}L|W|{}^{n+1}L\rangle = (1/2)n(n-1)(A-8B)-(3/2)\left|L(L+1)-\Sigma_k^n l_k(l_k+1)\right|B$$

$$= (1/2)n(n-1)(A-8B)-(3/2)\left|L(L+1)-6n\right|B \qquad \text{....(6.14)}$$

This therefore, represents the total electrostatic repulsion energy i.e., the total perturbation energy equivalent to $\sum_{k<l}[J(kl)-K(kl)]$ over all pairs of electrons of the corresponding $({}^{n+1}L)$ term of a (d^n) system. It is known that the values of the highest spin multiplicities of the terms increase from 2 to 6 as we mount from d^1 to d^5. Thereafter, it goes on dropping. Using (6.14) we find the electrostatic perturbation energies of ^{4}F, ^{4}P of (d^3), ^{5}D of (d^4) and ^{6}S of (d^5).

$$\left.\begin{aligned}\varepsilon\left({}^{4}F,d^{3}\right)&=3A-15B\\ \varepsilon\left({}^{4}P,d^{3}\right)&=3A\end{aligned}\right] \quad ....(6.15)$$

$$\varepsilon\left({}^{5}D,d^{4}\right)=6A-21B \quad ....(6.16)$$

$$\varepsilon\left({}^{6}S,d^{5}\right)=10A-35B \quad ....(6.17)$$

The perturbation energies of the highest spin states of (d^6), (d^7), (d^8) will be the same as of (d^4) (Eq (6.16), of d^3) (Eq (6.15) and of (d^2) (Eq (6.12) respectively. The actual energies however, will include a separate constant additive factor each for d^6, d^7 and d^8.

Energy levels of f^n system, which are of interest in the rare earth complexes, generally fall beyond the scope of our treatment. Nevertheless, it may be mentioned that Racah [5,6] and others published papers listing the ground term energies of different f^n system. Assuming a new set of linear combinations of Slater Condon parameters to define another four new parameters viz.

$$\left.\begin{aligned}E^{0}&=F_0-10F_2-33F_4-286F_6\\ E^{1}&=(7/9)(10F_2+33F_4+286F_6)\\ E^{2}&=(1/9)(F_2-3F_4+7F_6)\\ E^{4}&=(1/3)(5F_2+6F_4-91F_6)\end{aligned}\right\} \quad ....(6.18)$$

the ground term perturbation energies were listed as follows

$$\left.\begin{aligned}\varepsilon\left({}^{2}F,f^{1}\right)&=0\\ \varepsilon\left({}^{3}H,f^{2}\right)&=E^{0}-9E^{3}\\ \varepsilon\left({}^{4}I,f^{3}\right)&=3E^{0}-21E^{3}\\ \varepsilon\left({}^{5}I,f^{4}\right)&=6E^{0}-21E^{3}\\ \varepsilon\left({}^{6}H,f^{5}\right)&=10E^{0}-9E^{3}\\ \varepsilon\left({}^{7}F,f^{6}\right)&=15E^{0}\\ \varepsilon\left({}^{8}S,f^{7}\right)&=21E^{0}\end{aligned}\right\} \quad ....(6.19)$$

6.4 Evaluation of Racah parameters :

Racah parameters, A, B, and C, as defined earlier (Eq 2.83), are some specific linear combinations of F_k's the Slater Condon parameters. For studying p^n systems, evaluations of F_k's are necessary, but for d^n complexes. Racah parameters, being more convenient, are evaluated. Some new parameters are also defined for study of f^n systems (sec 6.3). The way the Racah parameters are assigned numerical values depends upon our way of fitting the theoretical equations for the energies of the multiplets to the experimentally observed values of the atom or ion. To ensure a best possible fit of the equations, recourse is taken to minimizing the mean square deviation. This naturally demands that the parameters satisfy some energy equations. The solutions of the equations yield the optimal values of the parameters. We shall briefly outline the process here described in more detail by Griffith[7] for Zr I, Zr III, VIII. The neutral Zr atom [Zr I in Stpectroscopists' notation] has the electronic configuration, $1s^2 2s^2 p^6 3s^2 p^6 d^{10} 4s^2 p^6 d^2 5s^2$ and is thus a (d^2) system. The electrostatic perturbation energies of (d^2) system [cf Table $T_{4.6}$].

$$\varepsilon(^3F) = A - 8B \qquad \varepsilon(^1G) = A + 4B + 2C$$

$$\varepsilon(^3P) = A + 7B \qquad \varepsilon(^1S) = A + 14B + 7C \qquad \text{....(6.20)}$$

$$\varepsilon(^1D) = A - 3B + 2C$$

If we write $\varepsilon(^3F) = a$, the Eqs (6.20), can be rewritten as

$$\left.\begin{array}{lll} \varepsilon(^3F) = a & \varepsilon(^1D) = a + 5B + 2C & \varepsilon(^1S) = a + 22B + 7C \\ \varepsilon(^3P) = a + 15B & \varepsilon(^1G) = a + 12B + 2C & \end{array}\right] \qquad \text{....(6.21)}$$

The general form of Eqs (6.21) is

$$z_i = a + x_i B + y_i C \qquad \text{....(6.22)}$$

where z_1, z_2 ..... z_5 represented by z_i stand for the experimental energy, $\varepsilon(^3F)$ $\varepsilon(^3P)$, $\varepsilon(^1D)$, $\varepsilon(^1G)$, and $\varepsilon(^1S)$ respectively. The experimental energy values of the multiplets always include the spin orbit interactions and any given term splits into levels each of which is characterized by a J value. A ^3F splits into $^3F_4$, $^3F_3$ and $^3F_2$ levels. The experimental value of the multiplet ^3F which we should accept for a fit with the theoretical equation should be a weighted mean of the level energies wherein the statistical weight factor is (2J+1).

Thus the experimental energy

$$\varepsilon(^3F) = (9/21)\varepsilon(^3F_4) + (7/21)\varepsilon(^3F_3) + (5/21)\,\varepsilon(^3F_2) \qquad \text{....(6.23)}$$

In the manner of (6.23), the experimental energies of the other terms to be used for calibrations of the theoretical equations are to be duly weighted with their respective degeneracies before averaging. Now using Eq (6.22) we choose a function 'g' as the weighted square deviation where the weighting factor $f_i = (2s_i + 1)(2L_i + 1) =$ total degeneracy viz.,

$$g = \Sigma_i f_i(z_i - a - x_i B - y_i C) \quad \text{....(6.24)}$$

and minimize it by equating

$$\partial g / \partial a = \partial g / \partial B = \partial g / \partial C = 0 \quad \text{....(6.25)}$$

From this set of three optimizing equations, we can finaly arrive at values of a, B and C that would minimize 'g'. Since 'a' is a common component of the energy expressions of all the terms, the mutual differences in the positions of the different terms on the energy axis, which are indicative of the extent of excess repulsion energy over their terms, are dictated by B and C values. One can thus formulate, from the set of equations emerging basically from (6.25), for B and C only, remembering that an equation can also be drawn up for 'a', if desired.

$$B = (3/50)(z_2 - z_4) + (1/70)(z_4 - z_3) \quad \text{....(6.26)}$$

$$C = -(1/10)(2z_1 + 3z_2) + (1/14)(3z_3 + 4z_4) \quad \text{....(6.27)}$$

There are equivalent alternative expressions if one uses the energy level z_5 of the high ^1S states which are not always experimentally obtainable easily for different atoms. In using Eqs. (6.26 and 6.27) numerical values of z_i's are generally computed from an equation like (6.23) using experimental data for spin orbit levels. A comprehensive collection of such data for atomic energy levels have been complied by Moore[8] and published in National Bureau of Standards (1949, 1952, 1958). Other sources also exist from which atomic energies can be obtained or calculated. Numerical values of B and C given by different researchers sometimes do not agree. But this is largely due to different source materials of uncertain experimental reliability. Griffith9 has drawn up a table listing the values of repulsion parameters B and C

which seem dependable. These are given in Table $T_{6.1}$ for atoms in different states of ionization indicated in spectroscopists' notations as MI, MII, MIII, MIV, MV which cover the range from neutral to quadruply ionized atoms.

Table $T_{6.1}$

(Repulsion parameters B and C in cm^{-1})

(First Transition Series Elements)

		Ti	V	Cr	Mn	Fe	Co	Ni
B	I	560	578	790	720	806	798	1025
	II	682	659	710	873	869	878	1037
	III	718	766	830	960	1058	1115	1084
	IV	-	861	1030	1140	-	-	-
	V	-	-	1039	-	1144	-	-
C	I	1840	2273	2520	3087	3506	4167	4266
	II	2481	2417	2790	3130	3638	3828	4314
	III	2629	2855	3430	3325	3901	4366	4831
	IV	-	4165	3850	3675	-	-	-
	V	-	-	4238	-	4459	-	-

Griffith examined whether or not there exists any correlation between atomic numbers, in the states of ionization (MI, MII etc.) of the transitional metals and their Racah parameters. Two such correlations were formulated, one for B and another for C, from the graphical plots of the Racah parameter values in different states of ionization of the elements against their corresponding atomic numbers. If m represents the states of ionization (i.e. for MI, m = 0 for MII, m = 1 etc.) and n the position of the element in its period of transition series (i.e. n = Z -18 where Z is the atomic number) then the relations are

$$B = 145 + 80n + 95m \qquad \text{....(6.28)}$$

And $C = 520 + 370n$ for $m = 0, 1$

$$C = 1400 + 300n \text{ for } m = 2 \qquad \text{....(6.29)}$$

while (6.28) is fairly good, (6.29) is somewhat a rough estimate.

Values of B and C, given by Eqs (6.28) and (6.29) for MI and MII states of ionization of the first transition series elements appear in Table $T_{6.2}$ These may be compared with the corresponding ones of Table $T_{6.1}$.

Table $T_{6.2}$

B and C Parameters (cm^{-1}) via Eqs (6.28) & (6.29)

		Ti	V	Cr	Mn	Fe	Co	Ni
MI	B	465	545	625	705	785	865	945
	C	2000	2370	2740	3110	3480	3850	4220
MII	B	560	640	720	800	880	960	1040
	C	2000	2370	2740	3110	3480	3850	4220

6.5 Condon Slater Parameters — Worked out Examples

Emission spectral data of free atoms provide scopes for calculations of parameters like F_2 and ζ the spin orbit coupling constant. A few of the sources from among several others, for emission energy levels and ζ occur in the lists provided by Moore[8], Griffith[9] and Golding[10]. The following exercises are planned on the basis of data available in such sources.

Example 1.

The F_2 value of C in $2p^2$ configuration is 1500 cm^{-1}. The spin-orbit coupling constant in this state is 30 cm^{-1}. If the ground state energy of C in its $^3P_0$ be taken as zero level of energy, find its energy levels in $^3P_1$, $^3P_2$, $^1D_2$ and $^1S_0$ optical states.

From chapter 4 we learn that electrostatic perturbation levels of $2p^2$ state of C are

$^3P = F_0 - 5F_2 = 0$ (zero level of energy)

$^1D = F_0+F_2$

$^1S = F_0+10F_2$

$$\varepsilon(^1D)-\varepsilon(^3P)=6F_2 = 6 \times 1500 \text{ cm}^{-1} = 9000 \text{ cm}^{-1}$$

$$\varepsilon(^1S)-\varepsilon(^3P)=15F_2 = 15 \times 1500 \text{ cm}^{-1} = 22500 \text{ cm}^{-1}$$

Therefore, $\varepsilon(^1D)=9000\text{ cm}^{-1}$ and $\varepsilon(^1S)=22500\text{ cm}^{-1}$

When the spin-orbit effect within a level is taken into account, then, using Lande interval rule, we have $\varepsilon(^3P_2)-\varepsilon(^3P_1)=\lambda J=(\zeta/2).2=15\,\text{cm}^{-1}$

$$\varepsilon(^3P_1)-\varepsilon(^3P_0)=(\zeta/2).1=15\,\text{cm}^{-1}$$

Therefore, $\varepsilon(^3P_2)=30\,\text{cm}^{-1}$ and $\varepsilon(^3P_1)=15\,\text{cm}^{-1}$.

Example 2

The analysis of emission spectra of Si with configuration $1s^2\,2s^2\,p^6\,3s^2\,p^2$ leads to the assignment of the following energy levels for $^1S_0$ and $^1D_2$ states (relative to $^3P_0$ as zero level) 15394 cm^{-1} and 6300cm^{-1} respectively. Find the Condon Slater parameter F_2.

$$\varepsilon(^1S_0)-\varepsilon(^3D_2)=9094cm^{-1}$$

or, $(F_0+10F_2)-(F_0+F_2)=9094cm^{-1}$

or, $9F_2=9094cm^{-1}$

Therefore, $F_2 = 1010.4\ \text{cm}^{-1}$

Exercises

1. Utilise Eq. (6.25) to establish the Eqs. (6.26) and (6.27) [Hint. Write the full form of g-function in terms of Eq. (6.21) and proceed to obtain the equations for the optimal values of B and C]
2. Elaborate the arguments underlying the conversion of Eq. (6.13) into Eq (6.14).

References

1. P.A.M. Dirac : See Sokolov, Ternov, Zhukovskii: *Quantum Mechanics* (Mir Publishers).
2. J.H. Van Vleck : Phys. Rev., 45, 416, 1934.
3. 7,9, J. S. Griffith : *The Theory of Transition Metal Ions* (Cambridge University Press)
4. F. Hund : *Linienspektren*, P. 124, 1927
5. G. Racah : *Phys. Rev.* 62. 438, 1942; also 61, 186, 1942.
6. a. J.P. Elliot, G. Racah : *Phys. Rev.*, 76, 1352, 1949.

   b. B. R. Judd and W. A Runeiman : Proc. Roy Soc., A 240, 509, 1957.
7. C.E. Moore : *U.S. National Bur. Std. Circ.* 467, 1949
8. R. M. Golding : *Applied Wave Mechanics*. (D. Van Nostrand)

Chapter Seven

Crystal Field Hamiltonians and Their Matrix Elements

General

The idea of crystal field was first mooted by Bethe[1] and van Vleck[2] to account for the perturbational effect of the electrostatic field exuded by an environment of ions arranged in the crystal lattice on the electronic orbitals of some central ion. Such an idea of perturbation was borrowed from the crystal science for application to individual molecules of transition metal complexes. It is assumed that a d^n metal ion of a complex in an environment of similar or mostly similar ligands is subjected to what is called a crystal field perturbation. Each ligand, irrespective of its atomic build-up, is assumed to serve as a negative point charge or as a point dipole with its negative end turned towards the central ion. Such an array of point charges or point dipoles exerts electrostatic repulsions on the d-electrons of the central ion causing a partial or total lifting of the degeneracies of the d-orbitals. In such a purely electrostatic treatment, the metal ion is deemed to be a distinct entity with no covalent overlap of its orbitals with those of the ligands, Conditions for these deficiencies are amended in the ligand field theory in an empirical manner and more or less quantitatively in the molecular orbital treatment. We shall, however, confine ourselves to the purely electrostatic viewpoint and assume the ligands to be locations of point charges, -Ze each, for similar ligands, where, Z is not necessarily unity.

Recall the relations (2.25 and 2.26) for the hamiltonian of an atom. Likewise we can write for the hamiltonian of a d^n metal ion present in a surrounding of ligands as

$$\mathscr{H}_{ion} = \mathscr{H}_{cent} + \sum_{k<l} e^2 / r_{kl} + \sum_k \xi\, l_k . s_k + \mathscr{H}_{crys} \qquad \text{.........(7.1)}$$

Wherein a new perturbation term is now introduced, viz.,

$$\mathscr{H}_{crys} = \sum_i^n h_{crys(i)} = h_{crys(1)} + h_{crys(2)} + \cdots + h_{crys(n)} \qquad \text{.........(7.2)}$$

Here $h_{crys(i)}$ is the crystal field hamiltonian experienced by the ith d-electron. For a d^1-metal ion complex

$$\mathscr{H}_{ionv} = \mathscr{H}_{cen} + \xi\, l.s + h_{crys} \qquad \text{.........(7.3)}$$

We shall examine in the chapters to follow how new situations or states evolve when $\mathscr{H}_{crys}/h_{crys}$ as in Eqs. (7.1) or (7.3) is added to the list of perturbations. As a prelude to all these, quantitative formulations $\mathscr{H}_{crys}/h_{crys}$ in different symmetry environments of ligands will first be made and will be followed by revealing the technique of evaluation of their matrix elements in this chapter. This will form the fundaments on which further explorations of the subsequent chapters will rest.

7.1 Crystal Field Hamiltonian – General Form.

We shall now address ourselves to the tasks, viz.,

1. Finding the general form of $\mathscr{H}_{crys}$
2. Sizing down the general form according to the requirements of geometry and symmetry of the environment.
3. Selecting mathematical method for evaluation of the matrix elements of $\mathscr{H}_{crys}/h_{crys}$.

1. General form of the crystal field hamiltonian

   Consider a ligand point charge, - Ze, of the pth ligand and an electron (say a d-electron) at a distance R_p from the former. The potential energy of repulsion is Ze^2/R_p. If we now use a coordinate system with the origin at O such that the coordinates of, p are $(a_p, \theta_p, \varphi_p)$ and those of the electron (a, θ, φ) it is then possible to express $1/R_p$ in terms of the Legendre's polynomial in (cos γ) [cf. Eq.1.26] where γ is the angle between the radial vectors a_p and r (see adjacent figure).

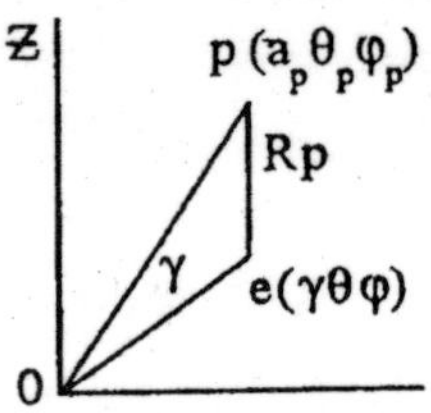

$$1/R_p = 1/a_p \sum_{k=0}^{\infty} (r/a_p)^k P_k(\cos\gamma) \quad ......(7.4)$$

It is possible to express Legendre's polynomials in terms of spherical harmonics by addition theorem (Eq. 1.28)

$$P_k(\cos\gamma) = \sum_{m=-k}^{k} (4\pi/2k+1) Y_{km}*(\theta_p, \varphi_p) Y_{km}(\theta, \varphi) \quad ......(7.5)$$

wherein the variables are measured with respect to the coordinate system set up at O, the z-axis serving as the polar axis. The potential due to the pth ligand and a single d-electron is

$$Ze^2/R_p = \sum_{k=0}^{\infty} \sum_{m=-k}^{k} (4\pi/2k+1) Ze^2 \{r^k/a_p^{k+1}\} Y_{km}*(\theta_p, \varphi_p) Y_{km}(\theta, \varphi) \quad ...(7.6)$$

Now in a complex ion, if there be N number of proxy charges for the ligands arranged around the central ion O in a particular symmetry (such as O_hT_d etc.) with each a_p (p=I to N) >r the total repulsion energy experienced by the d-electron due to all such proxy charges, i.e. the crystal field hamiltonian for a single d-electron, (using 7.6) is

$$h_{crys} = \sum_{k=0}^{\infty} \sum_{m=-k}^{k} \sum_{p=1}^{N} \{Ze^2/a_p^{k+1}\} (4\pi/2k+1) r^k Y_{km}*(\theta_p, \varphi_p) Y_{km}(\theta, \varphi) \quad ...(7.7)$$

Since for known geometry and symmetry of the distribution of the ligand proxy charges $\sum_{p=1}^{N}\{Ze^2/a_p^{k+1}\}Y_{km}*(\theta_p,\varphi_p)$ can be evaluated for a given k, m and Z, we may rewrite (7.7) as

$$h_{crys}=\sum_{k=0}^{\infty}\sum_{m=-k}^{k}A_k^m(4\pi/2k+1)r^kY_{km}(\theta,\varphi) \quad \text{.........(7.8)}$$

where $A_k^m=\sum_{p=1}^{N}\{Ze^2/a_p^{k+1}\}Y_{km}*(\theta_p,\varphi_p)$ ....(7.9)

which is the general expression for the crystal field hamiltonian for a single d-electron. When a multi –d electron (d^n) complex is considered, the general form of the crystal field Hamiltonian becomes

$$\mathscr{H}_{crys}=\sum_i^n h_{crys(i)}=\sum_{i=1}^{n}\{\sum_{k=0}^{\infty}\sum_{m=-k}^{k}A_k^m(4\pi/2k+1)r_i^kY_{km}(\theta_i,\varphi_i)\} \quad \text{.........(7.10)}$$

2. Sizing down the hamiltonian

The expression (7.7) shows that h_{crys} consists of an infinite number of terms. But the demands of environmental symmetry causes a severe clipping of terms in the hamiltonian to delimit its content to just an effective three or four term expression. The latter depends on the symmetry of the legand environment and the nature of the matrix elements of h_{crys} . Symmetry intervenes and helps in the simplification of the process in the following ways:

a) h_{crys} (and hence $\mathscr{H}_{crys}$) must be totally symmetric in the point group of the perturbed system (i.e., O_h, T_d etc.)

b) Any matrix element of $\mathscr{H}_{crys}$ / h_{crys} will resolve ultimately into a number of integrals consisting of bras and kets of d-electron orbital (in case of d^n complexes) with the operator $\mathscr{H}_{crys}$ / h_{crys} in between. The d-electron orbitals are of the same parity (in O_h, D_{4h}) and may belong to different IR's in other point groups (T_d, D_{3d}, C_{4v}). The matrix elements of the odd order spherical harmonics between two d-orbitals of even parity or between two f- orbitals of odd parity do not survive. Since the odd order spherical harmonics in $\mathscr{H}_{crys}$(or h_{crys}) [i.e., with k = 1, 3, 5 etc.] do not contribute, the effective crystal field hamiltonian should thus be shorn of all the odd order harmonics. The permissible values of k should thus be restricted to k = 0, 2, 4, 6. Similar conclusion applies to single or multi –f electron complexes of the rare earth series.

c) Gaunt's condition (Eq. 2.81) for the survival of the matrix element. $<\Psi'|h_{crys}|\Psi>$, where l' and l are the orbital angular momentum quantum numbers associated with the d-electron $<\Psi'|and|\Psi>$, is $|l-l'|\le k\le l+l'$. This therefore, restricts the permissible even values of k to 0, 2 and 4 when d-electron complexes are under consideration and 0, 2,4 and 6 and f-electron complexes.

d) The number of spherical harmonics in $\mathscr{H}_{crys}$ / h_{crys} with allowed values of k=0,2,4 (d-electron case) may further be curtailed depending on environmental symmetries. Thus while for D_{4h} symmetry k = 0, 2, 4 are to be considered, for O_h and T_d symmetries, spherical harmonic component with k = 2 in $\mathscr{H}_{crys}$ / h_{crys} is not permitted. This final weeding out of some specific even k value in some environmental symmetries emerges from the basic requirement laid down in (a) We shall find the specific forms of $\mathscr{H}_{crys}$ for different symmetries in later sections (secs. 7.3 – 7.9).

3. Selecting the mathematical methods for evaluation of matrix elements :

The matrix elements of the crystal field perturbation operator consist of integrals of the triple product of spherical harmonics besides the radial parts. Gaunt[3] carried out such integrations over the triple product of spherical harmonics and gave a general expression for the result of integration. The value of the component integrals over the product of the Associated Legendre's polynomials are available in the form of c^k table (Table $T_{2.1}$). We shall adopt this method of using the c^k values as we have done in Chaps. 4 and 5. Details concerning the actual evaluation of the matrix elements in the basis set of d-wavefunctions appear later (sec. 7.10).

Alternative method [4,5,6] have also been developed on the basis that the angular momentum operators, l_x, l_y, l_z, l^2 and their different linear combinations can substitute x , y, z and their powers in the matrix element expressions leading to integral values proportional to actual results.

7.2 Crystal Field Hamiltonian in Different Environmental Symmetries Choice of Axial Systems.

Having now learnt (sec. 7.1) that only spherical harmonics with k = 0, 2, and 4 may, in general, occur in $\mathscr{H}_{crys}$ of d-electron complexes and the these need to combine linearly in such a way that combination forms a basis of the totally symmetric IR of the point group of the complex, the immediate problem before us is two-fold.

a) Are all the harmonics with k = 0, 2, 4 present in the $\mathscr{H}_{crys}$ of a given point group symmetry or is any of them missing?

b) For any given symmetry of the complex, how is one to hit upon the concrete form of $\mathscr{H}_{crys}$ conforming to the totally symmetric IR?

We shall try to answer these two queries for the specific cases of environments of O_h, T_d, cubic (O_h), D_{4h}, D_{3d} and $D_{\infty h}$ symmetries around the d-electron metal ion of the complex in the next few forthcoming sections. Before this can be done one has to make a choice about the relative orientations of essentially two axial systems with the common origin O located at the central ion. The spherical harmonics of the d- orbital functions and those present in $\mathscr{H}_{crys}$ are essentially defined in terms of variables having the Z axis at O serving also as the quantizing axis. When a symmetry is developed by placing ligands (proxy charges) around the central ion, an additional axial system is needed at the origin O to define the locations of the symmetry axes and the reflection planes. As is the custom in the symmetry theory, the principal rotational axis or any one of the several highest fold symmetry axes is chosen to be the Z- axis of this second symmetry defining axial system. It is, however, possible to arrange the ligands of the environment about the central ion in such a way as to make

i) the highest fold symmetry axis and the quantizing axis collinear; thus the two axial systems at O become coincident and merge into one another.

or, ii) the symmetry axis non-collinear with the quantizing axis; thus two distinct axial systems with different orientations exist at O.

We shall stick to the case of coincidental axial systems in our treatment here postponing discussion on $\mathscr{H}_{crys}$ with choice of separate axial systems until Sec. 7.7 of this chapter.

7.3 Crystal Field Hamiltonian in Octahedral Field $\mathscr{H}_{crys}(0_h)$

i) While the spherical harmonic with k = 0 consists of a single member Y_{00} those with k = 2 and k = 4, comprise five functions Y_{2m} (m=2,1,0,-1,-2) and nine functions Y_{4m} (m=4,3,2,1,0,-1,-2,-3,-4) respectively. Since Y_{00} of any d-electron is constant (appendix Chap-2), it naturally forms a basis of the totally symmetric IR, A_{1g} of the 0_h point group and is thus a component of $\mathscr{H}_{crys}(0_h)$.

To check whether the Y_{2m} functions also span the A_{1g} of the 0_h point group, it is necessary to ascertain the types of IR subspaces present in a reducible representation of the 0_h point group (or even the 0 point group) drawn up in a basis set of the five spherical harmonics. Using the 0 point group (a subgroup of 0_h) and the relations for the characters of the rotation matrices (Eqs. 2.87, 2.88) the following values are obtained.

	E	$6C_4$	$6C_4^2$	$6C_2^/$	$8C_3$
χ^{red}	5	-1	1	1	-1

......(7.11)

Employing these character values in Eq. (2.84) of Sec. 2.10 along with the requisite data of the character table of the 0 point group, it is found that

$$n_{A1} = l/g \sum_R \chi_{A1}*(R)x^{red}(R) = 0$$

This indicates that the spherical harmonics, Y_{2m}'S do not have any component belonging to A_{1g} of the 0_h point group and are thus ruled out from participating in the expression of $\mathscr{H}_{crys}(O_h)$.

A similar exercise with the nine spherical harmonics of Y_{4m}(k=4) and the rotational characters of the 0 point group leads to a reducible character set

	E	$6C_4$	$6C_4^2$	$6C_2^/$	$8C_3$
χ^{red}	9	1	1	1	0

……(7.12)

These data, along with the appropriate ones from the character table of the point group 0, when fed into Eq. (2.84) for checking the presence of A_1. IR, provide an affirmative answer. Hence $\mathscr{H}_{crys}(O_h)$ contains some harmonic components of Y_{4m} along with that of Y_{00} already stated before. Thus for O_h symmetry environment the permissible values of k are just 0 and 4 in the relations (7.8, 7.9, 7.10), the value of k = 2 being ruled out. This answers the query (a) raised in sec. 7.2.

For an answer to the query (b) of sec. 7.2 we first write for O_h symmetry

$$\mathscr{H}_{crys}(O_h) = \mathscr{H}^{/}{}_{crys}(O_h) + \mathscr{H}^{//}{}_{crys}(O_h) = \Sigma_i\, h^{/}{}_{crys\,(i)}\,(0_h) + \Sigma_i\, h^{//}{}_{crys\,(i)}\,(0_h)$$

where $\mathscr{H}^{/}{}_{crys}(O_h) = \Sigma_i\, h^{/}{}_{crys\,(i)}\,(0_h)$ and $h^{/}{}_{crys\,(i)}\,(0_h)$, as given by Eq. (7.10), is $A^0{}_0 4\pi\, Y_{00}(\theta_i,\varphi_i)$ which is a constant as shown afterwards. For h $^{//}{}_{crys,}$ also given by Eq.(7.10) in the multi electron form, let us choose its component $\mathscr{H}^{//}{}_{crys(i)}$ for the ith electron. The $\Sigma_k \Sigma_m Y_{km}(\theta_i\varphi_i)$ part in $h^{//}{}_{crys(i)}$ has to be substituted by an A_{1g} symmetry adapted linear combination of some functions of the Y_{4m} set with proper values of m, k being equal to the fixed value 4. To find what these "some functions" of $Y_{4m}(\theta_i,\varphi_i)$ are, we adopt the technique enunciated in sec. 2.10. The results of some selected operators of the 0_h point group upon members of the $Y_{4m}(\theta_i,\varphi_i)$ yield some piecemeal information coordination of which enables us to arrive at the desired linear combination, symmetry adapted to A_{1g}, of some specific members of the Y_{4m} set. To apply the selected operations, it will be convenient to remember how x_i, y_i, z_i of the polynomial functions change under the various rotational operators of the 0 point group and also how θ_i and φ_i alter under such operators.

i) Effect of rotational operators under the 0 point group (C_4^z axis as the quantizing axis)

$$C_4^z\begin{vmatrix}x\\y\\z\end{vmatrix}=\begin{vmatrix}y\\-x\\z\end{vmatrix}\quad C_2^z\begin{vmatrix}x\\y\\z\end{vmatrix}=\begin{vmatrix}-x\\-y\\x\end{vmatrix}\quad C_2^/\begin{vmatrix}x\\y\\z\end{vmatrix}=\begin{vmatrix}x\\y\\-z\end{vmatrix}\quad C_3\begin{vmatrix}x\\y\\z\end{vmatrix}=\begin{vmatrix}y\\z\\x\end{vmatrix} \quad \text{………(7.13)}$$

With the results of (7.13) already available, let us operate by C_4^z upon the individual members of the Y_{4m} set (appendix Chap.2) in cartesian forms.

$$\left.\begin{array}{lll} C_4Y_{44} = Y_{44} & C_4Y_{40} = Y_{40} & C_4Y_{4-4} = Y_{4-4} \\ C_4Y_{43} = Y_{43}.i & & C_4Y_{4-3} = -Y_{4-3}.i \\ C_4Y_{42} = -Y_{42} & & C_4Y_{4-2} = -Y_{4-2} \\ C_4Y_{41} = -Y_{41}.i & & C_4Y_{4-1} = Y_{4-1}.i \end{array}\right\} \quad \text{.........(7.14)}$$

Since $\mathscr{H}_{crys}(0_h)$ [and hence $\mathscr{H}^{//}{}_{crys}(0_h)$] is to be totally symmetric, it is necessary that $C_4\ \mathscr{H}_{crys}(0_h) = \mathscr{H}_{crys}(0_h)$ and this, when extended to the ith d-electron, means $C_4h^{//}{}_{crys(i)}(0_h) = h^{//}{}_{crys(i)}(0_h)$. This demands that the $Y_{4m}(\theta_i,\varphi_i)$ members, when operated upon by C_4, should satisfy this invariance condition. Relations (7.14) tell us the $Y_{40}(\theta_i,\varphi_i)$, $Y_{44}(\theta_i,\varphi_i)$ and $Y_{4-4}(\theta_i,\varphi_i)$ of the Y_{4m} set retain this property. Hence, the A_{lg}- adapted linear combination for $h^{//}{}_{crys(i)}(0_h)$ is

$$Y_{40} + a\,Y_{44} + b\,Y_{4-4} \quad \text{...........(7.15)}$$

the spherical harmonics being functions of θ_i, φ_i of the ith electron

ii) the second piecemeal information is obtained by operating on (7.15) by one $C_2^{/}$ of the 0_h point group to insure the protection of the invariance property.

$$Y_{40} + a\,Y_{44} + b\,Y_{4-4} = C_2^{/}\left(Y_{40} + a\,Y_{44} + b\,Y_{4-4}\right) = Y_{40} + a\,Y_{4-4} + b\,Y_{44} \quad \text{.........(7.16)}$$

since $C_2^{/}Y_{40} = Y_{40}$, $C_2^{/}Y_{44} = Y_{4-4}$ and $C_2^{/}Y_{4-4} = Y_{44}$.

Relation (7.16) may represent an invariant property only if a =b. This makes the A_{lg} symmetry adapted linear combination in $h^{//}{}_{crys(i)}(0_h)$ to be

$$Y_{40} + a\left(Y_{44} + Y_{4-4}\right) \quad \text{...........(7.17)}$$

iii) The third useful information is obtained by demanding the retention of the invariance property when C_3 operates on (7.17) This will settle the magnitude of a. Using the functional forms of Y_{40}, Y_{44} and Y_{4-4} form appendix (Chap.2) in the cartesian forms and making use of (7.13) it is possible to write.

$$\sqrt{(9/4\pi)}\sqrt{(1/64)}1/r^4\left\{35z^4 - 30z^2r^2 + 3r^4\right\} + a\sqrt{\left(9/4\pi\right)}\sqrt{\left(35/128\right)}\left(1/r^4\right)\left\{(x+iy)^4 + (x-iy)^4\right\}$$

$$= C_3\sqrt{\left(9/4\pi\right)}\sqrt{\left(1/64\right)}1/r^4\left\{35z^4 - 30z^2r^2 + 3r^4\right\} + a\sqrt{\left(9/4\pi\right)}\sqrt{\left(35/128\right)}\left(1/r^4\right)\left\{(x+iy)^4 + (x-iy)^4\right\}$$

$$= \sqrt{\left(9/4\pi\right)}\sqrt{\left(1/64\right)}\,1/r^4\left\{35x^4 - 30x^2r^2 + 3r^4\right\} + a\sqrt{\left(9/4\pi\right)}\sqrt{\left(35/128\right)}\left(1/r^4\right)\left\{(y+iz)^4 + (y-iz)^4\right\} \quad \text{......(7.18)}$$

Since the invariance property is to be retained by (7.18), the latter becomes an identity relation. We can thus equate the coefficients of z^4 from the LHS and the RHS of (7.18). This yields (note the ticked terms on RHS and LHS involve z^4 terms)

$$\sqrt{(1/64)}\{35-30+3\}=\left(3/\sqrt{64}\right)+a\sqrt{(35/128)}\{1+1\}$$

$$\therefore a=\sqrt{(5/14)} \qquad \text{.........(7.19)}$$

Thus considering (7.17) and (7.19) together we find that $\Sigma_k \Sigma_m Y_{km}(\theta_i,\varphi_i)$ part in

$$h''_{crys(i)}(O_h) \text{ of (7.10) is } Y_{40}+\sqrt{(5/14)}(Y_{44}+Y_{4-4}) \qquad \text{.........(7.20)}$$

The full form of $h_{crys(i)}(O_h)$ can thus be written as [see Eq. 7.8 or 7.9]

$$h_{crys(i)}(O_h) = h'_{crys(i)}(O_h) + h''_{crys(i)}(O_h)$$

$$= A_0^{\,0}4\pi Y_{00}(\theta_i,\varphi_i)+\left\{(4\pi/9)A_4^{\,0}r_i^{\,4}Y_{40}+(4\pi/9)\sqrt{(5/14)}\left(A_4^{\,4}Y_{44}+A_4^{\,-4}Y_{4-4}\right)r_i^{\,4}\right\} \ldots\text{(7.20a)}$$

Thus once we evaluate A_0^0, A_4^0, A_4^4 and A_4^{-4} using Eq. (7.9), the expression for $h_{crys(i)}(O_h)$ will be fully revealed. This in turn will give the full expression for $\mathscr{H}_{crys}(O_h)$.

II. Specific values of A_0^0, A_4^0, A_4^4 and A_4^{-4}

Consider an octahedron of similar ligand charges (each-Ze) and a_p= a, the ligands being equidistant from the origin. If the proxy charges be labeled as 1, 2, 3, 4 in the XY -plane and 5, 6 for the axial positions, the spherical polar coordinates of the charges 1, 2, 3 .... 6 are

$$\left(a,\pi/2,0\right),\left(a,\pi/2,\pi/2\right),\left(a,\pi/2,\pi\right),\left(a,\pi/2,3\pi/2\right),\left(a,0,\pi/2\right) \text{and} \left(a,\pi,\pi/2\right) \quad \text{respectively.}$$

Using Eq. (7.9) for the six point charges,

$$A_0^{\,0} = A^0{}_{0(1)} + A^0{}_{0(2)} + A^0{}_{0(3)} + A^0{}_{0(4)} + A^0{}_{0(5)} + A^0{}_{0(6)}$$

$$=\left(Ze^2/a\right)\left[Y_{00}*(\pi/2,0)+Y_{00}*\left(\pi/2,\pi/2\right)+Y_{00}*\left(\pi/2,\pi\right)+\right.$$

$$\left.Y_{00}*\left(\pi/2,3\pi/2\right)+Y_{00}*\left(0,\pi/2\right)+Y_{00}*\left(\pi,\pi/2\right)\right]$$

$$=\left(1/\sqrt{4\pi}\right)6\,Ze^2/a \qquad \text{.........(7.21)}$$

$$A_4^{\,0} = \left(Ze^2/a^5\right)\left[Y_{40}*\left(\pi/2,0\right)+Y_{40}*\left(\pi/2,\pi/2\right)+Y_{40}*\left(\pi/2,\pi\right)+\right.$$

$$\left.Y_{40}*\left(\pi/2,3\pi/2\right)+Y_{40}*\left(0,\pi/2\right)+Y_{40}*\left(\pi,\pi/2\right)\right]$$

$$= \left(9/128\right)^{1/2}(2\pi)^{-1/2}.28\left(Ze^2/a^5\right) \qquad \text{.........(7.22)}$$

The specific values of the respective spherical harmonics are obtained by substituting the coordinates in the expressions for spherical harmonics (appendix Chap. 1).

We may evaluate A_4^4 and A_4^{-4} indirectly from (7.22) and the symmetry based conclusion leading to (7.20). The total symmetric behaviour of $h''_{crys(i)}(0_h)$ demands through (7.20) that the linear combination $\left\{Y_{40} + \sqrt{(5/14)}(Y_{44} + Y_{4-4})\right\}$ be present in the final expression. It is feasible if the coefficients associated with both $\sqrt{(5/14)}Y_{44}$ and $\sqrt{(5/14)}Y_{4-4}$ be each equal to the coefficient of Y_{40}. If c be the common coefficient viz. $C = (4\pi/9)A^0{}_4\, r_i^4$, (cf. Eq. 7.20a), then the symmetry adapted linear combination in

$h''_{crys(i)}$ is $h''_{crys(i)} = C\left[Y_{40} + \sqrt{(5/14)}(Y_{44} + Y_{4-4})\right]$ ....(7.23)

Thus (7.20a), (7.21) (7.22) and (7.23), when appropriately used in Eq. (7.8) leads to

$$
\begin{aligned}
h_{crys(i)}(0_h) &= h'_{crys(i)}(0_h) + h''_{crys(i)}(0_h)\\
&= 4\pi\left(1/4\pi\right)^{1/2}\left(6Ze^2/a\right)Y_{00}(\theta_i,\varphi_i) + \left(4\pi/9\right)\left(9/128\right)^{1/2}(2\pi)^{-1/2}\,28\left(Ze^2/a^5\right)r_i^4\\
&\quad x\left\{Y_{40}(\theta_i,\varphi_i) + \sqrt{(5/14)}(Y_{44} + Y_{4-4})(\theta_i,\varphi_i)\right\}\\
&= 6Ze^2/a + (7/3)\sqrt{\pi}\left(Ze^2/a^5\right)r_i^4 Y_{40}(\theta_i,\varphi_i) +\\
&\quad \left(35/18\right)^{1/2}\sqrt{\pi}\left(Ze^2/a^5\right)r_i^4(Y_{44} + Y_{4-4})(\theta_i,\varphi_i) \qquad \text{...........(7.24)}
\end{aligned}
$$

When one is dealing with a multi-d-electron complex, i.e., d^n complex, one has to use $\mathscr{H}_{crys}(0_h)$ Eq. (7.10) in which (7.24) has to be summed over all the n electron interactions with the ligand charges, viz.,

$$
\begin{aligned}
\mathscr{H}_{crys}(0_h). &= \Sigma_{i=l}^{n} h_{crys(i)}(O_h) = \Sigma_{i=l}^{n} h'_{crys(i)}(O_h) + \Sigma_{i=l}^{n} h''_{crys(i)}(O_h)\\
&= n6\,Ze^2/a + \Sigma_{i=l}^{n}\left[(7/3)\sqrt{\pi}\left(Ze^2/a^5\right)r_i^4 Y_{40}(\theta_i,\varphi_i) +\right.\\
&\quad \left.\left(35/18\right)^{1/2}\sqrt{\pi}\left(Ze^2/a^5\right)r_i^4(Y_{44} + Y_{4-4})(\theta_i,\varphi_i)\right] \qquad \text{...........(7.25)}
\end{aligned}
$$

Some authors[8], however, have preferred to shun the symmetry arguments and have calculated all the $A^m{}_k$'s in arriving at the final expression for $h_{crys(i)}(0_h)$. In such an approach a very large number of individual $A^m{}_k$ calculations become necessary. The labour saved by shunning symmetry based method pops up in a magnified form during the individual calculations of all the $A^m{}_k$'s.

7.4. Crystal Field Hamiltonian in Tetrahedral Environment $\mathscr{H}_{crys}(T_d)$

If we exclusively confine ourselves to purely d-electron complexes, then all the symmetry based arguments, employed in the previous section devoted to 0_h symmetry, apply equally well to the T_d environment of the same ligands around the central ion. In this case, one has to select only the group operators pertinent to T_d symmetry to arrive at a concrete expression for $\mathscr{H}_{crys}(T_d)$ or $h_{crys\ (i)}(T_d)$. A slight modification will, however, creep in the expression for $\mathscr{H}_{crys}(T_d)$ when the complex is not purely a d-electron complex but includes a slight amount of d-p mixing. This will be attended to later on.

Consider a cube, measuring $2l$ each side, with the metal ion at its centre O and the Z-directional axis (quantizing axis) coinciding with the C_4^z symmetry axis of the cube (Fig. 7.1).

In a tetrahedral complex the proxy ligand charges occupy the corners 1, 3, 6 and 8. The spherical polar coordinates of the point charges are $\left(\sqrt{3}l, \theta, \pi/4\right)$ $\left\{\sqrt{3}l, \theta, \pi + \pi/4\right\}$ $\left\{\sqrt{3}l, (\pi-\theta), \left(\pi - \pi/4\right)\right\}$ $\left\{\sqrt{3}l, (\pi-\theta), \left(2\pi - \pi/4\right)\right\}$ respectively where $\cos\,\theta = \sqrt{3/3}$ and $\cos\,(\pi-\theta) = -\cos\theta = -\sqrt{3/3}$. Symmetry based arguments being similar both for 0_h and T_d environments, we have to first evaluate A_0^o and A_4^o (cf. Eq. 7.9) to obtain the hamiltonian $h_{crys(i)}(T_d)$.

$$A_0^0 = A_{0(1)}^0 + A_{0(3)}^0 + A_{0(6)}^0 + A_{0(8)}^0$$

$$= Ze^2/\left(\sqrt{3}l\right)Y_{00}x4 = \left(4Ze^2/b\right)\,1/\sqrt{4\pi} \qquad \text{.........(7.26)}$$

Where $\sqrt{3}l = b =$ distance of the proxy charges from the metal ion at the centre.

$$A_4^0 = A_{4(1)}^0 + A_{4(3)}^0 + A_{4(6)}^0 + A_{4(8)}^0$$

$$= (2\pi)^{-1/2} Ze^2/b^5\left[\Theta_{40}(\cos\theta) + \Theta_{40}(\cos\theta) + \Theta_{40}\{\cos(\pi-\theta)\} + \Theta_{40}\{\cos(\pi-\theta)\}\right]$$

$$= (2\pi)^{-1/2}(9/2)^{1/2}(1/64)^{1/2}\left[35\cos^4\theta - 30\cos^2\theta + 3\right]4Ze^2/b^5$$

$$= -(2\pi)^{-1/2}\,(9/128)^{1/2}(28/9)4\,Ze^2/b^5 \qquad \text{.........(7.27)}$$

$$= -4/9\ \{A_4^0 \text{ of } O_h \text{ group with b substituting a}\}$$

Here $\Theta_{40}(\cos\theta)$ represents the normalized Associated Legendre's polynomial, the normalized Φ function being $(2\pi)^{-1/2}$, since m=0 in each case. We need not evaluate A_4^4 and A_4^{-4} separately, since on symmetry grounds, like in 0_h of Sec. 7.3 each of them should be related to A_4^o. Effectively it means that the coefficient of $\sqrt{(5/14)}\{Y_{44} + Y_{4-4}\} = c = A_4^o.\dfrac{4\pi}{9}.r_i^4$ Thus using (7.8) we can write for the tetrahedral crystal field hamiltonian for the d-electron

$$h_{crys(i)}(T_d) = A_0^o 4\pi Y_{00}(\theta_i,\varphi_i) + (4\pi/9)A_4^o\left[Y_{40}(\theta_i,\varphi_i) + \sqrt{(5/14)}(Y_{44} + Y_{4-4})(\theta_i,\varphi_i)\right]r_i^4$$

$$= 4Ze^2/b - 4/9\Big[(7/3)\sqrt{\pi}\left(Ze^2/b^5\right)r_i^4 Y_{40}(\theta_i,\varphi_i) +$$

$$\left(35/18\right)^{1/2}\sqrt{\pi}\left(Ze^2/b^5\right)r_i^4\left(Y_{44}+Y_{4-4}\right)(\theta_i,\varphi_i)\Big] \quad \ldots\ldots\ldots\ldots(7.28)$$

wherein, to arrive at (7.28), Eqs. (7.8, 7.26, 7.27) have been employed. The expression for $\mathscr{H}_{crys}(T_d)$ is obtained from (7.28) by summing over all the n-electron coordinates. As in $h_{crys(i)}(O_h)$, the expression for $h_{crys(i)}(T_d)$ can be slit into two parts, viz.

$$h_{crys(i)}(T_d) = h'_{crys(i)}(T_d) + h''_{crys(i)}(T_d)$$

where $h'_{crys(i)}(T_d)$ is $4Ze^2/b$ and $h''_{crys(i)}(T_d)$ represents the rest of (7.28). Of the two component parts of $h_{crys(i)}$, $h'_{crys(i)}$ is a constant being $6Ze^2/a$ for O_h symmetry and $4Ze^2/b$ for T_d symmetry and is thus independent of the stereodispositions of the ligands $(\theta_p$ *and* $\varphi_p)$. $h''_{crys(i)}$ part of the hamiltonian is expressive of the interactions of the d-electron dependent on the stereodispositions of the ligands and hence on the environmental symmetry O_h or T_d. As we shall see afterwards this $h''_{crys(i)}$ causes the splitting of the energy levels relative to the new baricentre generated by $h'_{crys(i)}$.

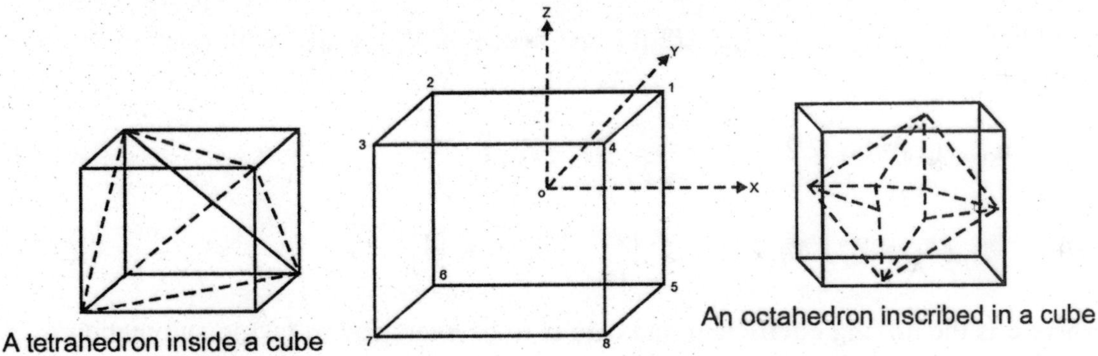

Fig. 7.1 Four-fold symmetry axis as the quantising axis

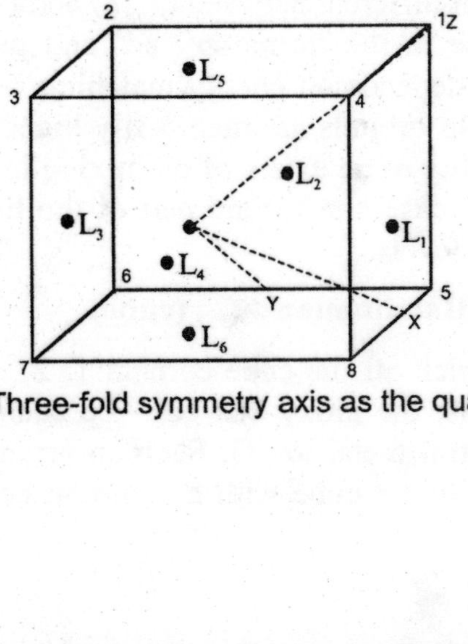

Fig. 7.2 Three-fold symmetry axis as the quantising axis

A scrutiny of Eqs. (7.24 & 7.28) reveals that when b=a, i.e., when the similar ligands are at the same distance from the central ion, both in 0_h and T_d environments, it is possible to write

$$\{h''_{crys(i)}(T_d)\}/\{h''_{crsys(i)}(0_h)\}=-4/9$$

i.e., $$h''_{crys(i)}(T_d)=-4/9\, h''_{crys(i)}(0_h) \qquad ....(7.29)$$

The application of this relation will be clarified in the next chapter

There exists a small aberration in our arguments for inducting the whole of the theory of $\mathscr{H}_{crys}$ of 0_h symmetry for constructing the $\mathscr{H}_{crys}$ of T_d symmetry. Tetrahedral systems do not have any inversion symmetry element whereas the 0_h systems do. Hence $\mathscr{H}_{crys}(T_d)$ needs be A_1 symmetry adapted and not A_{lg}. Additionally it is seen that a specific linear combination of odd order harmonics, viz., $\left(1/i\sqrt{2}\right)\{Y_{32}-Y_{3-2}\}$ behaves like A_1 under T_d group operators. Logically, therefore, $\left(1/i\sqrt{2}\right)\{Y_{32}-Y_{3-2}\}$ should also from a part of $h_{crys\,(i)}(T_d)$ [and hence of $\mathscr{H}_{crys}(T_d)$] with other even order harmonics.

$$h_{crys(i)}(T_d)=\text{Eq. 7.28}+c\left(1/i\sqrt{2}\right)\{Y_{32}-Y_{3-2}\} \qquad ....(7.30)$$

and $$\mathscr{H}_{crys}(T_d)=\Sigma_i^n\left[Eq.7.28+c\left(1/i\sqrt{2}\right)\{Y_{32}-Y_{3-2}\}\right] \qquad ....(7.31)$$

where c is the mixing coefficient and care is to be exercised in taking summation over the dummy index i leaving the imaginary quantity i unmolested. When we are considering purely d-electron complexes, the matrix elements of $\left(1/i\sqrt{2}\right)\{Y_{32}-Y_{3-2}\}$ vanish. There is then no material different between (7.28) and (7.30). Theoretically, the tetrahedral hybrids are possible in the forms sp^3, sd^3 and pd^3. Whichever of these materializes in a real situation depends on energy matching requirements. In the first transition series elements, the 4p orbitals are marginally higher in energy than the 3d orbitals. There is thus a possibility of an extent of pd mixing in the basis set of orbitals to be chosen. When such is the case, the second part of the Eq. (7.30) contributes its share to the crystal field energy levels.

7.5 Cubic (0_h) Environment : Hamiltonian $\mathscr{H}_{crys}$ (cube).

Consider Fig. 7.1 in which all the cube corners 1, 2, ... 7, 8 contain similar ligands or in crystal field terms, the proxy charges –Ze each and each such charge being at a distance b from central metal ion O. Such an arrangement is effectively a placement of two tetrahedra within a cube with a common central ion and a relative displacement of $(\pi/4)$ rotation about the C_4^z axis. The spherical coordinates of the point charges 1 through 8 are (cf. Sec. 7.4 for a tetrahedron):

$$(b,\theta,\pi/4)\{b,\theta,\pi-\pi/4\}\{b,\theta,\pi+\pi/4\}\{b,\theta,2\pi-\pi/4\},$$

$$(b,\pi-\theta,\pi/4)\{b,\pi-\theta,\pi-\pi/4\}\{b,\pi-\theta,\pi+\pi/4\}\{b,\pi-\theta,2\pi-\pi/4\}$$ where

$\cos\theta=\sqrt{3/3}$ and $\cos(\pi-\theta)=-\cos\theta=-\sqrt{3/3}$

It is easily comprehended that A_0^o and A_4^o terms should be twice as large as these are in the tetrahedral case of sec. 7.4 and the coefficients each of $Y_{44}(\theta_i,\varphi_i)$ and $Y_{4-4}(\theta_i,\varphi_i)$ are $\sqrt{(5/14)}$ times that of Y_4^o. Thus skipping the details to avoid repetition, we may write

$h_{crys(i)}$ (cube)= $h'_{crys(i)}$(cube) + $h''_{crys(i)}$(cube)

$$=8Ze^2/b-8/9\left[(7/3)\sqrt{\pi}(Ze^2/b^5)r_i^4Y_{40}(\theta_i,\varphi_i)+(35/18)^{1/2}\sqrt{\pi}(Ze^2/b^5)r_i^4(Y_{44}+Y_{4-4})(\theta_i,\varphi_i)\right]$$

.........(7.32)

where the spherical harmonics are functions of the angular coordinates θ_i,φ_i of the ith electron. The first term belongs to $h'_{crys(i)}$ (cube) and the rest of (7.32) represents $h''_{crys(i)}$(cube). For the same values of both a and b and of the charges –Ze in octahedral 0_h field and in cubic field.

$$\{h''_{crys(i)}(cube)\}/\{h''_{crys(i)}(0_h)\}=-8/9,$$ ....(7.33)

a result which is consistent with our expectation.

7.6 D$_{4h}$ environment – Crystal Field Hamiltonian, $\mathscr{H}_{crys}$ (D$_{4h}$).

I. D$_{4h}$ (sq. planner) Environment of d-Electron Metal Ion. The five symmetry based arguments make it possible to arrive at the final forms of $\mathscr{H}_{crys}(0_h)$ and $\mathscr{H}_{crys}(T_d)$ from the general forms given by Eqs. (7.8, 7.9, 7.10). To repeat, the five symmetry arguments are (1) Totally symmetric hamiltonian (2) Exclusion of odd order harmonics from (7.8) and (7.10) (3) Gaunt's condition, viz., finite number of even order harmonics (4) Selection of specific even values of k's out k=0, 2, 4 and of specific Y_{km} members for each selected k (5) Piecemeal operational effects of the point group operators to decide the final linear combination of the harmonics.

Arguing in the previous manner we start in the present case directly from the permissible values of k. On analysis (via 2.84) of the reducible representation of the D_4 group in the basis sets of Y_{2m}'s and Y_{4m}'s [Y_{00} being constant] it is found that the permissible values of k are 0, 2 and 4 for the D_{4h} point group. We now try to decide upon the linear combinations of Y_{2m}'s and Y_{4m}'s by coordinating the observations of the piecemeal effects of selected group operators.

Looking up for the functional forms of Y_{2m}'s (appendix Chap.2) and operating upon these by C_4, we find

$$C_4 \begin{bmatrix} Y_{22} \\ Y_{21} \\ Y_{20} \\ Y_{2-1} \\ Y_{2-2} \end{bmatrix} = \begin{bmatrix} -Y_{22} \\ iY_{21} \\ Y_{20} \\ iY_{2-1} \\ -Y_{2-2} \end{bmatrix} \qquad \text{...........(7.34)}$$

It is seen that the Y_{20} component of the Y_{2m} set can only be totally symmetric since the others fail to respond in this way and thus Y_{2m} must have totally symmetric components as stated before. As to the permissible members of the Y_{4m} set, and their linear combination, we can proceed in exactly the same manner as in the 0_h symmetry (sec. 7.3) trying the selected operators C_4 and $C_2^{//}$ to arrive at a possible linear combination $\{Y_{40} + d(Y_{44} + Y_{4-4})\}$ Such a linear combination behaves as an A_{1g} base function of D_{4h} with proper values or d inserted. To find the full linear combination involving $Y_{00}, Y_{20}, Y_{40}, Y_{44}, Y_{4-4}$ as a basis of A_{1g} symmetry composing $h_{crys(i)}(D_{4h})$ (sq.pl), we may shun the symmetry path now and seek to finds the A_k^m coefficients of Eq. (7.9). Thus the coordinates of the proxy charge are

$$(a, \pi/2, 0)(a, \pi/2, \pi/2)(a, \pi/2, \pi)(a, \pi/2, 3\pi/2)$$

$$A_0^0 = A_{0(i)}^0 + A_{0(2)}^0 + A_{0(3)}^0 + A_{0(4)}^0$$

$$= (Ze^2/a)\left[Y_{00}*(\pi/2, 0) + Y_{00}*(\pi/2, \pi/2) + Y_{00}*(\pi/2, \pi) + Y_{00}*(\pi/2, 3\pi/2)\right]$$

$$= (4Ze^2/a)\, x\, (1/4\pi)^{1/2} \qquad \text{...........(7.35)}$$

$$A_2^0 = (Ze^2/a^3)\,\Theta_{20}(\pi/2)(2\pi)^{-1/2}\{e^{-1.0.0} + e^{-1.0.\pi/2} + e^{-i.0.\pi} + e^{-i.0.3\pi/2}\}$$

$$= -(Ze^2/a^3)(5/8)^{1/2}(2\pi)^{-1/2}\, x4 \qquad \text{...........(7.36)}$$

Similarly, we find

$$A_4^o = (2\pi)^{-1/2}(81/8)^{1/2}(Ze^2/a^5) \qquad \text{...........(7.37)}$$

$$A_4^4 = A_4^{-4} = (\pi)^{-1/2}\left(9/2\right)^{1/2}\left(35/8\right)^{1/2}\left(Ze^2/a^5\right) \qquad \text{...........(7.38)}$$

Using (7.35) – (7.38) properly in Eq. (7.8) the required hamiltonian is found to be

$$\begin{aligned} h_{crys(i)}(D_{4h})\text{ (sq. pl)} &= 4Ze^2/a - \left(4/\sqrt{5}\right)\sqrt{\pi}\left(Ze^2/a^3\right)r_i^2 Y_{20} + \sqrt{\pi}\left(Ze^2/a^5\right)r_i^4 Y_{40} \\ &\quad + (35/18)^{1/2}\sqrt{\pi}\left(Ze^2/a^5\right)r_i^4\left(Y_{44}+Y_{4-4}\right) \\ &= h'_{crys(i)}(D_{4h})(Sq.pl) + h''_{crys(i)}(D_{4h})(Sq.pl) \qquad \text{...........(7.39)} \end{aligned}$$

Where $h'_{crys(i)}(D_{4h})$ represents the constant first term in (7.39) and the spherical harmonics of (7.39) are all functions of θ_i and φ_i of the ith electron. To find $\mathscr{H}_{crys}$ (D_{4h}) (sq.pl), it is just $\sum_{i=l}^{n}$ $h_{crys(i)}(D_{4h})$ (sq.pl).

II. D_{4h} (tetragonal) Environment of d-electron metal ion.

For calculation of $h_{crys(i)}(D_{4h})$ (Tetrag) two additional ligand charges have to be inserted at positions 5 and 6 with coordinates (b, 0,0) and (b, π,0) and their contributions have to be added to (7.39) to yield the crystal field hamiltonian. Here b, which is different from a, is the distance of each of the proxy charges in position 5 and 6 from the central ion.

The conclusions derived from symmetry considerations in the square planar case are also applicable in the present case so far as the crystal field hamiltonian is concerned. Additionally the Eqs. (7.35) through (7.38) are all valid for the present symmetry environment. Only we have to compute further $\left\{A^0_{0(5)} + A^0_{0(6)}\right\}$, $\left\{A^0_{2(5)} + A^0_{2(6)}\right\}$, $\left\{A^0_{4(5)} + A^0_{4(6)}\right\}$, $\left\{A^4_{4(5)} + A^4_{4(6)}\right\}$ and $\left\{A^{-4}_{4(5)} + A^{-4}_{4(6)}\right\}$ for the coordinates (b, 0, 0) and (b, π,0) of locations 5 and 6 respectively. Now,

$$\left\{A^0_{0(5)} + A^0_{0(6)}\right\} = \left(2Ze^2/b\right)\left(1/4\pi\right)^{1/2} \qquad \text{...........(7.40)}$$

$$\left\{A^0_{2(5)} + A^0_{2(6)}\right\} = 4\left(Ze^2/b^3\right)\left(5/8\right)^{1/2}(2\pi)^{-1/2} \qquad \text{...........(7.41)}$$

$$\left\{A^0_{4(5)} + A^0_{4(6)}\right\} = \left(Ze^2/b^5\right)(18)^{1/2}(2\pi)^{-1/2} \qquad \text{...........(7.42)}$$

$$\left\{A^0_{4(5)} + A^4_{4(6)}\right\} = \left\{A^{-4}{}_{4(5)} + A^{-4}_{4(6)}\right\} = 0 \qquad \text{...........(7.43)}$$

Using the relations (7.40) through (7.43) properly in Eq. (7.8) and adding these to Eq. (7.39) we have

$$h_{crys(i)}(D_{4h})(\text{tetrag}) = h'_{crys(i)}(D_{4h})(\text{tetrag}) + h''_{crys(i)}(D_{4h})(\text{tetrag})$$

$$= \left(4Ze^2/a + 2Ze^2/b\right) + \left\{\left(4/\sqrt{5}\right)\sqrt{\pi}Ze^2 r_i^2\left(1/b^3 - 1/a^3\right)Y_{20}\right.$$

$$+\sqrt{\pi}Ze^2r_i^4\left(4/3b^5+1/a^5\right)Y_{40}+\left(35/18\right)^{1/2}\sqrt{\pi}\left(Ze^2/a^5\right)r_i^4\left(Y_{44}+Y_{4-4}\right)\} \quad \text{.........(7.44)}$$

wherein the spherical harmonics are functions of θ_i and φ_i summing over all the d-electron coordinates in a d^n complex gives

$$\mathscr{H}_{crys}(D_{4h})(tetrag) = \Sigma_{i=1}^{n}\{h'_{crys(i)}(D_{4h})(tetrag)+h''_{crys(i)}(D_{4h})(tetrag)\}$$
$$= \Sigma_{i=1}^{n}\{Eq.(7.44)\} \quad \text{............(7.44a)}$$

7.7 Crystal Field Hamiltonian with C_3 Symmetry Axis as the Quantising Axis I. $\mathscr{H}_{crys}^*(O_h)$ II. $\mathscr{H}_{crys}^*(D_{3d})$

I. $\mathscr{H}_{crys}^*(O_h)$:

Attention is drawn to Sec. 7.2 where the different orientations of two axial systems (symmetry defining and Y_{lm}'s defining) were mentioned. We have already treated the cases where the two axial systems were coincident (secs.7.3-7.6). Presently an expression for $\mathscr{H}_{crys}^*(O_h)$ will be derived with the quantizing axis (Z-axis) coincident with C_3 axis and not with C_4^z symmetry axis. (The asterisk on $\mathscr{H}_{crys}$ implies this situation).

Here the general expression for the crystalfield Hamiltonian given by (7.8) and (7.9) is still valid. So also are the symmetry related conditions of Sec. 7.3 except the part dealing with the operational effects of selected group operators of the 0 point group deciding the nature of the linear combination of the harmonics. The requisite linear combination of the permitted even order harmonics has to be found out in the present case.

Consider a cube (Fig. 7.2) with upper corners numbered 1,2,3,4 and the lower ones as 5,6,7, and 8 with O as its centre. Let X and Y be the midpoints of the sides 5-8 and 5-6 respectively. Then the axial system OX, OY and OZ representing the positive directions of X, Y and Z axes, is one in which the the Z-axis, incidentally passing through the corners 1 and 7 and coincident with one C_3 symmetry axis, is the quantizing axis. Let L_1, L_2, ... L_5, L_6 be the six homo legends of an octahedral complex occupying the centres of the six faces of the cube with the metal ion at O.

Now to find the harmonics in the $h_{crys(i)}*(O_h)$, it is evident that Y_{00}, being a constant, is one of them. Of the Y_{4m} set, operation by C_3 upon the individual members (appendix Chap. 2) show that only Y_{40},Y_{43}, and Y_{4-3} remain invariant. This prescribes a linear combination ($Y_{40}+aY_{43}+bY_{4-3}$), a trial one still to be subjected to further selective operator actions. A C_2' operation (YOY$'$ axis is a C_2' axis) results in

$$C_2'\left(Y_{40}+aY_{43}+bY_{4-3}\right)=Y_{40}-aY_{4-3}-bY_{43}=\left[Y_{40}+a\left(Y_{43}-Y_{4-3}\right)\right] \quad \text{...........(7.45)}$$

if a = – b, a condition which must be enforced if invariance of the linear combination is to be protected. A further use of the group operator, C_4, upon the RHS of (7.45) divulges that coefficient a $=\left(10/7\right)^{1/2}$ with the linear combination $\left[Y_{40}+\left(10/7\right)^{1/2}\left(Y_{43}-Y_{4-3}\right)\right]$ retaining the invariance property under the transformation.

Using the coordinates of the six ligands we can find A_0^0 and A_4^0 from Eq. (7.9). The form of the linear combination suggests that whatever be the coefficient of Y_{40} in the crystal filed hamiltonian, that accompanying (Y_{43}-Y_{4-3}) should be $\left(10/7\right)^{1/2}$ times the former. It is found that.

$$A_0^0=\left(6Ze^2/a\right)\left(1/4\pi\right)^{1/2} \text{ and } A_4^0=\left(49/2\right)^{1/2}\left(2\pi\right)^{-1/2}\left(Ze^2/a^5\right) \qquad \text{.............(7.46)}$$

Utilizing these values suitably in (7.8), we have

$$h^*_{crys(i)}(O_h) = h'^*_{crys(i)}(O_h)+h''^*_{crys(i)}(O_h)$$

$$=6Ze^2/a-\left(196/81\right)^{1/2}\sqrt{\pi}\left(Ze^2/a^5\right)r_i^4Y_{40}-\left(280/81\right)^{1/2}\sqrt{\pi}\left(Ze^2/a^5\right)r_i^4\{Y_{43}-Y_{4-3}\}$$

$$\text{.............(7.47)}$$

$$\mathcal{H}_{crys^*}(O_h)=\Sigma_i\, h^*_{crys(i)}\left(O_h\right) \qquad \text{.............(7.47a)}$$

II. $\mathcal{H}_{crys^*}(D_{3d})$ in D_{3d} Environment

Consider, as in sandwich complexes, two equilateral triangles of equal area in parallel plains juxtaposed in perfectly staggered orientation with respect to each other. The vertices 1, 2, 3 of one triangle are the antipodes of $1', 2', 3'$ respectively of the other (Fig. 7.3). Let O be the midpoint of the line joining the centroids of the two triangles. The X, Y and Z axes are indicated in the Fig. 7.3 with O as origin situated in between the two planes and ZOZ$'$ being vertical to the planes of the two triangles. This construction has the symmetry D_{3d}. Six homoligands placed at the six vertices with a d- electron metal ion at O will constitute a complex of that symmetry. The six proxy charges have the coordinates $(a,\theta,0)_1$, $(a,\theta,2\pi/3)_2$; $(a,\theta,4\pi/3)_3$ $(a,\pi-\theta,\pi)'_1$, $(a,\pi-\theta,\pi+2\pi/3)'_2$ and $(a,\pi-\theta,2\pi/6)'_3$ respectively.

Arguing on the symmetry basis, viz. analysis of the reducible representations of D_3 point group with the spherical harmonis, Y_{00}, Y_{2m}, Y_{4m} as separate basis demands that Y_{00}, Y_{2m}, Y_{4m} components occur in the hamiltonian. Further examination of selected operations by C_3 upon Y_{2m}'s and by C_3 and C_2' on Y_{4m}'s reveals that Y_{20} and the linear combination $\{Y_{40}+a(Y_{43}-Y_{4-3})\}$ be present in $h^*_{crys(i)}(D_{3d})$. One can now evaluate A_k^m's from Eq. (7.9). The values are given below

$$\left.\begin{aligned} A_0^0 &= \left(6Ze^2/a\right)\cdot(1/4\pi)^{1/2} \\ A_2^0 &= \left(Ze^2/a^3\right)\cdot(2\pi)^{-1/2}(5/8)^{1/2}6\left(3\cos^2\theta-1\right) \\ A_4^0 &= \left(Ze^2/a^5\right)\cdot(2\pi)^{-1/2}(9/128)^{1/2}6\left(35\cos^4\theta-30\cos^2\theta+3\right) \\ A_4^3 &= -A_4^{-3} = -\left(Ze^2/a^5\right)\left(2\pi\right)^{-1/2}(315/32)^{1/2}6\sin^3\theta\cos\theta \end{aligned}\right\} \quad \text{.........(7.48)}$$

Subsituting the values of (7.48) in (7.8) we have

$$\begin{aligned} h^*_{crys(i)}(D_{3d}) &= h'^*_{crys(i)}(D_{3d}) + h''^*_{crys(i)}(D_{3d}) \\ &= 6Ze^2/a + \left[(2\pi)^{1/2}\left(18/5\right)^{1/2}\left(Ze^2/a^3\right)r_i^2\left(3\cos^2\theta-1\right)Y_{20}\right. \\ &\quad + (2\pi)^{1/2}\left(1/8\right)^{1/2}\left(Ze^2/a^5\right)r_i^4\left(35\cos^4\theta-30\cos^2\theta+3\right)Y_{40} \\ &\quad \left.-(2\pi)^{1/2}(35/2)^{1/2}\left(Ze^2/a^5\right)r_i^4\sin^3\theta\cos\theta\left(Y_{43}-Y_{4-3}\right)\right] \quad \text{.........(7.49)} \end{aligned}$$

where the spherical harmonics are functions of the angular coordinates of the electron i and the specific value of θ depends on the geometry of the ligand positions. As before, we can also write.

$$\mathcal{H}_{crys}^*(D_{3d}) = \Sigma_i^n h^*_{crys(i)}(D_{3d}) \quad \text{...........(7.49a)}$$

7.8 Perturbation of perturbation Hamiltonian. Tetragonal and Trigonal Distortion of Regular Octahedral Crystal Field.

It will be understood that in a complex of D_{4h} tetragonal symmetry, energy calculations of the crystal field terms of the metal ion must start with the formation of the matrix elements of the perturbation operator. $h_{crys(i)}(D_{4h})$(tetrag), of Eq.(7.44) and hence finally of Eq. (7.44a) between the unperturbed d- electron orbitals or, depending on situations, between its unperturbed LS states. But with the aid of Eq. (7.25) for $h'_{crys(i)}(D_{4h})$ (O_h), Eq. (7.44), can be rewritten, omitting baricentre raising constant parts, $h'_{crys(i)}(O_h)$ of Eq. (7.25) and $h'_{crys(i)}(D_{4h})$ (tetrag) of Eq. (7.44), as

$$h'_{crys(i)}(D_{4h})\,(\text{tetrag}) = h''_{crys(i)}\,(O_h) + Ar_i^2Y_{20} + Br_i^4Y_{40} \quad \text{...........(7.50)}$$

$$\text{where } A = \left(4/\sqrt{5}\right)\sqrt{\pi}Ze^2\left(1/b^3 - 1/a^3\right) \text{ and } B = \left(4/3\right)\sqrt{\pi}Ze^2\left(1/b^5 - 1/a^5\right) \quad \text{........(7.51)}$$

Eq. (7.50) enables us to reinterpret that $h''_{crys(i)}(D_{4h})$ (tetrag) as a perturbation $\left(Ar_i^2Y_{20} + Br_i^4Y_{40}\right)$ imposed on a bigger perturbation hamiltonian $h''_{crys(i)}$ (O_h). In fact this is so if we disregard the difference in the baricentre raising energy components, viz. $h'_{crys(i)}(D_{4h})$ (tetrag) and $h'_{crys(i)}$ (O_h). The nature of splitting of the energy levels in tetragonal d- electron complex is what one can expect of the splitting of levels in octahedral environment when the perturbation operation $\left(Ar_i^2Y_{20} + Br_i^4Y_{40}\right)$ is applied subsequent to it.

Arguing in a similar vein, it can be shown that $h^*_{crys(i)}(D_{3d})$ with C_3 as the quantizing Z-axis can be expressed as a perturbation of the crystal field hamiltonian $h^*_{crys(i)}(0_h)$ of Eq. (7.47) if we apply the perturbation $\left(A'r_i^2Y_{20} + B'r_i^4Y_{40}\right)$ subsequently. The perturbation, $\left(A'r_i^2Y_{20} + B'r_i^4Y_{40}\right)$, is to be applied on the eigenfunctions of $h^*_{crys(i)}(0_h)$ and not upon the unperturbed wavefunctions of $\mathscr{H}_{cent}$.

$$h^*_{crys(i)}(D_{3d}) = h^*_{crys(i)}(0_h) + \left(A'r_i^2Y_{20} + B'r_i^4Y_{40}\right) \qquad \text{............(7.52)}$$

These considerations can be easily extended to $\mathscr{H}_{crys}^*(D_{3d})$ and $\mathscr{H}_{crys}^*(0_h)$ for multi-d-electron systems

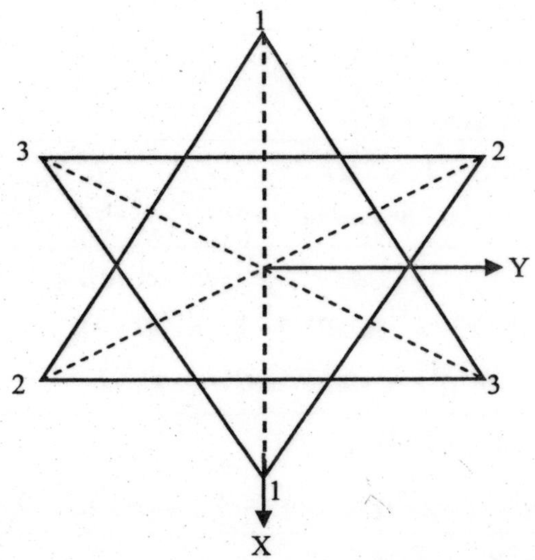

[Fig. 7.3 Axial set up in D_{3d} symmetry system]

Table $T_{7.1}$

Crystal Field Hamiltonian in Different symmetries for (d^1) electron systems.

Environment Symmetry	$h_{crys(i)}$	Ref. to Eqs.
O_h (Octahedral)	$6Ze^2/a+\sqrt{\pi}Ze^2\left(r_i^4/a^5\right)\left\{(7/3)Y_{40}+(35/18)^{1/2}\left(Y_{44}+Y_{4-4}\right)\right\}$	7.25
O_h^* (Octahedral) C_3- quant. axis	$6Ze^2/a-\sqrt{\pi}Ze^2\left(r_i^4/a^5\right)\left\{(196/81)^{1/2}Y_{40}+(280/81)^{1/2}\left(Y_{43}-Y_{4-3}\right)\right\}$	7.47
O_h (Hexahedral-Cube)	$8Ze^2/a-\sqrt{\pi}Ze^2\left(r_i^4/a^5\right)(8/9)\left\{(7/3)Y_{40}+(35/18)^{1/2}\left(Y_{44}+Y_{4-4}\right)\right\}$	7.32
T_d (Tetrahedral)	$4Ze^2/a-\sqrt{\pi}Ze^2\left(r_i^4/a^5\right)(4/9)\left\{(7/3)Y_{40}+(35/18)^{1/2}\left(Y_{44}+Y_{4-4}\right)\right\}$	7.28
D_{4h} (sq. planar)	$4Ze^2/a-\left(4/\sqrt{5}\right)\sqrt{\pi}Ze^2\left(r_i^2/a^3\right)Y_{20}+\sqrt{\pi}Ze^2\left(r_i^4/a^5\right)$ $\left\{Y_{40}+(35/18)^{1/2}\left(Y_{44}+Y_{4-4}\right)\right\}$	7.39
D_{4h} (tetragonal)	$\left(4Ze^2/a+2Ze^2/b\right)-\left(4/\sqrt{5}\right)\sqrt{\pi}Ze^2r_i^2\left(1/a^3-1/b^3\right)Y_{20}+$ $\sqrt{\pi}Ze^2\,r_i^4\left(1/a^5+4/3b^5\right)Y_{40}+(35/18)^{1/2}\sqrt{\pi}Ze^2\left(r_i^4/a^5\right)\left(Y_{44}+Y_{4-4}\right)$	7.44
D_{3d}^* C_3 –quant. axis	$6Ze^2/a+(2\pi)^{1/2}(18/5)^{1/2}\,Ze^2r_i^2/a^3\left(3\cos^2\theta-1\right)Y_{20}+$ $(2\pi)^{1/2}(1/8)^{1/2}Ze^2r_i^4/a^5\left(35\cos^4\theta-30\cos^2\theta+3\right)Y_{40}-$ $(2\pi)^{1/2}(35/2)^{1/2}Ze^2r_i^4/a^5\sin^3\theta\cos\theta\left(Y_{43}-Y_{4-3}\right)$	7.49
D_{4h} (tetragonal) as distorted Octahedron	$h_{crys(i)}\left(O_h\right)+Ar_i^2Y_{20}+Br_i^4Y_{40}$ where $A=-\left(4/\sqrt{5}\right)\sqrt{\pi}Ze^2\left(1/a^3-1/b^3\right), B=-4/3\sqrt{\pi}\ Ze^2\left(1a^5-1/b^5\right)$	7.50 7.51
D_{4h} (sq. planar) as distorted O_h	$h_{crys(i)}\left(O_h\right)+Ar_i^2Y_{20}+Br_i^4Y_{40}$ where $A=-\left(4/\sqrt{5}\right)\sqrt{\pi}Ze^2/a^3$ and $B=-4/3\sqrt{\pi}\ Ze^2/a^5$	cf 7.39 7.39a
D_{3d}^* as distorted O_h, with C_3 as quantising axis)	$h_{crys(i)}\left(O_h\right)+Ar_i^2Y_{20}-Br_i^4Y_{40}$ Y_{40} is taken with minus sign to stabilize the d_{z2} orbital	7.52
$D_{\infty h}(linear)$	$\left(4/\sqrt{5}\right)\sqrt{\pi}Ze^2r_i^2/b^3+(4/3)\sqrt{\pi}Ze^2\left(r_i^4/b^5\right)Y_{40}+2Ze^2/b$	cf 7.51

7.9. General Technique of Evaluation of Matrix Elements

With the forms of the different crystal field Hamiltonians already available (Table $T_{7.1}$), it is necessary to evaluate the matrix elements of the perturbation hamiltonian. To this end we shall employ the basis sets of five complex d-electron kets, $|d_m>'s, m=\pm 2, \pm 1, 0$ and/or five real d-electron kets, $|d_z^2>, |d_{x^2-y^2}>$ etc. which are some specific linear combinations of the former set (see appendix Chap.1). One should, however, remember that the different types of initial basis sets for d-electron complexes, such as free ion wavefunctions, wavefunctions including spin, determinantal wavefunctions are sometimes necessary for the construction of the perturbation matrix. But in all these cases, the forms of the matrix elements are reducible to those involving single d-electron kets either in complex $|d_m>$ forms or in real forms. It is therefore, useful to have, ready on hand, lists of values of matrix elements of the different crystal field Hamiltonians in the basis sets of the single d-electron kets.

The general form of a normalized d-electron ket will be written as

$$|d_{ml}> = R_{n2}Y_{2ml} = R_{n2}\Theta_{2ml}\Phi_{ml}$$

where R_{n2} is the mormalised radial function, Θ_{2ml} the normalized associated Legendre polynomial with l=2 and $\Phi_{ml} = (1/2\pi)^{1/2} e^{iml\varphi}$, a normalized function.

A. Matrix element of $h_{crys}(O_h)$ single d- electron kets as basis:

$$< d_{ml}|h_{crys}(O_h)|d_{ml}'> = <d_{ml}|h'_{crys}(O_h)|d_{ml}'> + <d_{ml}|h''_{crys}(O_h)|d_{ml}'>$$

$$= < R_{n2}\Theta_{2ml}\Phi_{ml}\left|6Ze^2/a + (7/3)\sqrt{\pi}(Ze^2/a^5)r^4Y_{40} + (35/18)^{1/2}\sqrt{\pi}(Ze^2/a^5)r^4(Y_{44}+Y_{4-4})\right|$$

$$R_{n2}\Theta 2_{ml}'\Phi_{ml}' >$$

$$= (6Ze^2/a)\delta_{ml,ml'} + (7/3)\sqrt{\pi}(Ze^2/a^5)\overline{r^4} < \Theta_{2ml}|\Theta_{40}|\Theta_{2ml}'> < \Phi_{ml}|\Phi_0|\Phi_{ml}'> \delta_{ml,ml'}$$

$$+ (35/18)^{1/2}\sqrt{\pi}(Ze^2/a^5)\overline{r^4} < \Theta_{2ml}|\Theta_{44}|\Theta_{2ml}'> < \Phi_{ml}|\Phi_4|\Phi_{ml}'> \delta_{ml,(ml'+4)}$$

$$+ (35/18)^{1/2}\sqrt{\pi}(Ze^2/a^5)\overline{r^4} < \Theta_{2ml}|\Theta_{4-4}|\Theta_{2ml}'> < \Phi_{ml}|\Phi_{-4}|\Phi_{ml}'> \delta_{ml,(ml'-4)}$$

...........(7.53)

where $< R_{n2}|(Ze^2/a^5)r^4|R_{n2}> = (Ze^2/a^5)\overline{r^4} = \alpha_4$(say), and $\overline{r^4}$ is the mean fourth power of the radial distance of the electron from the central ion and $h_{crys}(O_h)$ represents

$h_{crys(i)}(O_h)$ of Eq. (7.24) from which the suffix i has been dropped.

Proper use of Eq. (7.53) yields the values of the various matrix elements in terms of the parameter α_4 and the constant $6Ze_2/a$. Scrutiny of the kronecker δ factors indicates that the diagonal elements emerge from the first two expressions on the RHS of Eq. (7.53) and the off diagonal ones from the last two terms.

Writing thc matrix elements as $d'_{00}, d'_{+1+1,} d'_{-1-1}, d'_{+2+2}$ and so on, the Θ integrals in the forms $|l\, ml>$, one gets from (7.53), remembering that (cf. Sec. 2.8).

$< \Theta_{2ml} | \Theta_{km} | \Theta_{2ml}{}' > = \sqrt{\left(2k + \tfrac{1}{2}\right)}\, c^k\left(m_1, m_1'\right)$ the following results.

The diagonal elements:

$$d'_{oo} = 6Ze^2/a + \left(\tfrac{7}{3}\right)\sqrt{\pi}(1/2\pi)^{1/2}\ \alpha_4 < 20|40|20 > = 6Ze^2/a + \alpha_4$$

$$= 6Ze^2/a + 6Dq, \text{ on using}$$

C^k Table $T_{2.1}$ and putting $\alpha_4 = 6Dq = \left(Ze^2/a^5\right)\overline{r^4}$

$$d'_{+1+1} = d'_{-1-1} = 6Ze^2/a + \left(\tfrac{7}{3}\right)\sqrt{\pi}\left(\left(\tfrac{1}{2\pi}\right)^{1/2}\right)\alpha_4 < 2^+{-1}|40|2^+{-1} > = 6Ze^2/a - \left(\tfrac{2}{3}\right)\alpha_4$$

$$= 6Ze^2/a - 4Dq \qquad \text{........(7.54)}$$

$$d'_{+2+2} = d'_{-2-2} = 6Ze^2/a + \left(\tfrac{1}{6}\right)\alpha_4 = 6Ze^2/a + Dq$$

The off diagonal elements

$$d'_{+2-2} = d'_{-2+2} = \left(\tfrac{5}{6}\right)\alpha_4 = 5Dq$$

$$d'_{0+2} = d'_{0-2} = d'_{0+1} = d'_{0-1} = 0; also\ d'_{+1+2} = d'_{-1-2} = d'_{+1-2} = d'_{-1+2} = 0 \qquad \text{...........(7.54)}$$

$$and\ d'_{+1-1} = d'_{-1+1} = 0$$

The first constant term, $6Ze^2/a$, in the diagonal elements of (7.54) indicates that the energy level of the original degenerate d- functions is raised by this amount to form the new baricentre of the energy level in the crystal field. The second terms, involving Dq's in the diagonal elements, represent the values of the energies of the split levels relative to the new baricentre taken as zero energy, i.e., $\epsilon_d^o + 6Ze^2/a = 0$ where ϵ_d^0 is the degenerate energy of the unperturbed state.

Now we turn to the real d- orbitals as the basis set, the forms of which are given in the appendix (Chap.1).

$$d'_{z^2z^2} = d'_{00} = 6Ze^2/a + \alpha_4 = 6Ze^2/a + 6Dq \qquad \text{...........(7.55)}$$

$$d'_{xz\,xz} = \tfrac{1}{2}\{<(d_{-1}-d_1)|h_{crys}(O_h)|(d_{-1}-d_1)>\}$$

$$= 6Ze^2/a + \tfrac{1}{2}\{<d_{-1}|h''_{crys}(0h)|d_{-1}> - <d_{-1}|h''_{crys}(0_h)|d_1>$$

$$- <d_1|h''_{crys}(O_h)|d_{-1}> + <d_1|h''_{crys}(O_h)|d_1>\}$$

$$= 6Ze^2/a - 4Dq \text{ (using Eq. 7.54)} \qquad \text{............(7.56)}$$

$$d'_{yz\,yz} = \tfrac{1}{2}\{<i(d_{-1}+d_1)|h_{crys}(O_h)|i(d_{-1}+d_1)>\}$$

$$= \tfrac{1}{2}\{<(d_{-1}+d_1)|h_{crys}(O_h)|(d_{-1}+d_1)>\} \qquad \text{note i in bra = -i}$$

$$= 6Ze^2/a - 4Dq \text{ (using Eq. 7.54)} \qquad \text{............(7.57)}$$

and similarly

$$d'_{xy\,xy} = 6Ze^2/a - 4Dq$$

$$d'_{(x^2-y^2)(x^2-y^2)} = 6Ze^2/a - 4Dq$$

Since our knowledge of group theory tells us that d_z^2, $d_{(x^2-y^2)}$ belong to the IR, E_g, and d_{xz}, d_{yz} and d_{xy} form basis of the IR T_{2g}, of the O_h point group, there can be no mixing of the former set (IR E_g) with any of the latter under the crystal field perturbation, so that

$$d'_{z^2\,xy} = d'_{z^2\,xz} = d'_{z^2\,yz} = 0$$

$$d'_{(x^2-y^2)xy} = d'_{(x^2-y^2)xz} = d'_{(x^2-y^2)yz} = 0 \qquad \text{............(7.57a)}$$

B. $h_{crys}(O_h)$ (cube) and h_{crys} (T_d) – Matrix elements (d^1-complex) with d- electron kets as basis set.

The matrix elements and the energy values are respectively –(8/9) and –(4/9) times the figures found already for the $h_{crys}(O_h)$ case. Only the constant terms raising the baricentres for the O_h (cube) and the T_d symmetries are $8Ze^2/a$ and $4Ze^2/a$ respectively.

C. Matrix elements of $h_{crys}(D_{4h})$ between single d- electron kets – sq. planar case.

Looking at the form of the Hamiltonian in Table 7.1, it is easy to write, in the manner of Eq. (7.53), the general form of the matrix elements.

$$<d_{ml}|h_{crys}(D_{4h})(sq.pl)|d_{ml}'> = (4Ze^2/a)\delta_{ml\cdot ml'} -$$

$$(4/\sqrt{5})\sqrt{\pi}\,\alpha_2 <\Theta_{2ml}|\Theta_{20}|\Theta_{2ml'}><\Phi_{ml}|\Phi_0|\Phi'_{ml}>\delta_{ml\,,ml'}+$$

$$\sqrt{\pi}\alpha_4 < \Theta_{2ml}|\Theta_{40}|\Theta_{2ml}{}' >< \Phi_{ml}|\Phi_0|\Phi_{ml}{}' > \delta_{ml\cdot ml}{}' +$$

$$\left(35/18\right)^{1/2}\sqrt{\pi}\ \alpha_4\{< \Theta_{2ml}|\Theta_{44}|\Theta_{2ml}{}' >< \Phi_{ml}|\Phi_4|\Phi_{ml}{}' > \delta_{ml\cdot(ml'+4)} +$$

$$< \Theta_{2ml}|\Theta_{4-4}|\Theta_{2ml}{}' >< \Phi_{ml}|\Phi_{-4}|\Phi_{ml}{}' > \delta_{ml\cdot(ml'-4)}\} \quad \text{...........(7.58)}$$

$$\alpha_{2=} = Ze^2\overline{r^2}/a^3 =< R_{n2}|Ze^2r^2/a^3|R_{n2} > and\ \alpha_4 = Ze^2\overline{r^4}/a5 =< R_{n2}|Ze^2r^4/a^5|R_{n2} >$$

Remembering that the surviving integrals over Φ functions yield $(1/2\pi)^{1/2}$ [cf. sec. 2.8] and those over the product of Θ functions are ascertainable from the c^k table ($T_{2.1}$), it is now a simple matter to write down the actual values of the matrix elements.

Basis set of complex d- functions: matrix elements.

The diagonal elements are

$$d'_{00} = 4Ze^2/a - \left(4/\sqrt{5}\right)\sqrt{\pi}\alpha_2 < 20|20|20 > (1/2\pi)^{1/2} + \sqrt{\pi}\alpha_4 < 20|40|20 > (1/2\pi)^{1/2}$$

$$= 4Ze^2/a - \left(4/7\right)\alpha_2 + \left(3/7\right)\alpha_4$$

$$d'_{+1+1} = d'_{-1-1} = 4Ze^2/a - \left(2/7\right)\alpha_2 - \left(2/7\right)\alpha_4$$

$$d'_{+2+2} = d'_{-2-2} = 4Ze^2/a + \left(4/7\right)\alpha_2 + \left(1/14\right)\alpha_4$$

The off-diagonal elements, obtainable from either the fourth or the fifth terms of (7.58), are ....(7.59)

$$d'_{+2-2} = d'_{-2+2} = (5/6)\,\alpha_4$$

and the rest being zeros.

Basis set of real d-orbitals : matrix elements

$d'_{z^2z^2}$ is the same d'_{00} found above. It is seen d_1 has the same energy as d_{-1} Eq. (7.59). Therefore, any specific IR -symmetry adapted linear combination of the degenerate wavefuncitons, d_1 and d_{-1}, among which d_{xz} and d_{yz} figure, in the same sq. planar perturbation field has the same degenerate energy value as of d_1 or of d_{-1}, thus the three diagonal elements are

$$d'_{z^2z^2} = 4Ze^2/a - (4/7)\alpha_2 + (3/7)\alpha_4$$

$$d'_{xz\,xz} = d'_{yz\,yz} = 4Ze^2/a - (2/7)\alpha_2 - (2/7)\alpha_4 \quad \text{...........(7.60)}$$

The orbitals $d_{(x^2-y^2)}$ and d_{xy} are each some suitable linear combination of d_2 and d_{-2}, the latter two being degenerate themselves in the sq. planar perturbation field. But unlike d_{xz}, d_{yz} degenerate pair, these two are not degenerate. They will mix under D_{4h} sq. planar field giving off-diagonal matrix elements which modulate their respective energies. Here one has to solve for energy from the 2×2 order matrix. Omitting the constant term $4Ze^2/a$ from the energy of d_2 and d_{-2}, the matrix is

$$\begin{bmatrix} \left(\frac{4}{7}\right)\alpha_2 + (1/14)\alpha_4 & \left(\frac{5}{6}\right)\alpha_4 \\ \left(\frac{5}{6}\right)\alpha_4 & (4/7)\alpha_2 + \left(\frac{1}{14}\right)\alpha_4 \end{bmatrix}$$

which gives the perturbational

Energy eigenvalus $\in_1 = \left(\frac{4}{7}\right)\alpha_2 + \left(\frac{19}{21}\right)\alpha_4$ associated with the eigenfunction $\left(\frac{1}{\sqrt{2}}\right)\{d_{-z} + d_{-z}\}$ and hence with $d_{(x^2-y^2)}$ orbital, and $\in_2 = \left(\frac{4}{7}\right)\alpha_2 - \left(\frac{16}{21}\right)\alpha_4$ associated with the eigenfunction $\left(\frac{1}{\sqrt{2}}\right)\{d_2 - d_{-2}\}$ and hence with $\left(\frac{1}{\sqrt{2}}\right)\{d_{-2} - d_2\}$ i.e., with d_{xy}. The matrix elements are

$$\left.\begin{aligned} d'(x^2-y^2)(x^2-y^2) &= 4Ze^2/a + \left(\frac{4}{7}\right)\alpha_2 + \left(\frac{19}{21}\right)\alpha_4 \\ d'_{xy\,xy} &= 4Ze^2/a + \left(\frac{4}{7}\right)\alpha_2 - \left(\frac{16}{21}\right)\alpha_4 \end{aligned}\right] \qquad \text{.................(7.61)}$$

D. h_{crys} (D_{4h}) (tetrag): Tetragonal distortion of an octahedral field. Matrix elements with the d-orbitals as basis.

Turning to Eqs. (7.50, 7.51) of Table $T_{7.1,}$ we find the matrix elements of the tetragonal distortion of O_h crystal field level e_g are provided in the matrix

$$\begin{bmatrix} d'_{z^2\,z^2} & d'_{z^2(x^2-y^2)} \\ d_{(x^2-y^2)z^2} & d'(x^2-y^2)(x^2-y^2) \end{bmatrix}$$

where,

$d'_{z^2(x^2-y^2)} = \langle d_z^2 | Ar^2Y_{20} + Br^4Y_{40} | d_{(x^2-y^2)} \rangle$ and so on for the rest. Such a matrix has to be solved for energy and eigenvectors to lead to perturbed energy levels and eigenfunction. We shall indicate, without working out in full, how this can be done.

$$\begin{aligned} d'_{z^2(x^2-y^2)} &= \langle d_o | Ar^2Y_{20} + Br^4Y_{40} | (1/\sqrt{2})\{d_{-2} + d_2\} \rangle \\ &= \langle d_o | P | (1/\sqrt{2}) d_{-2} \rangle + \langle d_o | P | (1/\sqrt{2}) d_2 \rangle, \end{aligned}$$ where P stands for the distortion

perturbation. Each matrix element has to be found from the general expression

$$< d_{ml}|P|d_{ml}' > = -(4/\sqrt{5})\sqrt{\pi}[\alpha_2(eq)-\alpha_2(ax)] < \Theta_{2ml}|\Theta_{20}|\Theta_{2ml}' > < \Phi_{ml}|\Phi_o|\Phi_{ml}' > \delta_{ml',ml}'$$

$$-(4/3)\sqrt{\pi}[\alpha_4(eq)-\alpha_4(ax)] < \Theta_{2ml}|\Theta_{40}|\Theta_{2ml}' > < \Phi_{ml}|\Phi_o|\Phi_{ml}' > \delta_{ml',ml}' \quad \text{............(7.62)}$$

It is evident that only the diagonal matrix elements can exist. Hence both d_z^2 and $d_{(x^2-y^2)}$ are still eigenfunctions under tetragonal distortion. The energy values are obtained from (7.62) via c^k table ($T_{2.1}$).

$$\left.\begin{aligned} d'_{z^2z^2} = d'_{00} &= -(4/7)[\alpha_2(eq)-\alpha_2(ax)] - (12/21)[\alpha_4(eq)-\alpha_4(ax)] \\ d'_{(x^2-y^2)(x^2-y^2)} &= (1/2)\{d'_{-2-2}+d'_{-22}+d'_{2-2}+d'_{22}\} \\ &= d'_{22} = (4/7)[\alpha_2(eq)-\alpha_2(ax)] - (2/21)[\alpha_4(eq)-\alpha_4(ax)] \end{aligned}\right\} \quad \text{......(7.63)}$$

The degeneracy of the octahedral e_g orbitals is thus lifted under distortion perturbation. The parameters α_2 and α_4 appearing in (7.62) and in succeeding relations are defined as

$$< R_{n2}|Ze^2r^2\{(1/a^3)-(1/b^3)\}|R_{n2} > = [\alpha_2(eq)-\alpha_2(ax)]$$

and $< R_{n2}|Ze^2r^4\{(1/a^5)-(1/b^5)\}|R_{n2} > = [\alpha_4(eq)-\alpha_4(ax)]$

where (eq.) and (ax) indicate parameters associated with equatorial

and axial values of the respective averages. On introducing two new parameters,

$$Ds = (2/7)[\alpha_2(eq)-\alpha_2(ax)] \text{ and } Dt = (2/21)[\alpha_4(eq)-\alpha_4(ax)] \quad \text{............(7.64)}$$

it follows,

$$\left.\begin{aligned} &\in_{dz^2} = -2Ds-6Dt \\ \text{And} \quad &\in_{d(x^2-y^2)} = -2Ds-Dt \end{aligned}\right] \quad \text{............(7.65)}$$

These represent the derivations of the splitting of an octahedral degenerate level, e_g, into two nondegenerate levels (7.65) when the octahedral field is subjected to tetragonal distortion. Referred to the baricentre of the energy levels in O_h field taken as zero, the energies of the split levels are

$$\left.\begin{aligned} &\in_{dz^2} = 6Dq-2Ds-6Dt \\ \text{and} \quad &\in_{d(x^2-y^2)} = 6Dq+2Ds-Dt \end{aligned}\right] \quad \text{............(7.66)}$$

In the D_{4h} (tetragonal) field although the baricentre of energy levels may be a new one, the nature of splitting and the energy gap between the split levels will remain the same.

A similar procedure followed for the triply degenerate (d_{xy}, d_{xz}, d_{yz}), i.e., the t_{2g} level of O_h field subjected to tetragonal distortion leads to a two-fold degenerate (d_{xz}, d_{yz}) and a nondegenrate (d_{xy}) levels turning up.

$$\left.\begin{aligned} \in_{dxy} &= -4Dq + 2Ds - Dt \\ \in_{dxz} &= -4Dq - Ds + 4Dt \\ \in_{dyz} &= -4Dq - Ds + 4Dt \end{aligned}\right\} \qquad \text{............(7.67)}$$

It is possible to extend this treatment of perturbation of a perturbation (i.e., two successive perturbations) to d –electron system in D_{4h} (sq. pl.) field starting from 0_h field results or from D_{4h} (tetrag-distortion) results.

E. Matrix elements of D_{4h} (sq. pl.) treated as infinite tetragonal distortion of 0_h with d-electron orbitals as basis.

The procedure here is the same as in the foregoing finite tetragonal distortion of the 0_h. field. The matrix with d_{z^2} and $d_{(x^2-y^2)}$ as basis is to be built using the general form of the matrix element of the perturbation operator.

$$< d_{ml}|P|d'_{ml} > = -\left(4\big/\sqrt{5}\right)\sqrt{\pi\alpha_2} < \Theta_{2ml}|\Theta_{20}|\Theta_{2ml'} >< \Phi_{ml}|\Phi_0|\Phi_{ml}' > \delta_{ml,ml'}$$

$$-\left(4\big/3\right)\sqrt{\pi\alpha_4} < \Theta_{2ml}|\Theta_{40}|\Theta_{2ml}' >< \Phi_{ml}|\Phi_0|\Phi_{ml'} > \delta_{ml,ml'}$$

where P is substituted by (7.39a) of Table $T_{7.1}$. The matrix turns out be diagonal with different energy values for d_{z^2} and $d_{(x^2-y^2)}$ in the final perturbation field.

Similarly perturbation of the t_{2g} orbitals of the O_h field can be explored by forming the matrix in the basis d_{xy}, d_{xz} and d_{yz} with P as perturbation operator. There will result a diagonal matrix indicating d_{xy}, d_{xz}, d_{yz} to be individual eigenfunctions in the final perturbation field. There will be one energy value for d_{xy} orbital and another one for both d_{xz}, d_{yz} orbitals, the latter two forming the base function of E_g IR in D_{4h} (sq. pl) field. Of the two parameters, Ds Dt defined by taking only the first parts of the equations (7.64), Ds does not figure in the final energy expression and the latter viz., Dt appearing in the energy expression is substituted in terms of Dq parameter.

F. $h^*_{crys}(0_h)$Matrix elements with d-electron wavefunctions as basis. (C_3 symmetry axis is the quantizing axis in 0_h field).

Referring to Eq. (7.47) of Table $T_{2.1}$ one can write the general expression for the matrix element.

$$< d_{ml}\left|h^{*}_{crys}(O_h)\right|d'_{ml} >=$$

$$(6Ze^2/a)\delta_{ml',ml}{}' - (196/81)^{1/2}\sqrt{\pi}\alpha_4 < \Theta_{2ml}|\Theta_{40}|\Theta_{2ml}{}' >< \Phi_{ml}|\Phi_o|\Phi'_{ml} > \delta_{ml',ml}{}'$$

$$-(280/81)^{1/2}\sqrt{\pi}\alpha_4 < \Theta_{2ml}|\Theta_{43}|\Theta_{2ml}{}' >< \Phi_{ml}|\Phi_3|\Phi_{ml}{}' > \delta_{ml',ml}{}'_{+3}$$

$$+(280/81)^{1/2}\sqrt{\pi}\alpha_4 < \Theta_{2ml}|\Theta_{4-3}|\Theta_{2ml}{}' >< \Phi_{ml}|\Phi_{-3}|\Phi_{ml}{}' > \delta_{ml',ml}{}'_{-3} \quad ...(7.68)$$

Utilising (7.68), the c^k table ($T_{2.1}$) and taking $\alpha_4 = 6Dq$, one may derive the following nonvanishing diagonal and off-diagonal matrix elements.

$$\left.\begin{aligned} d'_{oo} &= 6Ze^2/a - (2/3)\alpha_4 = 6Ze^2/a - 4Dq \\ d'_{+1+1} &= d'_{-1-1} = 6Ze^2/a + \left(4/9\right)\alpha_4 = 6Ze^2/a + \frac{8}{3}Dq \\ d'_{+2+2} &= d'_{-2-2} = 6Ze^2/a - (1/9)\alpha_4 = 6Ze^2/a - (2/3)Dq \\ d'_{+2-1} &= d'_{-2+1} = d'_{+1-2} = d'_{-1+2} = \pm(10/18)\sqrt{2}\alpha_4 = \pm(10/3)\sqrt{2}Dq \end{aligned}\right\} \quad ....(7.69)$$

where the minus preceding the last result applies to the two middle matrix elements. The other possible matrix elements are all zeros. As an illustration for detail, we work out the 1st matrix element.

$$d'_{+2-1} = < d+2\left|h^{*}_{crys}(O_h)\right| d_{-1} >$$

$$= -\left(280/81\right)^{1/2}\sqrt{\pi}\alpha_4 < \Theta_{22}|\Theta_{43}|\Theta_{2-1} >< \Phi_2|\Phi_3|\Phi_{-1} >$$

$$= -\left(280/81\right)^{1/2}\sqrt{\pi}\alpha_4 < 22|43| < 2-1 > \left(1/2\pi\right)^{1/2}$$

$$= -\alpha_4\left(280/81\right)^{1/2}\left(1/2\right)^{1/2}\left(9/2\right)^{1/2}\left\{-(35)^{1/2}/21\right\}$$

$$= \left(10/3\right)\sqrt{2}Dq$$

G. $h^{*}_{crys}(D_{3d})$ -Trigonal distortion of O_h field. Matrix elements with d-electron orbitals as basis.

We shall employ here the distortion Hamiltonian (7.52 of Table $T_{7.1}$) as the perturbation applied on O_h crystal field and use the basis eigenfunctions of $h^{*}_{crys}(O_h)$ of Eq. (7.47) with C_3 as the quantizing axis, to form the matrix elements. The basis eigenfunctions are provided by Eqs. (8.4, 8.5) of the next chapter. The choice of a negative sign for the coefficient of Y_{40}, i.e., $Ar^2Y_{20} - Br^4Y_{40}$ is intended to stabilize the

d_z^2 orbital on elongation along the trigonal C_3 axis.

Since the real forms of the d-orbitals emanate from linear combinations of the complex d- wavefunctions, the general form of the matrix element of the perturbation operator in this basis is shown below.

$$<d_{ml}|(Ar2Y_{20} - Br4Y_{40})|d'_{ml}> =$$

$$<R_{n2}|Ar^2|R_{n2}><\Theta_{2ml}|\Theta_{20}|\Theta 2'_{2ml}><\Phi_{ml}|\Phi_0|\Phi_{ml}'>\delta_{ml,ml'}$$

$$-<R_{n2}|Br^4|R_{n2}><\Theta_{2ml}|\Theta_{40}|\Theta_{2ml}'><\Phi_{ml}|\Phi_0|\Phi_{ml}'>\delta_{ml,ml'} \quad \text{......(7.70)}$$

It is apparent that only the diagonal matrix elements will survive. Writing I_2 and I_4 for the two radial integral parts of (7.70), we have the following non-zero matrix elements. The evaluations involve the use of c^k table ($T_{2.1}$).

$$<d^{+}_{-2}|\left(Ar^2Y_{20} - Br^4Y_{40}\right)|d^{+}_{-2}>$$

$$= I_2\left(1/2\pi\right)^{1/2} <2^{+}_{-}2|20|2^{+}_{-}2> - I_4\left(1/2\pi\right)^{1/2} <2^{+}_{-}2|40|2^{+}_{-}2>$$

$$= -I_2\left(1/2\sqrt{\pi}\right)\left(2\sqrt{5}/7\right) - I_4\left(1/2\sqrt{\pi}\right)\left(1/7\right)$$

$$= -\left(1/2\sqrt{\pi}\right)\left(2\sqrt{5}/7\right)I_2 - \left(1/2\sqrt{\pi}\right)\left(1/7\right)I_4 \quad \text{......(7.71)}$$

Similarly the other non-zero matrix elements are

$$<d_o|\left(Ar^2Y_{20} - Br^4Y_{40}\right)|d_o> = \left(1/2\sqrt{\pi}\right)\left(2\sqrt{5}/7\right)I_2 - 6I_4\left(1/2\sqrt{\pi}\right)\left(1/7\right) \quad \text{......(7.72)}$$

$$<d^{+}_{-1}|\left(Ar^2Y_{20} - Br^4Y_{40}\right)|d^{+}_{-1}> = +\left(1/2\sqrt{\pi}\right)\left(\sqrt{5}/7\right)I_2 + 4I_4\left(1/2\sqrt{\pi}\right)(1/7) \quad \text{......(7.73)}$$

Now introducing two parameters $D\sigma$ and $D\tau$ defined as

$$D_\sigma = -\left(1/2\sqrt{\pi}\right)\left(\sqrt{5}/7\right)I_2$$

$$\text{And } D\tau = \left(1/2\sqrt{\pi}\right)\left(1/7\right)I_4 \quad \text{......(7.74)}$$

The foregoing Eqs. (7.71, 7.72, 7.73) can be expressed without the common constant terms, as

$$d'_{+2,+2} = d'_{-2,-2} = 2D\sigma - D\tau \qquad \text{......(7.75)}$$

$$d'_{+1,+1} = d'_{-1,-1} = -D\sigma + 4D\tau \qquad \text{......(7.76)}$$

$$d'_{0,0} = -2D\sigma - 6D\tau \qquad \text{......(7.77)}$$

Viewed against the background of the O_h field being distorted by a trigonal component to constitute an effective D_{3d} field, the matrix elements of the distortion Hamiltonian are then wanted in the t_{2g} and e_g basis eigenfunctions of $h^*_{crys}(O_h)$ [C_3 as the quantizing axis]. These basis eigenfunctions as well as the desired matrix elements are derived in the next chapter (secs 8.2, 8.4 III) wherein the Eqs. (7.75-7.77) come in handy for direct use.

Exercises

1. Find the crystal field hamiltonian perturbing a metal ion placed at the centre of a rhombohedron in which the angles in the equatorial plane of the corresponding rhombus are 120^0's and 60°'s respectively.
2. A hypothetical homoligand compelex has its metal ion situated at the centre of a triangular bipyramid. Assuming the five proxy charges (-ze each) to be equidistant from the central ion, find its perturbation hamiltonian.
3. Find the diagonal elements of the Hamiltonian (Ex.2) in the basis set of single-electron complex d-wave functions.
4. Three point charges are at locations $(r, \pi/2, \pi/3), (r, \pi/2, \pi)$ and $(r, \pi/2, 2\pi - \pi/3)$. what will be the crystal field potential for a metal ion placed at the centroid of the triangle formed by joining the three points?
5. What will be the form of the hamiltonian (Ex.4) if the molecule be subjected to a rotation of $2\pi/3$ about the Z-axis?
6. Use group theoretical arguments to ascertain which of the spherical harmonics Y_{lm}'s would be present in the Hamiltonian of Ex. 4.
7. Establish the relations shown in Eq. (7.60)

References

1. H.A. Bethe : *Ann. Physik.* 3, 133, 1929.
2. J.H. Van Vleck: See ref. 7. chap. 8
3. J.A. Gaunt : *Trans, Roy, Soc. A228, 194, 1929; Phil Trans. A228, 151, 1929*
4. R.M. Golding : *Applied Wave Machanics* (Addison Wesley)
5. 7.C.J. Ballhausen : *Introduction to Ligand Field Theory* (Mc. Graw-Hill)
6. K.W.H. Stevens : *Proc. Roy. Soc. A219, 542, 1953; Proc. Phys. Soc. A65, 209, 1952*
7. B.N. Figgis : *Introduction to legend Fields (Wiley Eastern)*

Chapter Eight

Splitting of Energy Levels, Symmetries, Relative Order, Energy Eigenvalues and Eigenfunctions of d^1 - Electron Metal Ion Complexes - Cubic Symmetry System, Perturbations of Cubic Symmetry. Lower Symmetries.

The applications of crystal field lie mainly in interpreting the electronic spectra and the magnetic properties of the transition metal complexes. To do this one must have, to the extent possible, a qualitative and quantitative knowledge of the items listed in the caption of the chapter. This knowledge is achieved through a joint use of group theoretical reasonings for the first three and a mathematical analysis, related to the symmetry and perturbations, of the last two items, viz., energy values and eigenfunctions. Of course, this mathematical analysis incidentally provides direct confirmation of the conclusion as regards the first three items predicted by the group theory.

We shall make a start with d^1 complexes of cubic symmetry and then discuss such complexes of lower symmetry.

8.1 d^1 Metal Ion Complex with Octahedral Crystal Field

The five d - orbitals of a free metal ion in unperturbed state are degenrate. For a d^1- electron system , one perurbation term out of three in Eq (7.1) drops out, so that

$$\mathscr{H}(d^1) = \mathscr{H}_{cent} + \xi(r)\, l.s + h_{crys}(O_h) \qquad \text{.............(8.1)}$$

We shall assume that h_{crys} $(0_h) >> \xi(r)\, l.s$ in the present treatment and leave the case, for a future discussion, when both the perturbations of (8.1) are to be considered in either order viz., $h_{crys} > \xi(r) l.s$ or $\xi(r) l.s > h_{crys}$

(a) Role of group theory

The splitting of energy levels, i.e., the nature of lifting the degeneracy of the d-orbitals and the symmetries of the split levels in O_h crystal field can immediately be gathered from the character table of the O_h point group. The basis states d_{z2} and $d_{(x2-y2)}$ belong to the IR. E_g and the rest, viz., d_{xy}, d_{xz}, d_{yz} to T_{2g} of the O_h point group. Therefore, the split levels are twofold and threefold degenerate in the crystal field.

With the C_4^z axis serving as the quantizing axis, the ligands (hence the assumed proxy charges) face directly the d_{z2} and $d_{(x2-y2)}$ orbitals. Hence a d electron, when present in any of these two orbitals, is more strongly repelled by the proxy charges than when present in d_{xy}, d_{xz} or $d_{yz.}$ Thus one gains the idea that the e_g orbitals (i.e., d_{z2}, $d_{(x2-y2)}$ possess higher energy than the t_{2g} orbitals in a relative energy sequence order.

(b) The energy eigenvalues and eigenfunctions are obtained from a solution of the O_h crystal field perturbation matrix. The matrix elements of the perturbation operator have already been worked out in Sec (7.9) (Chap 7) in the two basis sets (i) d_o, d_1, d_{-1}, d_2, d_{-2} and (ii) d_{z2}, $d_{(x2-y2)}$, d_{xy}, d_{xz}, d_{yz}

Leaving out the constant, $6ze^2/a$, the energy baricenter raising component, from the diagonal term, the perturbation matrix in the complex d-wavefunction basis set is (cf. Eq 7.54)

$$\begin{bmatrix} 6Dq & & & & \\ & -4Dq & & & \\ & & -4Dq & & \\ & & & Dq & 5Dq \\ & & & 5Dq & Dq \end{bmatrix}$$

The energy eigenvalues of which are 6Dq, -4Dq, -4Dq, 6Dq and -4Dq respectively. While the eigenvectors, associated with the first four are +1 each, the two eigenvectors emerging from the last (2×2) matrix are $\begin{bmatrix} 1/\sqrt{2} \\ 1/\sqrt{2} \end{bmatrix}$ and $\begin{bmatrix} 1/\sqrt{2} \\ -1/\sqrt{2} \end{bmatrix}$.

Thus energywise, the eigenfunctions, degeneracies and symmetries are, (relative to the baricenter energy taken as zero)

Energy	Eigenfunction	Symmetry	Degeneracy	Orbital Notation
6Dq	d_0			e_g^0
6Dq	$(d_{-2}+d_2)/\sqrt{2}$	E_g	2-fold	e_g^+
-4Dq	d_1			t_{2g}^+
-4Dq	d_{-1}	T_{2g}	3-fold	t_{2g}^-
-4Dq	$-(d_{-2}-d_2)/\sqrt{2}$			t_{2g}^o

The + - and 0 in orbital symmetry notation are just arbitrary and are used only for distinguishing the individual members of the doublets and triplets in case a necessity arises for doing so. These superscripts may as well be substituted by other symbols or letters such as a,b,c etc. The symmetries of the eigenfunctions, already revealed by group theory, may be confirmed by subjecting the functional forms to transformations under the operators of the O_h point group.

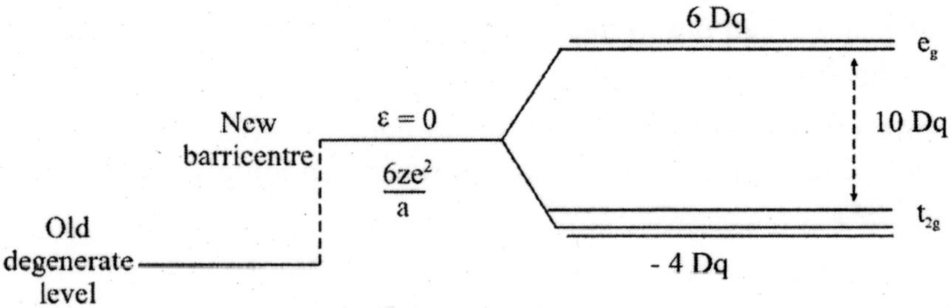

Fig. 8.1: d-orbital splitting in octahedral d^1 complex.

The energy difference between the split levels is thus 10 Dq where Dq= $\frac{1}{6}\alpha_4 = Ze^2r^4/6a^5$. As already stated, it is a parameter to be ascertained experimentally for the complex under study.

If we use the real d-orbitals i.e. d_{z2}, d_{xy}, d_{xz}, d_{yz}, $d_{(x2-y2)}$ as the basis set , the perturbation matrix of $h_{crys}(O_h)$, scooped from Eqs (7.55, 7.56, 7.57) without the baricenter raising component, is the following diagonal written briefly in a row.

(6Dq, -4Dq, -4Dq, -4Dq, 6Dq) ....(8.2)

with the same energy values for the two-fold and three-fold degenerate wavefunctions as in the complex d-basis functions. Thus d_{z2} and $d_{(x2-y2)}$ are the e_g orbitals with energy 6Dq and d_{xy}, d_{xz}, and d_{yz} form the t_{2g} orbitals with energy -4Dq , the energies being relative to the baricenter level = 0

8.2 d^1 Metal Ion Complex with Octahedral Crystal Field (C_3 as quantizing axis)

. Let us now examine how the results are modified or appear to be so, when the basis d functions are defined taking C_3 symmetry axis as the quantizing axis.

(a) Role of group theory

For the nature of lifting degeneracy and the distribution of the d-orbitals under different IR's of the point group O_h, we cannot here directly refer to the character table. The conventional character table shows the distribution of basis functions. (viz., the spherical harmonics, d-orbitals) defined with respect to C_4^z axis as the quantizing axis. So in the present case a reducible representation of the O-point group in the basis set Y_{2ml}. spherical harmonics with variable θ defined with respect to C_3 axis, is first constructed and then resolved into IR's via Eq (2.84) using the conventional character table of O point group of data on characters. The IR's so found are E and T_2 and hence for the O_h point group these happen to be E_g and T_{2g}. It thus transpires from symmetry that the degeneracy will be lifted partially to form a pair of two-fold degenerate orbitals (the e_g set) and a set of three orbitals with 3-fold degeneracy (the t_{2g} set) with no idea gained uptil now of their corresponding basis functions comprising the d-orbitals. Nor do we have any qualitative idea as to which energy level, e_g or t_{2g}, will be higher. The answers to these two latter properties are available only after we solve for energy eigenvalues and eigenvectors of the perturbation matrix.

(b) Eigenvalues, Eigenfunctions

The matrix elements of $h_{crys}{}^*(O_h)$, shown in Eqs (7.69), suggest that the crystal field, while allowing d_0 to remain pure, scrambles d_2 with d_{-1} and also d_{-2} with d_1. Using these basis functions in the sequence as mentioned here, the (5×5) perturbation matrix, directly compilable from (7.69), written in units of Dq with the constant baricenter raising term $6Ze^2/a$ dropped, is found to be block diagonal. It consists of one (1×1) matrix with basis function d_0, one (2×2) matrix with basis functions d_2 and d_{-1} and another (2×2) matrix with d_{-2} and d_1 as the basis functions. The total matrix is

$$\begin{bmatrix} -4 & & & & \\ & -2/3 & 10\sqrt{2}/3 & & \\ & 10\sqrt{2}/3 & 8/3 & & \\ & & & -2/3 & -10\sqrt{2}/3 \\ & & & -10\sqrt{2}/3 & 8/3 \end{bmatrix} \qquad \text{..............(8.3)}$$

After solving all the three component matrices for their eigenvectors and eigenvalues, the following table is compiled:

It is interesting to note that with the shift of the quantising axis from the C_4^z symmetry axis to C_3, the $d_{z^2}(\equiv d_0)$ orbital, defined in the new orientation, becomes a member of the triply degenerate set, t_{2g}, shedding its symmetry tag, e_g, of the previous situation (sec 8.1).

Table 8.1

Matrix from (8.3)	Energy Eigenvalues	Eigenvector Matrix	Eigenfunctions	Symmetry
(1×1)	-4Dq	(1)	d_o	T_{2g}
First (2×2)	-4Dq	$\begin{bmatrix}\sqrt{2/3}\\ -1/\sqrt{3}\end{bmatrix}$	$\sqrt{2/3}d_2 - 1/\sqrt{3}d_{-1}$	T_{2g}
	6Dq	$\begin{bmatrix}1/\sqrt{3}\\ \sqrt{2/3}\end{bmatrix}$	$1/\sqrt{3}d_2 + \sqrt{2/3}d_{-1}$	E_g
Second (2×2)	-4Dq	$\begin{bmatrix}\sqrt{2/3}\\ 1/\sqrt{3}\end{bmatrix}$	$\sqrt{2/3}d_{-2} + 1/\sqrt{3}d_1$	T_{2g}
	6Dq	$\begin{bmatrix}1/\sqrt{3}\\ -\sqrt{2/3}\end{bmatrix}$	$1/\sqrt{3}d_{-2} - \sqrt{2/3}d_1$	E_g

The symmetry tags, shown in the last column against the corresponding eigenfunctions, are only ensured by subjecting the double orbilals with energy 6Dq and the triple orbitals with energy -4Dq to specific symmetry transformations of the point group O which lead to the conclusions on symmetry noted in the last column of table $T_{8.1}$. The transformation process, though straightforward, is a bit laborious and not shown here.

If we want the eigenfunctions in terms of real d-orbital set, we can easily achieve these from those expressed in Table $T_{8.1}$. Writing the t_{2g} and the e_g orbitals together as separate sets one has

T_{2g} set

$$d_o \quad \text{.............(8.4a)}$$

$$\sqrt{2/3}\, d_{+2} - 1/\sqrt{3}\, d_{-1} \quad \text{.............(8.4b)}$$

$$\sqrt{2/3}\, d_{-2} + 1/\sqrt{3}\, d_{+1} \quad \text{.............(8.4c)}$$

e_g set

$$1/\sqrt{3}d_2 + \sqrt{2/3}\, d_{-1} \quad \text{.............(8.5b)}$$

$$1/\sqrt{3}\, d_{-2} - \sqrt{2/3}\, d_1 \quad \text{.............(8.5c)}$$

Thus taking the t_{2g} set, adding (8.4b) and (8.4c) and multiplying the whole linear combination by $1/\sqrt{2}$ we get a t_{2g} orbital

$$\sqrt{2/3}\, d_{(x2-y2)} - 1/\sqrt{3}\, d_{xz} \qquad \text{.............(8.6)}$$

Similarly by subtracting (8.4c) from (8.4b) and multiplying $1\sqrt{2}i$ we have another t_{2g} orbital

$$\sqrt{2/3}\, d_{xy} + 1\sqrt{3}\, d_{yz} \qquad \text{.............(8.7)}$$

and the final t_{2g} orbital is $d_o = d_{z2}$ ....(8.8)

Repeating this technique for the e_g orbital set (8.5b, 8.5c) we have the following degenerate e_g pair

$$\left.\begin{array}{l} 1/\sqrt{3}\, d_{(x2-y2)} + \sqrt{2/3}\, d_{xz} \\ 1/\sqrt{3}\, d_{xy} - \sqrt{2/3}\, d_{yz} \end{array}\right\} \qquad \text{.............(8.9)}$$

We thus glean all the information about the effect of the O_h crystal field on the splitting of d-orbitals when C_3 axis is the quantizing axis. With a shift of the quantization axis to C_3, everything remains the same excepting the forms of the eigenfunctions.

8.3 d^1 Metal Ion Complex with Octahedral (Cube) and Tetrahedral Crystal Fields

As has been our practice, we employ group theory first to reveal some facets of splitting of d-orbitals in these two symmetry environments and follow these up by finding the eigenvalues and eigenfunctions.

d^1 complex - O_h, (cube) field

(a) Role of group theory:

We do not have anything to add here beyond that already described under O_h (regular octahedron) field in sec 8.1 (a) except to point out that the baricenter raising constant energy term is $8Ze^2/a$ and that ligand proxy charges being all located at the corners of the cube, cause a reversal of the relative ordering of the two-fold degenerate e_g and three-fold degenerate t_{2g} orbitals. The d_{z2} and $d_{(x2-y2)}$ orbitals are relatively further away than the orbital set d_{xy}, d_{xz}, d_{yz} from the proxy charges. A d-electron in any one of the former pair experiences less of electrostatic repulsion by the ligands than it does when present in any of the latter triplet set.

(b) Eigenvalues and Eigenfunctions

As to the energy eigenvalues (eigenfunctions are already known) we can bank on O_h field result via Eq (7.30) and need not solve the perturbation matrix separately.

Thus

$$\varepsilon(e_g) = -8(6Dq)/9$$

and $\varepsilon(t_{2g}) = 8(4Dq)/9$ where $Dq = 1/6\left(Ze^2 r^4 / a^5\right) = 1/6\,\alpha^4$

Thus the effective energy gap between the t_{2g} and c_g levels is

$$\Delta\varepsilon_{(t2g-eg)} = (8\times10/9)Dq$$

d^1 complex - T_d Symmetry

(a) Role of group theory:

Reference to T_d character table (Appendix Chap-2) suggests that the d-orbitals split into e and t_2 set. As in O_h (cube), geometry of the locations of the proxy charges indicates d_{xy}, d_{yz}, d_{xz} orbitals (t_2 set) be energetically higher than d_{z2}, $d_{(x2-y2)}$ (e set) orbitals

(b) Equation (7.29) enables us to skirt any actual solution of the perturbation matrix for the energy eigenvalues, the eigenfunctions being already known. The energy eigenvalues of t_2 and e sets are -4/9 times the corresponding energy values of t_{2g} and e_g in octahedral environments

$$\left.\begin{aligned}\varepsilon(e) &= -4(6Dq)/9\\ \varepsilon(t_2) &= -4(-4Dq)/9 = 4\times4Dq/9\end{aligned}\right\}$$ see Fig. 8.2 for different splitting

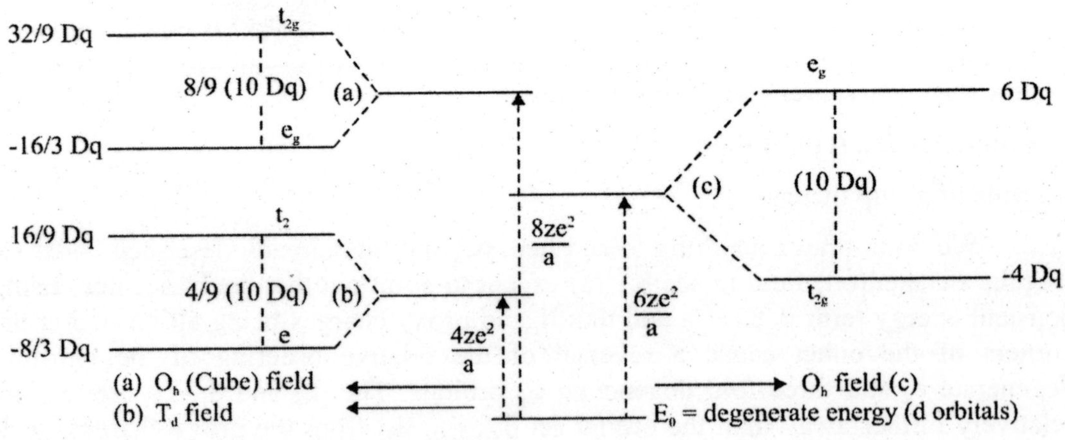

Fig. 8.2: Energy Level splitting (a) d^1 in O_h (cube) environment, (b) d^1 in T_d symmetry, (c) d^1 in (octahedral) symmetry

8.4 Crystal Fields of Loner Symmetries –d^1 Metal Ion Complex. Perturbations of Perturbed Systems.

Many relatively lower symmetry environments[1.3] (D_{4h} (tetragonal, sq. planar), D_{3d}, C_4, D_{3h}) may be deemed as the direct perturbed forms of an initially unperturbed spherically symmetric system. Alternatively, these may be deemed as the second stage perturbation effect of an already perturbed form (O_h, T_d) of an originally unperturbed completely symmetric systems. We shall try to illustrate here the approach based on the second viewpoint for some of the symmetry environments of a d^1 metal ion.

I. d^1 Metal Ion in D_{4h} (tctrag) Field. Distortion of O_h field

(a) Role of group theory

Instead of directly referring to the character table of D_{4h}, which would have been brief, neat and concise, one may as well indulge in some 'thought' experiment in order to fetch information otherwise available from the character table. Imagine a D_{4h} tetragonal complex ML_4V_2 having all the L ligands situated at two opposite locations on the X and Y axes in the equatorial plane and the two V ligands on the vertical ZOZ' axis. We thus know the initial dispositions and spread of all the d-orbitals of the metal ion at O. If we now toss the complex and catch it, we can still discern which is the Z-direction axis although we cannot tell the x axis from the y-axis. Thus after tossing, we can recognize the vertical and equatorial planes. Hence, after the toss, equivalent to a succession of symmetry operations of D_{4h}, we can still recognize the spreads of d_{z2}, $d_{(x2-y2)}$ orbitals and also of the d_{xy} orbital, the last two being in the equatorial plane. We can also discern the disposition of the pair of d_{xz} and d_{yz} orbitals though, unlike the former three, the latter two cannot be individually distinguished. It means d_{z2}, $d_{(x2-y2)}$, d_{xy} must form the base of one-dimensional IR's each whereas d_{xz}, d_{yz} belong to two-dimensional IR of D_{4h}. Under C_4 rotation, $z \rightarrow z$, $x \rightarrow y$, $y \rightarrow -x$ indicating d_{z2} belongs to A IR and $d_{(x2-y2)}$ and d_{xy} to B IR's. Under C_2' the changes are $z \rightarrow -z$, $x \rightarrow x$, $y \rightarrow -y$, hinting d_{z2} belongs to A_{1g}, $d_{(x2-y2)}$,to B_{1g} and d_{xy} to B_{2g} of D_{4h}. Of course d_{xz}, d_{yz} belong to E_g. The g suffix results from the fact that the d-orbitals are all gerade.

The totality of group theoretical information suggests that the degeneracies associated with the t_{2g} and e_g orbitals of O_h symmetry are lifted in the following way under a tetragonal distortion perturbation (such as elongation along z-axis due to substitution of two L-ligands by two V's reducing symmetry from O_h to D_{4h})

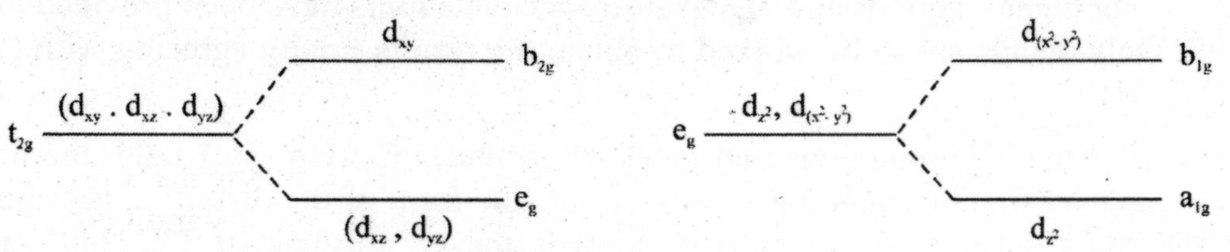

O_h symmetry $\rightarrow$ D_{4h} symmetry

(b) Eigenvalues and Eigenfunctions.

The constructions and solutions of two perturbation matrices of the distortion hamiltonian $Ar^2 Y_{20} + Br^4 Y_{40}$ (cf 7.50), one in the basis set of the e_g orbitals and the other in that of the t_{2g} orbitals, have already been accomplished in the discussion on the matrix elements (Sec 7.9D). Eq (7.62) suggests that only diagonal elements can emerge in these two basis sets. So all the individual orbitals, d_{z2}, $d_{(x2-y2)}$, d_{xy}, d_{xz}, d_{yz} are all eigenfunctions. We rewrite the results below (cf Sec 7.9D).

Eigenfunctions	IR	Energyeigenvalues (relative to baricentre energy=0)
d_{z2}	A_{1g}	$6dq - 2Ds - 6Dt$
$d_{(x2-y2)}$	B_{1g}	$6Dq + 2Ds - Dt$
d_{xy}	B_{2g}	$-4Dq + 2Ds - Dt$
d_{xz}, d_{yz}	e_g	$-4Dq - Ds + 4Dt$

……(8.10)

The Dq, Ds and Dt parameters have already been defined previously.

II. d^1 Metal Ion in D_{4h} (sq. pl) and C_{4v} (sq. pyramid) Fields as Distortions of O_h Field.

The distortion of O_h field through infinite elongation of the ligand bonds along the z-axis generates a D_{4h} (sq. pl) field around the metal ion. Similarly a finite and another infinite elongations respectively of the two ligand bonds along the z-axis give rise to a C_{4v} (sq. pyramid) field from distortion of an O_h environment. Using either the respective character tables or the 'thought experiments' or the technique of the branching of IR's on reduction in symmetry. we get to know the different IR's and their corresponding d-orbital base functions both for D_{4h} (sq. pl.) and C_{4v} (sq. py.) cases.

For eigenfunctions and eigenvalues, adequate hints have been provided in Sec 7.9E. Only details are to be worked to obtain the results finally agreeing with (7.60) and (7.61)

For the eigenfunctions and eigenvalues under C_{4v} (sq. py.) field, the matrix elements may first be found out using the suitable form of the distortion hamiltonian in the basis sets of the e_g orbitals and of the t_{2g} orbitals of O_h point group. Eigenfunctions and eigenvalues of energy follow from this perturbation matrix.

III. d^1 Metal Ion in D_{3d} Environment. Treated as Trigonal Distortion of O_h Field

a) Role of group theory:

The way the degeneracies of t_{2g} and e_g orbitals (Eqs 8.4 , 8.5) with C_3 as the quantizing axis is going to be affected may be ascertained by applying the principle of branching of the IR's, T_{2g} and E_g of the O_h point group when trigonal distortion reduces the symmetry to D_{3d} . Using this group theoretical technique the breakdown of the IR's in the present case is found to be

$T_{2g} \rightarrow A_{1g} + E_g$, $E_g \rightarrow E_g$

Furthermore, a look at the common components (d_{+-2}, d_{-+1}) of the eigenfunctions of the t_{2g} and e_g orbitals suggests that these are likely to remain scrambled when the symmetry is lowered to D_{3d}, but we cannot say anything definite beyond this likelihood. Actual transformations under symmetry operations of D_{3d} can establish this.

(b) Perturbation Matrix and Eigenvalues

We construct the perturbation matrix of the distortion hamiltonian (Ar^2Y_{20} - Br^4 Y_{40}) in the joint basis set provided by the t_{2g} set and the e_g set (Eqs 8.4 a, b, c and 8.5 b , c) The matrix elements to be found are

$$< t_{2g}^{a} | P | t_{2g}^{a} >, < t_{2g}{}^{b,c} | P | t_{2g}{}^{b,c} >, < t_{2g}{}^{b,c} | P | e_g{}^{b,c} >, < e_g{}^{b,c} | P | e_g{}^{b.c} >$$

where P stands for the perturbation hamiltonian. The other types of mixed matrix elements in the (5×5) matrix vanish as may be inferred directly from the composition of the t_{2g} and e_g orbitals vis a vis the results (Eqs 7.74 - 7.77 and 7.70) wherein too, the mixed matrix elements involving the d-wavefunctions are all zeros.

It follows from Eqs. (7.71 and 7.77)

$$< t_{2g}{}^{a} | P | t_{2g}{}^{a} > = < d_o | (Ar^2Y_{20} - Br^4Y_{40}) | d_o > = -2D_\sigma - 6D_\tau \quad ....(8.11)$$

Again using Eqs (8.4b, c and 7.74, 7.75)

$$< t_{2g}{}^{b,c} | P | t_{2g}{}^{b,c} > = < (\sqrt{2/3}\, d_{+-2} - +1/\sqrt{3}\, d_{-+1}) | Ar^2Y_{20} - Br^4Y_{40} | + (\sqrt{2/3}\, d_{+-2} - +1/\sqrt{3} d_{-+1}) >$$

$$= 2/3 < d_{+-2} | (Ar^2Y_{20} - Br^4Y_{40}) | d_{+-2} > + 1/3 < d_{-+1} + | (Ar^2Y_{20} - Br^4Y_{40}) | d_{-+1} >$$

$$= D\sigma + 2/3\, D\tau \quad ....(8.12)$$

$< e_g{}^{b,c} | P | e_g{}^{b,c} >$, on using (8.5b, c and 7.74 and 7.75),

$$=1/3 < d_{+-2}\left|\left(Ar^2Y_{20} - Br^4Y_{40}\right)\right|d_{+-2} > +$$

$$2/3 < d_{-1+1} + \left|\left(Ar^2Y_{20} - Br^4Y_{40}\right)\right|d_{-+1} >$$

$$= 7/3D\tau \qquad \text{.............(8.13)}$$

and finally

$$< t_{2g}{}^{b,c}\left|\left(Ar^2Y_{20} - Br^4Y_{40}\right)\right|e_g{}^{b,c} >$$

$$= \sqrt{2}D\sigma - 5\sqrt{2}/3\,D\tau \qquad \text{.............(8.14)}$$

Arranging the basis set in the sequence $t_{2g}^a, t_{2g}^b, e_g^b, t_{2g}^c, e_g^c$, there results, using (8.11-8.14) the following perturbation matrix

$$\begin{bmatrix} -2D\sigma - 6D\tau & & & & \\ & D\sigma + 2/3\,D\tau & \sqrt{2}D\sigma - 5\sqrt{2}/3\,D\tau & & \\ & \sqrt{2}D\sigma - 5\sqrt{2}/3\,D\tau & 7/3D\tau & & \\ & & & D\sigma + 2/3\,D\tau & \sqrt{2}D\sigma - 5\sqrt{2}/3\,D\tau \\ & & & \sqrt{2}D\sigma - 5\sqrt{2}/3\,D\tau & 7/3\,D\tau \end{bmatrix}$$

.............(8.15)

The matrix (8.15) is block-diagonal one, with two (2×2) submatrices in it identical. The two different roots from one (2×2) matrix will be the same as the couple obtained from the second. Altogether there will be a pair of E_g IR's with a pair of e_g orbitals associated with each E_g. The last orbital is one of A_{1g} symmetry having d_0 as the eigenfunction with perturbation energy $\varepsilon = -2D\sigma - 6D\tau$. To this must be added -4Dq (i.e, the unperturbed energy value of t_{2g}) to give the displacement of the non-degenerate level relative to the baricenter of the O_h level taken as zero.

Analytic solutions of (2×2) matrices are feasible if the values of $D\sigma$ and $D\tau$, indirectly gathered from other experimental sources, be inserted in the matrix. Alternatively, parametric solutions may be obtained by assuming a suitable proportionality between $D\sigma$ and $D\tau$ for the specific D_{3d} complex. Finally either a -6Dq or a 4Dq term has to be added.

8.5 d[1] Metal Ion with Two Perturbations (Spin orbit and Crystal Field). Double Groups

When the above two perturbations are considered for a d^1 system, the treatment naturally involves the application of the lesser perturbation subsequent to that of the relatively large one. Turning to Eq (8.1), one finds two distinct cases requiring attention.

A) $\xi(r)l.s >> h_{crys}$ B) $h_{crys} >> \xi(r)l.s$

To deal with such cases we have to make copious application of the principles and formulations of double groups with which, we assume, our readers are familiar.

It is known that an electronic wavefunction consists of a space part as well as a spin part. If the hamiltonian is spin-free, the symmetry of the wavefunction is decided by the IR of the space part. If, however, the hamiltonian also contains a spin dependent component, then the symmetry of the wavefunction is governed by its total behaviour under the group operations of the corresponding double group. In other words, the net symmetry will be indicated by an IR of the double group and to arrive at this IR, the direct product representation (dpr) of the double group plays a vital role in many cases.

Before we take up the question of perturbations and the symmetries of the wavefunctions, let us find the symmetries of the spin kets, $|1/2\,1/2>$, $|1/2\,-1/2>$ of an electron in the double groups $D_4^{/}, O^{/}$, and $D_3^{/}$. The spin kets are nothing but α and β spin functions and the question here is one of finding the symmetries of the spin functions in environments of $D_4^{/}, O^{/}$ and $D_3^{/}$ fields which are just hypothetical extensions of real environments of D_4, O and D_3 fields.[5,6]

As the spin kets indicate, the spin angular momentum quantum number s = 1/2 = j (no orbital motion of the electron is conveyed by spin kets). The characters of the representations of rotational operations on such a pair of spin functions are calculated from the general formula

$$\chi(C_\gamma) = sin\,(J+1/2)\gamma / sin\,\gamma/2 \qquad \text{.............(8.16)}$$

Thus for $D_4^{/}$ group, with J=j=1/2

$$\chi(C_4) = \sqrt{2} = \chi(C_4^3 R) : \chi(C_2^{/}) = \chi(C_2^{/} R) = 0$$

$$\chi(C_4^3) = \chi(C_4 R) = -\sqrt{2} : \chi(C_2^{//}) = \chi(C_2^{//} R) = 0$$

$$\chi(C_2) = \chi(C_2 R) = 0$$

The complete set of characters in tabular form is

$D_4^{/}$	E	R	$C_4^3 R$ C_4	$C_4 R$ C_4^3	$C_2 R$ C_2	$2C_2^{/} R$ $2C_2^{/}$	$2C_2^{//} R$ $2C_2^{//}$
$\chi(j=1/2)$	2	-2	√2	-√2	0	0	0

When checked against the character table of $D_4^{/}$ (App chap 2) the set is found to be identical with that appearing alongside $\Gamma_6^{/}$ (i.e. $E_2^{/}$). Hence the spin kets constitute a degenerate pair with symmetry $\Gamma_6^/$ .Since $D_4^/$ is a subgroup of $D_{4h}^/$, the kets will also be of same symmetry in $D_{4h}^/$ field.

Turning now to the environments of $O^/$, the set of characters in the basis set of spin kets, obtainable from (8.16), is tabulated below

$O^/$	E	R	$3C_4^3R$ $3C_4$	$3C_4R$ $3C_4^3$	$3C_2R$ $3C_2$	$4C_3^2R$ $4C_3$	$4C_3R$ $4C_3^2$	$6C_2^/R$ $6C_2^/$
$\chi(j=1/2)$	2	-2	√2	-√2	0	1	-1	0

The symmetry is thus identified with $\Gamma_6^/$ (i.e. $E_2^/$) of $O^/$ double group. This is also the expected IR of the spin kets in the $O_h^/$ field.

A similar follow up with $D_3^/$ double group indicates that α and β spin functions form a degenerate pair belonging to $\Gamma_4^/$ of $D_3^/$.

8.6 Successive Perturbations: $\xi(r)\,l.s >> h_{crys}$

We have already worked out in chapter 3 the eigenfunctions and eigenvalues when perturbation due to spin orbit interaction acts on a single valence /optical electron system including d[1]. We know there occur, for the latter, six fold degenerate states, $|j,m_j>$ with j = 5/2 and m_j running from 5/2 to-5/2 and also four fold degenerate states with j = 3/2 and $-3/2 \le m_j \le 3/2$. Thus the first of the successive perturbations causes a d[1] system i.e., a ^{2}D atomic or ionic state to split into two fine structure levels (j = 5/2, j = 3/2) with inherent degeneracies as stated above.

Let us now examine how these doublet levels split when the system is placed in a crystal field of D_4 environment. The physical realities of such a perturbing field are also realised if we suppose that D_4 symmetry field is substituted by $D_4^/$ field. This double group provides the additional theoretical advantages of dealing with symmetries of spin containing eigenfunctions.

Role of group theory:

The six degenerate eigenfunctions are with spin α (j = 5/2) and these are $|jm_j>\alpha$ with $-5/2 \le m_j \le 5/2$ *and* $j = 5/2$. Similarly the four degenerate spin orbit states with j = 3/2 and $-3/2 \le m_j \le 3/2$ are $|j,m_j>\beta$. We find the sets of characters of the group elements of $D_4^/$ double group in the two degenerate basis sets, viz., six of $|j\,m_j>\alpha$ and four of $|j\,m_j>\beta$. Putting j = 5/2 for the first set and j = 3/2 for the second, the characters of the rotational operators are obtained via Eq (8.16). Some simple calculations are shown below.

$$j=5/2, \chi(C_4)=\sin(5/2+1/2)\pi/2/\sin\pi/4=-\sqrt{2}$$

$$j=5/2, \chi(C_4^3)=\sin 9\pi/2/\sin\pi/4=\sqrt{2}$$

$$j=3/2, \chi(C_4)=\sin(3/2+1/2)\pi/2/\sin\pi/4=0$$

All such values collected separately lead to the following characters for D_4' point group

D_4'	E	R	(C_4, C_4^3R)	(C_4^3, C_4R)	(C_2, C_2R)	$(2C_2', 2C_2'R)$	$(2C_2'', 2C_2''R)$
$\chi(j=5/2)$	6	-6	-√2	√2	0	0	0
$\chi(j=3/2)$	4	-4	0	0	0	0	0

Each character set suggests that the corresponding representation is reducible. The IR's contained in each of them are found out by applying the relation (2.84) and using D_4' character table. For the set j = 5/2

$$\eta_{\Gamma_6'}=1/16\left[6\times2+6\times2+2\times\sqrt{2}\times(-\sqrt{2})+2\times\sqrt{2}\times(-\sqrt{2})\right]=1$$

$$\eta_{\Gamma_7'}=1/16\left[6\times2+6\times2+2\times(-\sqrt{2})\times(-\sqrt{2})+2\times\sqrt{2}\times\sqrt{2}\right]=2$$

$$\therefore \Gamma^{red}(j=5/2)=\Gamma_6'+2\Gamma_7' \quad \text{.............(8.17)}$$

Similarly for the set

$$\Gamma^{red}(j=3/2)=\Gamma_6'+\Gamma_7' \quad \text{.............(8.18)}$$

The interpretation of the results is that an ion with one d-electron has its spin orbit fine structure terms (first perturbation effect) further split into a number of crystal field terms when placed in an environment of D_4 or of D_{4h} We, however, content ourselves with the group theoretical findings and do not proceed further to obtain the final energy eigenvalues of the split levels.

When the system is placed in an environmental symmetry of O (or O_h) the characters of the reducible representation of O' double group for j = 5/2 and 3/2 are calculated by using Eq (8.16) and shown below

O'	E	R	$3C_4^3R$ $3C_4$	$3C_4R$ $3C_4^3$	$3C_2R$ $3C_2$	$4C_3^2R$ $4C_3$	$4C_3R$ $4C_3^2$	$6C_2'R$ $6C_2'$
$\chi(j=5/2)$	6	-6	$-\sqrt{2}$	$\sqrt{2}$	0	0	0	0
$\chi(j=3/2)$	4	-4	0	0	0	-1	+1	0

Application of resolution relation (2.84) leads to

$$\Gamma_{(j-5/2)^{red}} = \Gamma_7' + \Gamma_8' \text{ and } \Gamma_{(j=3/2)^{red}} = \Gamma_8' \qquad \text{.............(8.19)}$$

Collecting the results of (8.17), (8.18) and (8.19) we can have the following diagram for splitting of a d[1] system (^2D state) –Fig. 8.3.

Also employing techniques[7] similar to the foregoing ones, one arrives at the nature of splittings of p^1 (doublet P state) and of f^1 (Doublet F state) under spin orbit, O_h crystal field perturbations (Fig. 8.4).

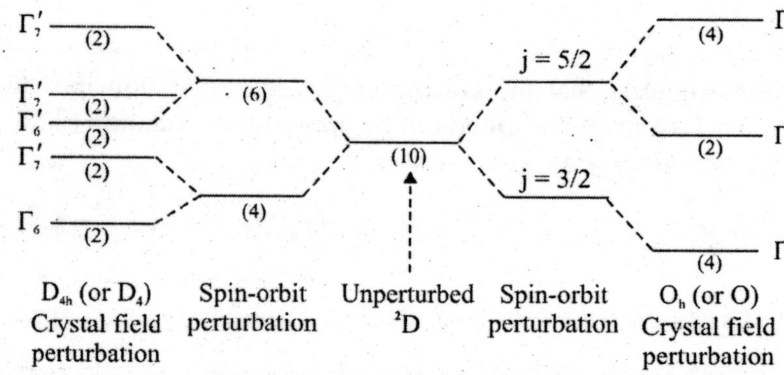

Fig. 8.3: Energy level splittings of a d[1] system under spin-orbit and crystal field perturbations. Figures within parentheses under each energy level indicate total intrinsic degeneracies including spin.

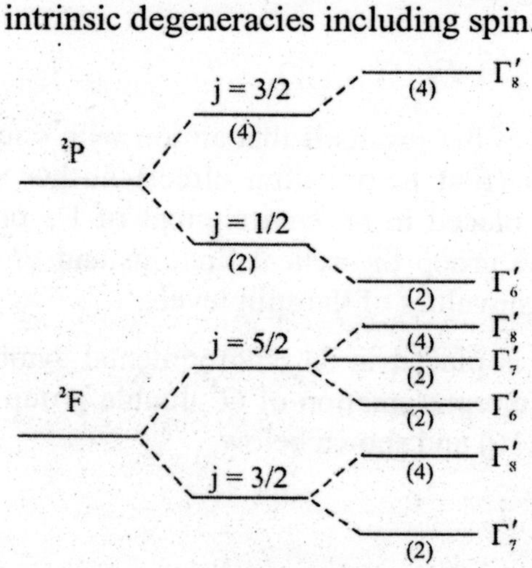

Fig. 8.4: Level splittings of p[1] and f[1] configurations under spin-orbit and O_h crystal field perturbations. Numbers within parenthesis indicate degeneracies.

8.7 d[1] Systems with Crystal Field and Spin Orbit Perturbations $h_{crys} >> \xi(r)l.s$

We now attend to the reverse case of the application of two successive perturbations, the one due to the crystal field being larger of the two. Consider a d^1 system (^2D state) placed in O_h field.

Role of group theory:

The lessons of Sec (8.1) tell us that the d-orbital space splits in O_h field into two subspaces, E_g and T_{2g} with basis functions d_{z2}, $d_{(x2-y2)}$ for the former and d_{xy}, d_{xz}, d_{yz} spanning the latter. The orbital motions of the electron are characterized by the symmetry properties of E_g and T_{2g} of O_h i.e., equivalently by the symmetry properties of $\Gamma_3^/$ and $\Gamma_5^/$ of the $O^/$ double group. Again the spin functions of the electron span the $\Gamma_6^/$ IR of the $O^/$ point group (Sec 8.5). Since the spin orbit perturbation represents interaction of spin and orbital motions, we have now to investigate how their joint wavefunctions in both the orbital and spin spaces are classified into IR's of the $O^/$ field. In other words we have to consider how the following two dpr's, viz.,

a) $\Gamma_3^/ \times \Gamma_6^/$ with the basis product functions $d_{z2}\alpha,\ d_{z2}\beta,\ d_{(x2-y2)}\alpha,\ d_{(x2-y2)}\beta$

and

b) $\Gamma_5^/ \times \Gamma_6^/$ with basis product functions $d_{xy}\alpha,\ d_{xz}\alpha,\ d_{yz}\alpha,\ d_{xy}\beta,\ d_{xz}\beta,\ d_{yz}\beta$ resolve into IR's of the O' double group.

Using the character table of $O^/$ (Appendix, Chap 2), the characters in the above two dpr's (see below) and the relation (2.84) we arrive at results noted against (8.20)

$O^/$	E	R	$4C_3^2R$ $4C_3$	$4C_3R$ $4C_3^2$	C_2R C_2	$3C_4^3R$ $3C_4$	$3C_4R$ $3C_4^3$	$6C_2^/R$ $6C_2^/$
$\chi(\Gamma_3^/ \times \Gamma_6^/)$	4	-4	-1	1	0	0	0	0
$\chi(\Gamma_5^/ \times \Gamma_6^/)$	6	-6	0	0	0	-√2	√2	0

$$\Gamma_3^/ \times \Gamma_6^/ = \Gamma_8^/ \quad \text{and} \quad \Gamma_5^/ \times \Gamma_6^/ = \Gamma_8^/ + \Gamma_7^/ \qquad \text{..............(8.20)}$$

The following diagram expresses the final results in $O^/$ field.

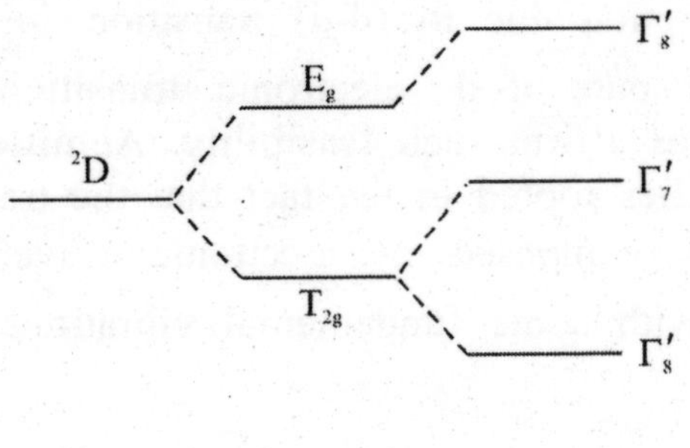

It is seen that the symmetries of the ultimate levels (cf Fig 8.3) are the same irrespective of the order of the applications of the two perturbations (cf Sec 8.6). This generalization covers only the ultimate symmetries and is not to be extended to the energy magnitudes.

An extension of this technique to D_{4h} field accompanied with spin orbit interaction can easily be made for a d^1 system. The character sets of the necessary direct products of the IR's of D_4' group are

D'	E	R	(C_4, C_4^3, R)	$(C_4^3, C_4 R)$	(C_2, C_2R)	$(2C_2', 2C_2'R)$	$(2C_2'', 2C_2''R)$
$\chi(\Gamma_3' \times \Gamma_6')$	2	-2	-√2	√2	0	0	0
$\chi(\Gamma_4' \times \Gamma_6')$	2	-2	-√2	√2	0	0	0
$\chi(\Gamma_5' \times \Gamma_6')$	4	-4	0	0	0	0	0
$\chi(\Gamma_1' \times \Gamma_6')$	2	-2	√2	-√2	0	0	0

Scrutiny of the character table of D_4' reveals that the final symmetry levels emerge from the following changeovers:

$$B_{1g} \rightarrow \Gamma_7'; \; B_{2g} \rightarrow \Gamma_7'; \; E_g \rightarrow \Gamma_8'; \; A_{1g} \rightarrow \Gamma_6'$$

This method of studying the effect of two successive perturbations can also be applied to $f^1(^2F)$ and p^1 (^{2}P) systems easily.

We wish to conclude this chapter with some remarks on some apparent, but not real discord between the crystal field results and the group theoretical demand. The evidence of lifting of degeneracy of the single-electron d-orbitals resulting from treating d^1-complex by crystal field theory, must leave its signature in the shape of spectral absorption due to (d-d) transition $(e_g \leftarrow t_{2g})$. But symmetry arguments forbid such pure d- d electronic transition. Contrarily, however, experimental observations affirm such feasibility. Admittedly, the transitions are weak. The explanation lies rooted in the fact that the transition moment integral involves wavefunctions, composed of electronic wavefunction, ψ_{eg} and ψ_{t2g} inseparably mixed up with some fundamental vibrational wavefunctions of the complex. Thus the transition moment integral is not given by $<\psi_{eg}|\mu|\psi_{t2g}>$, but by $<\psi_{eg}\varphi'|\mu|\psi_{t2g}\ \ \varphi''>$ where φ' and φ'' represent vibrational wavefunctions. Such transition moment integrals still continue to be responsible for the so called pure (d - d) transitions although, in reality, these are vibronic transitions.

This explanation, based on vibrational mixing, is also valid for transitions between crystal field levels originating from D/F or other atomic states of d^n complexes treated in chapters 9 and 10.

These types of vibralional contaminations of electronic wave functions occur whenever one or more of the IR's of the vibrational fundamentals happen to be the same as one or more of the symmetry types of the d^n complex. Further details of such simple calculations of vibronic transitions are presented in texts of group theory[5,6,7].

8.8 Workcd-out Examples

As an exercise for obtaining the Zeeman splitting of the quartet $\Gamma_8'(^2T_{2g})$ and of the doublet $\Gamma_7'(^2T_{2g})$ [Eq 8.20] we would in fact be dealing with the fourth perturbational operator (third perturbational operator in the case of single d-eleclron system) being applied in succession to the original unperturbed stale of the central field hamiltonian. The sequence of perturbations, in descending order of relative dominance, is the coulomb field, the crystal field, the spin-orbit interaction and, finally, the external magnetic field.

The external magnetic field splits a spin-orbit eigenstate into Zeeman levels and thus permits a calculation of "g"-factor of that level. While considering the "g"-factors and later magnetic susceptibilities and moments (chap 11) occasions arise when we have to consider the mixing of zero-order ground and excited states. One type of mixing is the spin-orbit coupling of the two states leading to newly formed actual ground state described by a new wave function with a first order correction. This is prior to and independent of whether or not any external magnetic field be set up to cause the subsequent Zeeman splitting of the spin-orbit level. The second type of mixing is the magnetic field - induced mixing of the ground spin-orbit level with a higher level (equivalent to selective second order Zeeman correction). This gives rise to additional paramagnetism, called TIP, i.e., temperature independent paramagnetism (see chap 11).

Example 1:

Obtain the "g"-factor for the ground spin-orbit stale $\Gamma_8^{/}\left({}^2T_{2g}\right)$ degenerate quartet (Eq.8.20) of a d^1 octahedral complex ion. Collect the necessary wave functions of $\Gamma_8^{/}\left({}^2T_{2g}\right)$ from Eq. 11.9a of chap. 11.

The wave function are (a) $\left|-1^+ >\right. = \left|d_{-1}^+ >\right.;$

b) $\left(1/\sqrt{3}\right)\left|-1^- > +\sqrt{(2/3}\right|\zeta_1^+ > \;=\; \left(1/\sqrt{3}\right)\left|d_1^- > +\sqrt{(2/3)}\right|1/\sqrt{2}\left(d_2^+ - d_{-2}^+\right)>$

c) $\sqrt{(2/3)}\left|\zeta_1^- > -1/\sqrt{3}\right|1^+ > \;=\; \sqrt{(2/3)}\left|1/\sqrt{2}\left(d_2^- - d_{-2}^-\right) > -1/\sqrt{3}\right|1^+ >$

d) $\left|1^- >\right. = \left|d_1^- >\right.$

Let the magnetic field be applied in the z direction so that the perturbation Hamiltonian is $\beta/\hbar(l_z + 2s_z)\,B_z$ .The perturbation energies are (remember that the kets are normalized and mutually orthogonal).

a) $\varepsilon^a = < d_{-1}^+ \left|\beta/\hbar\,(l_z + 2s_z)\right| d_{-1}^+ > B_z = \beta/\hbar\,(-1 + 2(1/2)\hbar B_z +$

$$< d_{-1}^+ \left| \; d_{-1}^+ > = 0 \text{ for } m = m_l + m_s = \pm 1/2\right.$$

$$\varepsilon^b = \left\{1/\sqrt{3} < d_{-1}^-\right| + \sqrt{(2/3)}1/\sqrt{2} < \left(d_2^+ - d_{-2}^+\right)\left|\;\right\}\beta/\hbar(l_z + 2s_z)\,B_z \times$$

$$\left\{1/\sqrt{3}\left|d_{-1}^- > + \sqrt{(2/3)}1/\sqrt{2}\right|\left(d_2^+ - d_{-2}^+\right)>\right\}$$

$$= 1/3\{-1 + 2(-1/2)\} < d_{-1}^- \left|d_{-1}^- > \beta B_z +\right.$$

$$1/3\beta B_z\left\{(2 + 2.(1/2)) < d_2^+\left|d_2^+ > + (-2 + 2.(1/2)) < d_{-2}^+\right| d_{-2}^+ >\right\}$$

$= 0$ for $m = 0$

A follow-up calculation of the other two diagonal matrix elements with wave functions (c) and (d) shows that both these spin –orbit levels remain unsplit like (a) and (b). Hence the splitting factor g = 0. It is however, to be noted that the spin -orbit interaction gives rise to $\Gamma_8^{/}\left({}^2E_g\right)$degenerate quartet in addition to $\Gamma_8^{/}\left({}^2T_{2g}\right)$ [Fig. In sec. 8.7]. There is thus a possibility of spin-orbit coupling of these two symmetry states giving rise to a different ground state level of $\Gamma_8^{/}$ symmetry. If these newly obtained and correct states of $\Gamma_g^{/}$ symmetry be taken into account in considering Zeeman splitting, we get the following final states - two non-degenerate and one doubly-degenerate.

$\varepsilon^e = (4\zeta_d / 10Dq)\beta B_z$ for m = 3/2, $\varepsilon^b = 0$ for m = 1/2

$\varepsilon^a = 0$ for m = -1/2, $\varepsilon^d = -(4\zeta_d / 10Dq)\beta B_z$ for m = -3/2

Therefore, the magnetic splitting factor

g = $(\varepsilon^c - \varepsilon^b)/\beta B_z = (\varepsilon^a - \varepsilon^d)/\beta B_z = 4\zeta_d / 10Dq$

Example 2.

Calculate the g-factor of the spin-orbit level $\Gamma_7'({}^2T_{2g})$of a cubic d[1] complex ion. Gather the requisite wave functions ψ^e and ψ^f from Eq. (1 1.9b) of chap. 11.

We find the diagonal elements of the operator $\beta/\hbar(l_z + 2s_z)$ B_z with respect to the wave functions ψ^e and ψ^f

$$\varepsilon^e = <\psi^e|\beta/\hbar\,(l_z + 2s_z)\,B_z|\psi^e> = 2/3[-1 + 2.(-1/2)\,]\beta B_z + 1/6\,[2 + 2.(1/2)]\beta B_z + 1/6[-2 + 2.(1/2)]\beta B_z$$

$$= -\beta B_z$$

Similarly, it is found that $\varepsilon^f = \beta B_z$. Thus $\Gamma_7'({}^2T_{2g})$level, in magnetic field, splits into two levels with g = $(\varepsilon^f - \varepsilon^e)/\beta B_z$ = z

Example 3

Compare the splitting factors, g's, of d[1] free ion and a cubic d[1]complex ion. Calculate also the spread (maximum energy gap) of the spin-orbit interaction states of these two types of d[1] ions.

We turn to chapter 3 for the expressions of Lande' g-factor and of spin-orbit interaction energy of a free ion.

$$g = 1 + [j(j+1) - l(l+1) + s(s+1)]/2j(j+1)$$

$$\varepsilon\,(\text{spin orbit}) = \zeta/2[j(j+1) - l(l+1) - s(s+1)].$$

Therefore $\Delta\varepsilon(max) = \varepsilon_1 - \varepsilon_2$

$$= \zeta\, s\,(2l+1)$$

(with $j_1 = l + s$ and $j_2 = |l - s|$)

For a free $d_1$ion, $g(j = 5/2) = 6/5$ and $g(j = 3/2) = 4/5$

$\Delta\varepsilon = 5\zeta/2$

For a cubic d^1 complex, $\Delta\varepsilon$ (max), due to spin-oibit interaction of any initial ground doublet $^2T_{2g}$ level, has been worked out in chap. 11 to be ζ and $-\zeta/2$ giving $\Delta\varepsilon = 3\zeta/2$. This can also be obtained by choosing two wave functions, one from Eq. (11.9a) and other from Eq. (l 1.9b) and working out the diagonal matrix elements of the operator $\zeta/\hbar^2\,(l.s)$.

Combining the present set of results with those available from examples 1 and 2, we have the following table for comparison.

	d^1 free ion	d^1 cubic complexion
g- factor	6/5 for j = 5/2 4/5 for j= 3/2	i) 0 for $\Gamma_8'\left({}^2T_{2g}\right)$ ii) $4\zeta\, d/10Dq$ for $\Gamma_8'\left({}^2T_{2g}\right)$ from a more correct initial wave function iii) 2 for $\Gamma_7'\left({}^2T_{2g}\right)$
$\Delta\varepsilon$ (max) for a given initial level	$5\zeta/2$ on splitting ^2D state of ion	$3\zeta/2$ on splitting of the $^2T_{2g}$ crystal field state.

Exercises

1. Find the transformations of the $Y_{2\ ml}$'s under rotations of the O-point group when the fuctions are defined with respect to one C_3 axis of a cube as the quantising axis. Which of the spherical harmonics mix among themselves ?
2. Using the lessons of chap. 7, find the form of $h_{crys}.(C_{4v})$ of a penta coordinated complex with a square base. What then will be the form of the distortion hamiltonian if the complex be deemed to originate from distortion of a regular octahedral complex ?
3. Apply the method of "tossing" (cf. sec. 8.1) to ascertain the types of degenerate and nondegenerate orbitals into which the five d-orbitals split when the metal ion is situated at the centre of the base of a square pyramidal crystal field.

References

1. C. J. Ballhausen : *Introduction to Ligand Field Theory* (Mc. Graw-Hill)
2. B. N. Figgis : *Introduction to Ligand Fields* (Wiley Eastern).

3. R. M. Golding : *Applied Wave Mechanics* (Addison Wesley)
4. L. E. Orgcl : *Ligand Field Theory* (Mcthucn and Co.)
5. M. Tinkharn : *Group Theory and Quantum Mechanics* (Mc. Graw-Hill)
6. S. C. Rakshit : *Molecular Symmetry Groups and Chemistry* (New Book Stall)
7. F. A. Cotton : *Chemical Applications of Group Theory* (Interscience)

Chapter Nine

Symmetries, Energyeigenvalues and Eigenfunctions of d^n Metal Ions In Weak Crystal Fields

If we look at the hamiltonian expression of a multi d-electron ion in a crystal field (Eq 7.1) we find that there are three perturbation terms affecting the central field hamiltonian. Depending on the comparative effects of the three preturbation hamiltonians, theoretical studies may be made under the following heads:

1. Weak crystal field cases, viz., $\sum e^2 / r_{kl} >> \mathcal{H}_{crys} > \sum_i \xi(r_i) l_i.s_i$

2. Strong crystal field cases, viz., $\mathcal{H}_{crys} >> \sum e^2 / r_{kl} > \sum_i \xi(r_i) l_i.s_i$

   as in the covalent complexes

3. Fields of intermediate strengths $\mathcal{H}_{crys} \sim \sum e^2 / r_{kl} > \sum_i \xi(r_i) l_i.s_i$

   examples of which depend upon the nature of liquids.

4. Spin orbit interactions larger than crystal field viz., $\sum_i \xi(r_i) l_i.s_i > \mathcal{H}_{cry}$ as in the heavier metal ion complexes including those of the rare earths.

We shall discuss the first three cases in some detail especially in O_h crystal fields. Of the three perturbation terms, the influences of the first two will be examined sequentially. Lastly of the different d^n systems, details will be worked out for the d^2 systems and to an extent partially of d^3 systems as well. The considerations will then be logically extended to the higher d^n systems. A great deal of analogy exists in the treatments of d^1 complex and of d^n systems, though more complexities creep in because of multiple number of electrons in the latter cases. On the whole, however, like in d^1 systems (Chaps 7,8), group theory is first applied to elicit as much of the desired information as possible. The evaluations of energyeigenvalucs of the split levels follow in its wake.

9.1 d^2 Metal Ion in O_h Crystal Field (Weak Field Case) General.

The hamiltonian written in conformity with (7.1) is

$$\mathcal{H}_{ion} = \mathcal{H}_{cent} + e^2 / r_{12} + \sum_k (r_k) l_k.s_k + \mathcal{H}_{crys}(O_h) \qquad \text{..........(9.1)}$$

where $\mathcal{H}_{crys}(O_h)$ is given by (7.10) in general and by (7.25) in particular for an O_h field. In the present weak field treatment, e^2/r_{12} is the main perturbation which is trailed by the weak crystal field perturbation $\mathcal{H}_{crys}(O_h)$. Spin orbit interaction will be neglected here.

Since the electrostatic repulsion is dominant , the two equivalent d-electrons couple, according to Russell Saunders' scheme, to yield the primary multiplets ^3F, ^3P and the singlets ^1D, ^1G and ^1S. (cf see 1.5.2). The secondary LS couplings in these initial multiplets are disregarded since the LS interactions are subordinate to crystal field interactions in the present perturbation scheme. It is therefore, clear that the wavefunctions, degeneracies and energyeigenvalues of the Russell Saunders' multiplets are the ones resulting from the application of the first perturbation Hamiltonian e^2/r_{12} to the central field of d^2 ion (cf chap 4). These states, already described in chap 4, form the new pedestal, i.e., the new initial states on which the effects of the second perturbation viz., $\mathscr{H}_{crys}(O_h)$are to be worked out.

As we know from the angular momentum operator principles, these S, P, D, F states have the general eigenfunction form $RY_{L\ ML}(\theta,\phi)$. These, being eigenfunctions of ($\mathscr{H}_{crys}+e^2/r_{12}$), are also eigenstates of e^2/r_{12}. Our preliminary basis set of functions are, therefore, D_2, D_1 , D_o, D_{-1}, D_{-2} for ^1D state, F_3, F_2, F_1, F_o, F_{-1}, F_{-2}, F_{-3} for the ^3F state and so on where the subscripts indicate the M_L values of the corresponding spherical harmonics. To find the effect of $\mathscr{H}_{crys}(O_h)$ (i.e., $\sum_i h_{crys}(O_h)$ we do not immediately construct the matrix of $h_{crys}(O_h)$ in the basis set of functions, viz., the D_{ML}'s or the F_{ML}'S. We however manage, in an indirect way, to derive the forms of final wavefunctions first of the foregoing Russell Saunders' multiplets in the O_h crystal fie

ld. This indirect way heavily leans on group theory.

The eigenkets of ($\mathscr{H}_{cent}$ $+e^2/r_{12}$) of a d^2 system are represented in the conventional notations, $|LSM_LM_S>$which denotes a spatial wavefunction of the type, as already mentioned, $R(r)Y_{L\ ML}(\theta,\phi)$ where θ and θ are the coordinates of an effective point mass having a total angular momentum $\sqrt{L(L+1)}\hbar$ and its z component value being $M_L\ \hbar$.

Within a given multiplet, say ^3F, we shall denote, for the sake of brevity, the foregoing atomic kets by ψ_{ML}'s, where M_L's range from +3 to -3. For ^1D, M_L covers the integral values +2 to -2 in ψ_{ML}'s. We now find the forms of wavefunctions when the crystal field is on. This is done in the next section.

9.2 d^2 Multiplets. Eigenfunctions and Energy eigenvalues in O_h Field

I. Crystal field eigenfunctions of ^3F states.

Role of group theory : Using the seven M_L's, viz., $\psi_3,\psi_2,\psi_1,....,\psi_{-3}$ as the basis functions, where each function contains a spherical harmonic $Y_{3ML}(\theta,\phi)$with the symmetry axis C_4^z of O_h field as the quantizing axis, the character set of the reducible representation of the O-point group may be found by applying Eqs (2.87, 2.88). The χ's are

O-group	E	$6C_4$	$3C_2$	$8C_3$	$6C_2^{/}$
χ^{red}	7	-1	-1	1	-1

The Ir's present in the reducible representation, found via Eq (2.84.) and the requisite character table, are A_2, T_1 and T_2. Since the single d-electron orbitals are basically gerade, the F set of states, comprising two d-electron each, classify themselves under symmetry types A_{2g}, T_{1g} and T_{2g} in the O_h field.

We now try to form the foregoing symmetry adapted basis functions with the seven members of ψ_{ML}'s of the F-multiplet. These will constitute the final wavefunctions (i.e. eigenfunctions) in the O_h field. To do this we avoid the general process of projection operator technique and instead consider the survival condition of the matrix elements of $\mathscr{H}_{crys}$ in the basis set of ψ_{ML}'s. Dropping the baricentcr raising component Y_{00} from $\mathscr{H}_{crys}$ (in the sense of Eq 7.24, since ψ_{ML}'s. are simple single entity i.e. atomic functions), it is seen that the remaining harmonic components in the hamiltonian, viz., Y_{40}, Y_{44} and Y_{4-4} permit us to draw the following conclusions.

i) Y_{40} ensures the survival of only the diagonal matrix elements, i.e., $< \psi_{ML} | Y_{40} | \psi'_{ML} > \delta_{ML \cdot ML}'$ for all M_L's ranging from 3 to -3

ii) Y_{44} and Y_{4-4} guarantee the survival of those off diagonal elements $< \psi_{ML} | (Y_{44} + Y_{4-4}) | \psi'_{ML} >$ that will have $\delta_{(mL/+-4)' ML} =$ unity

Considering the implications of the above two conclusions, it is evident that ψ_0, by itself, will form a basis function. The other six will be some linear combinations of $(\psi_1 \& \psi_{-3})$,$(\psi_{-1} \& \psi_3)$ and $(\psi_2 \& \psi_{-2})$ since each such pair satisfies the condition $\delta_{(ML/+-4)' ML} = 1$

Since all the wavefunctions are to be mutually orthogonal we may write the set as

a) ψ_0

b) $p\psi_1 + q\psi_{-3}$

c) $r\psi_{-1} + s\psi_3$

d) $u\psi_2 + v\psi_{-2}$ ....(9.2)

e) $q\psi_1 - p\psi_{-3}$

f) $s\psi_{-1} - r\psi_3$

g) $v\psi_2 - u\psi_{-2}$

The function $\psi_0 = f(r) 1/\sqrt{\pi}\sqrt{7/4} 1/2\, z(5z^2 - 3r^2)/r^3$ is seen to belong to the dpr $(T_{1g} \times A_{1g})$ of the O_h point group since z and ($5z^2$-$3r^2$) as basses of T_{1g} and A_{1g} respectively. The wavefunction ψ_0 is thus of T_{1g} symmetry. The other members of this triply degenerate IR, T_{1g} are obtainable by suitable rotational operation of the group upon the eigenfunction ψ_0

$$C_3\psi_0 = C_3 f(r) 1/\sqrt{\pi\sqrt{7/4}} 1/2z(5z^2 - 3r^2)/r^3$$

$= f(r)1/\sqrt{\pi\sqrt{7/4}}\ 1/2x(5x^2 - 3r^2)/r^3$, which is an independent real f state function

$$= 1/\sqrt{2}\{(\sqrt{3/8}\,\psi_{31} + \sqrt{5/8}\,\psi_{3-3}) - (\sqrt{3/8}\,\psi_{3-1} + \sqrt{5/8}\,\psi_{33})\}$$

$$= 1/\sqrt{2}\{(\sqrt{3/8}\,\psi_1 + \sqrt{5/8}\,\psi_{-3}) - (\sqrt{3/8}\,\psi_{-1} + \sqrt{5/8}\,\psi_3)\} \quad \text{..........(9.3)}$$

The penultimate step of Eq (9.3) follows from our acquaintance with practice in quantum chemistry of expressing the real f orbitals as linear combinations of complex f- wavefunctions. The relation also settles the values of the coefficients in Eqs. (9.2b, c, e, f), viz.,

$$p = r = \sqrt{3/8} \quad q = s = \sqrt{5/8}$$

The third member of the T_{lg} set may be obtained by C_3^2 operator acting on ψ_0.

$$C_3^2\,\psi_0 = C_3^2 f(r) 1/\sqrt{\pi}\sqrt{7/4}\ 1/2z(5z^2 - 3r^2)/r^3$$

$= f(r)1/\sqrt{\pi}\sqrt{7/4}\ 1/2y(5y^2 - 3r^2)/r^3$, a real independent F state function

$$= 1/\sqrt{2}\,i\{(\sqrt{3/8}\,\psi_1 + \sqrt{5/8}\,\psi_{-3}) + (\sqrt{3/8}\,\psi_{-1} + \sqrt{5/8}\,\psi_3)\} \quad \text{..........(9.4)}$$

Although (9.3) and (9.4) represent two degenerate wavefunction of T_{1g} symmetry in addition to the third one ψ_0, the two halves of either (9.3) or of (9.4) are themselves independent basis functions of T_{1g} symmetry generated by C_3 or C_3^2 operators. We intend to work with these two simple basis functions as eigenfunctions of T_{1g} symmetry rather than with their linear combinations denoted by (9.3) or (9.4).

We now try to write down an eigenfunction symmetry adapted to A_{2g}. Trial operators C_4 C_2, C_2' of O-group on $1/\sqrt{2}(\psi_2 - \psi_{-2})$ lead to character values in agreement with these under A_2 of the point group O. This also brings out the coefficient values u = v = 1√2 in Eqs (9.2.d,g).

The wavefunction $1/\sqrt{2}\,(\psi_2 + \psi_{-2})$ should, therefore, have the symmetry T_{2g} as the wavefunctions of the other symmetry types (T_{1g} and A_{2g}) have all been found out. The other two degenerate wavefunctions of T_{2g} IR are the respective orthogonal complements of each of the two halves of either (9.3) or (9.4). The net result of this elaborate discussion on group theoretical treatment for finding the nature of splitting and the form of eigenfunctions of the split levels is summarized in Table $T_{9.1}$.

Table $T_{9.1}$

Split Levels of ^3F state (d^2-configuration) in O_h crystal filed

Symmetry Type	Wavefunctions
a) A_{2g}	$1/\sqrt{2}(\psi_2 - \psi_{-2})$
b) T_{1g}	ψ_0
c) T_{1g}	$\left(\sqrt{3/8}\,\psi_1 + \sqrt{5/8}\,\psi_{-3}\right)$
d) T_{1g}	$\left(\sqrt{3/8}\,\psi_{-1} + \sqrt{5/8}\,\psi_3\right)$
e) T_{2g}	$1/\sqrt{2}(\psi_2 + \psi_{-2})$
f) T_{2g}	$\left(\sqrt{5/8}\,\psi_1 - \sqrt{3/8}\,\psi_{-3}\right)$
g) T_{2g}	$\left(\sqrt{5/8}\,\psi_{-1} - \sqrt{3/8}\,\psi_3\right)$

……(9.5)

II. Energy of split Levels of ^3F in O_h Crystal Field (d^2 system)

Let us choose the A_{2g} function (9.5a), one T_{1g} function (9.5b) and one T_{2g} function (9.5e) and evaluate their energies due to crystal field perturbation. To do this one ought to have the "single-electron spin orbitals containing" two-electron determinantal forms of chosen atomic F state wavefunctions. The components in these wavefunctions are ψ_2, ψ_0 and ψ_{-2}. Since the energy values are independent of M, in ^{3}F state we take $M_s = 1$ in all the components of atomic kets.

The only det form of the atomic ket $|3121> \;=\; \psi_2 = \left(2^+0^+\right)$ ……….(9.6)

Working upon the LHS with L- and equivalently on the RHS of (9.6) with ($l_{1-}+l_{2-}$),

$$\sqrt{10}|3111> \;=\; \sqrt{10}\psi_1 = 2\left(1^+0^+\right) + \sqrt{6}\left(2^+ -1^+\right)$$

$$\therefore \psi_1 \;=\; \sqrt{2/5}\left(1^+0^+\right) + \sqrt{3/5}\left(2^+ -1^+\right) \qquad \text{……….(9.7)}$$

A repeat operation upon ψ_1 by L. followed by rearrangement yields from (9.7)

$$|3101> \;=\; \psi_0 \;=\; 2/\sqrt{5}\left(1^+ -1^+\right) + 1/\sqrt{5}\left(2^+ -2^+\right) \qquad \text{……….(9.8)}$$

Like ψ_2, ψ_{-2} has also a single det form

$$|31-21>=\psi_{-2}=(0^+-2^+) \qquad \text{..........(9.9)}$$

The energy of ψ_{A2g} state can be obtained by substituting (9.6) and (9.9) in [9.5(a)] and finding the corresponding diagonal matrix element of $\mathscr{H}_{crys}$ (O$_h$) [i.e. $h_{crys(1)}(O_h)$+ $h_{crys(2)}(O_h)$], given by Eq [7.25].

$$\varepsilon(^3A_{2g};^3F,d^2) = 1/2<\{\psi_2-\psi_{-2}\} \mid \mathscr{H}_{crys}(O_h)|\{\psi_2-\psi_{-2}\}>$$

$$= 1/2\,[<\psi_2|\mathscr{H}_{crys}(O_h)|\psi_2> + <\psi_{-2}|\mathscr{H}_{crys}(O_h)|\,\psi_{-2}> - <\psi_2|\mathscr{H}_{crys}(O_h)|\,\psi_{-2}>$$
$$- <\psi_{-2}|\mathscr{H}_{crys}(O_h)|\,\psi_2>]$$

$$= 1/2\,[<(2^+0^+)|h_{crys(1)}(O_h)+ h_{crys(2)}(O_h)|(2^+0^+)> +$$
$$<(0^+-2^+)|h_{crys(1)}(O_h) + h_{crys(2)}(O_h)|(0^+-2^+)> -$$
$$<(2^+0^+)|h_{crys(1)}(O_h) + h_{crys(2)}(O_h)|(\,0^+-2^+)> -$$
$$<(0^+-2^+)|h_{crys(1)}(O_h) + h_{crys(2)}(O_h)|(2^+0^+)>] \qquad \text{.....(9.10)}$$

Changing the columns of the determinants (0^+-2^+) in the last two integrals, expanding the determinants and remembering that the inherent unwritten normalizing factors of the dets cause only one effective integral (i.e., one the product of the diagonal elements) to subsist, we have

$$\varepsilon(^3A_{2g};^3F,d^2) = 1/2[<d_2|h_{crys(1)}(O_h)d_2> + <d_o|h_{crys(2)}(O_h)\,d_o>$$
$$+ <d_o\,|h_{crys(1)}(O_h)d_0> + <d_{-2}|h_{crys(2)}(O_h)d_{-2}>$$
$$+ <d_2\,|h_{crys(1)}(O_h^{\cdot})d_{-2}> + <d_{-2}|h_{crys(2)}(O_h)d_2>] \quad \text{..(9.11)}$$

$$=12D_q \qquad \text{on using (7.54)} \qquad \text{..........(9.12)}$$

from which the baricentre raising term has been dropped.

$\varepsilon(^3T_{2g})$To calculate it, we find the diagonal matrix element of $\mathscr{H}_{crys}(O_h)$ of (7.25) with respect to the eigenfunction (9.5e) with (9.6) substituted in it. Like in (9.11), we will have a similar expression with the difference that the signs of the last two integrals are changed.

$$\varepsilon(^3T_{2g}) \quad = 1/2\,[< d_2|h_{crys(1)}(O_h)d_2> + <d_o|\,h_{crys(2)}(O_h)d_o>$$
$$+ <d_o\,|h_{crys(1)}(O_h)d_0> +< d_{-2}|h_{crys(2)}(O_h)d_{-2}>$$
$$+ <d_2\,|h_{crys(1)}(O_h)d_{-2}> +< d_{-2}|h_{crys(2)}(O_h)d_2> \text{..(9.13)}$$

$$= 2Dq, \text{ on using (7.54)} \qquad \text{..........(9.14)}$$

a value sans the baricentre raising energy.

$\varepsilon(^3T_{lg})$: To calculate this, we find the diagonal matrix element of the crystal field hamiltonian with respect [9.5(b)] as modified by (9.8) substituted in it.

$$\varepsilon(^3T_{1g},\ ^3F,d^2) = 4/5 < (1^+\text{-}1^+)|\mathscr{H}_{crys}(O_h)|(1^+\text{-}1^+)> + 1/5 <(2^+\text{-}2^+)|\mathscr{H}_{crys}(O_h)|(2^+\text{-}2^+)>$$
$$+ 2/5 < (1^+\text{-}1^+)|\mathscr{H}_{crys}(O_h)|(2^+\text{-}2^+)>$$
$$+ 2/5 <(2^+\text{-}2^+)|\mathscr{H}_{crys}(O_h)|(1^+\text{-}1^+)>$$

On expanding the determinant, taking care of the (unwritten) normalizing factor, orthogonality and orthonormality of the functions and remembering that $\mathscr{H}_{crys}(O_h)=\sum_i h_{crys(i)}(O_h)$, there finally results, without the baricenter-raising energy contribution,

$$\varepsilon(^3T_{1g};^3F,d^2) = -6Dq \qquad \text{..........(9.15)}$$

If we scoop the energy value due to primary electrostatic perturbation in forming the ^3F term of d^2 configuration from table ($T_{4\cdot6}$), the total perturbation energy of the split levels on dropping the baricenter raising term is,

$$\left.\begin{aligned}\varepsilon(^3A_{2g};^3F,d^2) &= F_0-8F_2+9F_4+12Dq = A-8B+12Dq\\ \varepsilon(^3T_{2g};^3F,d^2) &= F_0-8F_2+9F_4+2Dq = A-8B+2Dq\\ \varepsilon(^3T_{1g};^3F,d^2) &= F_0-8F_2+9F_4-6Dq = A-8B-6Dq\end{aligned}\right\} \qquad \text{..........(9.16)}$$

III. Energy Evaluation (Alternative method)

If any of the resulting crystal field wavefunctions in the form of its overall atomic kets (as in table $T_{9.1}$) is such that we can readily determine its energy from its det form, obtainable directly from Table $T_{4.3}$ or by applying shift operator (L_+ or L_-) on some cell in it, then this energy can be used in association with diagonal sum rule and the principle of nonshift of the baricenter to determine the energies of the other crystal field states.

To illustrate, the atomic ket form of the $\psi(^3T_{1g}) = \psi_0$ is given by $|3101|>$. The det form of this atomic ket is obtainable readily from repeated L_- operations starting from cell E of table $T_{4.3}$. So that finally one gets

$$\psi(^3T_{1g};^3F,d^2) = \psi_0 \equiv |3101> = 2/\sqrt{5}(1^+-1^+) + 1/\sqrt{5}(2^+-2^+) \qquad \text{..........(9.17)}$$

As already found, its energy, evaluated from (9.17) is (cf 9.15)

$$\varepsilon(^3T_{1g};^3F,d^2) = -6Dq \qquad \text{..........(9.18)}$$

Now the wave functions

$\psi\left({}^3A_{2g}\right)=1/\sqrt{2}(\psi_2-\psi_{-2})$ and $\psi\left({}^3T_{2g}\right)=1/\sqrt{2}\,(\psi_2+\psi_{-2})$ (In table T9.1)

contain the atomic kets $\psi_2=|3121>$ and $\psi_{-2}=|31-21>$ From table T4.3 these two atomic kets in det forms are (2^+0^+) and $(0^+\ -2^+)$ respectively. These two atomic kets do not occur in any other wavefunction. Hence the crystal field perturbation matrix in the basis set of these two atomic kets, viz., $|21>$ and $|-21>$ $\{M_LM_s notation\}$ involves in its trace the sum of energies $\varepsilon\left({}^3A_{2g}\right)$ and $\varepsilon\left({}^3T_{2g}\right)$. This is the principle of the diagonal sum rule.

$$\varepsilon\left({}^3A_{2g};{}^3F,d^2\right)+\varepsilon\left({}^3T_{2g};\ {}^3F,d^2\right)$$

$$=\varepsilon\left(2^+0^+\right)+\varepsilon\left(0^+-2^+\right) = \text{sum of diagonal elements of the dets}$$

$$= <d_2|h_{crys(1)}(O_h)|d_2>+<d_0\,|h_{crys(2)}(O_h)|\,d_0>+<d_0|h_{crys(1)}(O_h)|\,d_0>$$

$$+<d_{-2}|h_{crys\,(2)}(O_h)\,|d_{-2}>$$

$$= 14Dq, \text{ on using (7.54)} \qquad \text{..........(9.19)}$$

Since, the baricenter of the three energy levels of the symmetry states ${}^3A_{2g}, {}^3T_{2g}$ and ${}^3T_{2g}$ all emerging from ^{3}F remains unmoved after splitting by crystal field into the former states, we can write an equation

$$\varepsilon\left({}^3A_{2g}\right)+3\varepsilon\left({}^3T_{2g}\right)+3\varepsilon\left({}^3T_{1g}\right)=0 \qquad \text{..........(9.20)}$$

Now using the three eqs (9.18), (9.19) and (9.20) we have

$$\varepsilon\left({}^3A_{2g}\right) = 12Dq,\ \varepsilon\left({}^3T_{2g}\right) = 2Dq \text{ and } \varepsilon\left({}^3T_{1g}\right) = -6dq$$

9.3 Splitting of the Other Multiplets : ^{1}S, ^{3}P, ^{1}D, ^{1}G of the d^2 System in O$_h$ Field.

I. The role of group theory and other specific details were explained in arriving at the final wavefunctions of the split levels of the ${}^3F$ term in O_h crystal field. The following remarks are helpful in arriving at the forms of wavefunctions of the split levels or at least of the majority of the set rather quickly for any multiplet of d^n system in O_h crystal field.

The general process of obtaining symmetry adapted wavefunctions through the use of projection operator in O_h crystal field is, in so far as the labour is involved, no mean task. That is why in actual cases (cf ${}^3F$ of sec 9.2) indirect methods, aided by intelligent guesses, use of transformation properties of selected group operators and of suitable relations, are often employed to achieve the end.

The following is a recipe for working out in general many, in fact, most of the linear combinations constituting wavefunctions belonging to different symmetries in the crystal field. This is generally true for d^2, d^3 and other d^n functions.

R1.

Ascertain the IR's of the wavefunctions using the basis set of the corresponding spherical harmonics using Eqs (2.87 and 2.88) e.g., those which arise for the F set of states of d^n configurations , are A_{2g}, T_{1g}, and T_{2g} in O_h field (see sec 9.2).

R2.

Divide the basis functions into groups or batches (each batch may comprise either two or three functions and sometimes even one) using the logic of survival of the matrix elements of the crystal field hamiltonian occurring as integral of triple product of spherical harmonics.

R3.

(a) For any batch having $|M_L|$ the same in a pair, the normalized linear combinations viz., $1/\sqrt{2}(\psi_{ML} \pm \psi_{-ML})$ constitute a pair of wavefunctions suiting two of the intended symmetries found already by applying RI.

(b) For a batch having $M_L \neq M_L$, a.pair of acceptable wavefunctions satisfying the intended symmetry types are two linear combinations taken in proportions of "a" and "b" in one and "b" and "a" in another with + and - signs in between respectively where "a" and "b" are the coefficients of the normalized Y_{LML} and $Y_{LML/}$, say, in their cartesian forms (appendix chap 2) The linear combinations arc finally taken in normalized forms.

(c) For a batch comprising three functions in it as in G and H multiplets, while in some cases considerations of R3(b) will apply, constituting the complementary orthogonal member is not that simple. One has to consider additional features to arrive at the acceptable forms of wavefunctions. Generally speaking, no shortcut technique can be prescribed for a combination of three or more functions.

R4.

The final assignments of the symmetries found under R1. to the specific wavefunctions collected under R3(a) , (b) and (c) are completed by using selected principles of group theory, eg, dpr's , selected transformations under group operators or by referring to the already derived result. In many cases, it turns out that wavefunctions of symmetry type E of the split levels of d^n systems include one or the other of the pair of spherical harmonics of even order such as $Y_{22} \pm Y_{2-2}, Y_{44} \pm Y_{4-4}, Y_{42} \pm Y_{4-2}$ .This hint helps in picking up the proper form from R3(a). Subject those to trial transformations under selected group operators for confirmation or rejection of their claim for membership of the E symmetry club. Similarly forms coming under R3(b) constitutes wavefunctions of the triply degenerate T IR.

We now utilise the recipe to draw up a table of the forms of wavefunctions of the split levels of the multiplets of d^n systems in O_h, environment. The multiplets S P D F G and H are considered . For d^2 system, there will be no H multiplet. Although ^3F of d^2 has been treated in detail (Sec 9.2), this is included here for the sake of completeness of the table. While table $T_{9.2}$ shows the application of the recipe R1 to R3, in table $T_{9.3}$ there appear the individual symmetries of the wavefunctions (R4)

In order to differentiate the ψ symbols of the different multiplets, the corresponding L values of the states concerned will be subscripted to ψ alongwith the M_L's rendering the final form as ψ_{LML} in the tables $T_{9.2}$ and $T_{9.3}$ The L subscript will be withdrawn later in the follow-up discussions in this and the other chapters where there arises no possibility of confusion. Appendix (chap 2) is used for assignment of coefficient values a, b and c in table $T_{9.2}$.

Table $T_{9.2}$

Forms of wave functions of split-levels of atomic multiplets of d^n systems in O_h crystal field. (Application of recipe rules)

Atomic multiplets	R1 symmetries of wave functions	R2 Batches of ψ_{LM_L}'s forming acceptable wavefunctions in linear combinations	R3(a) Forms of wave-functions	R3(b) Forms of wave functions	R3 (c) Forms of wave-functions
S	A_{1g}	(ψ_{00})	(ψ_{00})		
P	T_{1g}	$(\psi_{10});(\psi_{11},\psi_{1-1})$	ψ_{10} $(\psi_{11}\pm\psi_{1-1})/\sqrt{2}$		
D	E_g, T_{2g}	$\psi_{20}\ (\psi_{21},\psi_{2-1});$ (ψ_{22},ψ_{2-2})	ψ_{20} $(\psi_{21}\pm\psi_{2-2})/\sqrt{2}$ $(\psi_{22}\pm\psi_{2-2})/\sqrt{2}$		
F	A_{2g}, T_{1g}, T_{2g}	$\psi_{30};(\psi_{32},\psi_{3-2});$ $(\psi_{31},\psi_{3-3});(\psi_{3-1}\psi_{33})$	ψ_{30} $(\psi_{32}\pm\psi_{3-2})/\sqrt{2}$	$\sqrt{3/8}\,\psi_{31}+\sqrt{5/8}\,\psi_{3-3}$ $\sqrt{3/8}\,\psi_{3-1}+\sqrt{5/8}\,\psi_{33}$ $\sqrt{5/8}\,\psi_{31}-\sqrt{3/8}\,\psi_{3-3}$ $\sqrt{5/8}\,\psi_{3-1}-\sqrt{3/8}\,\psi_{33}$	
G	A_{1g}, E_g T_{1g}, T_{2g}	$\psi_{40};(\psi_{42},\psi_{4-2});$ $(\psi_{44},\psi_{4-4});$ $(\psi_{41},\psi_{4-3});$ (ψ_{4-1},ψ_{43})	$(\psi_{42}\pm\psi_{4-2})/\sqrt{2}$ $(\psi_{44}\pm\psi_{4-4})/\sqrt{2}$	$\sqrt{7/8}\,\psi_{41}+\sqrt{1/8}\,\psi_{4-3}$ $\sqrt{7/8}\,\psi_{4-1}+\sqrt{1/8}\,\psi_{43}$ $\sqrt{1/8}\,\psi_{41}-\sqrt{7/8}\,\psi_{4-3}$ $\sqrt{1/8}\,\psi_{4-1}-\sqrt{7/8}\,\psi_{43}$	$\sqrt{7/12}\,\psi_{40}+\sqrt{5/24}\,(\psi_{44}+\psi_{4-4})$ $\sqrt{5/12}\,\psi_{40}-\sqrt{7/24}\,(\psi_{44}-\psi_{4-4})$**
H	E_g, T_{1g}, T_{1g}, T_{2g}	$\psi_{50};(\psi_{52},\psi_{5-2});$ $(\psi_{54},\psi_{5-4});$ $(\psi_{55},\psi_{51};\psi_{51},\psi_{5-3})$ $(\psi_{53},\psi_{5-1};\psi_{5-1},\psi_{5-3})$	ψ_{50} $(\psi_{52}\pm\psi_{5-2})/\sqrt{2}$ $(\psi_{54}\pm\psi_{5-4})/\sqrt{2}$		$\sqrt{126/256}\,\psi_{55}+\sqrt{60/256}\,\psi_{51}+\sqrt{70/256}\,\psi_{5-3}$ $\sqrt{126/256}\,\psi_{5-5}+\sqrt{60/256}\,\psi_{5-1}+\sqrt{70/256}\,\psi_{53}$ $\sqrt{10/256}\,\psi_{55}+\sqrt{84/256}\,\psi_{51}-\sqrt{162/256}\,\psi_{5-3}$ $\sqrt{10/256}\,\psi_{5-5}+\sqrt{84/256}\,\psi_{5-1}-\sqrt{162/256}\,\psi_{53}$ $-\sqrt{120/256}\,\psi_{55}+\sqrt{112/256}\,\psi_{51}+\sqrt{24/256}\,\psi_{5-3}$ $-\sqrt{120/256}\,\psi_{5-5}+\sqrt{112/256}\,\psi_{5-1}+\sqrt{24/256}\,\psi_{53}$

Table T9.3

(Symmetries of wave functions of split levels of d^n configuration in O_h field)

Multiplets	Symmetries of split levels	Wave functions of split levels (from table T9.2)
S	A_{1g}	ψ_{00}
P	T_{1g}	$\psi_{10}.(\psi_{11}+\psi_{1-1})/\sqrt{2}$, $(\psi_{11}-\psi_{1-1})/\sqrt{2}$
D	E_g	$\psi_{20},(\psi_{22}+\psi_{2-2})/\sqrt{2}$
	T_{2g}	$(\psi_{21}+\psi_{2-1})/\sqrt{2}$, $(\psi_{21}+\psi_{2-1})/\sqrt{2}$, $(\psi_{22}-\psi_{2-2})/\sqrt{2}$
F	A_{2g}	$(\psi_{32}-\psi_{3-2})/\sqrt{2}$
	T_{1g}	$\psi_{30}\cdot[(3/8)^{1/2}\psi_{31}+(5/8)^{1/2}\psi_{3-3}]\,[(3/8)^{1/2}\psi_{3-1}+(5/8)^{1/2}\psi_{33}]$
	T_{2g}	$(\psi_{32}+\psi_{3-2})/\sqrt{2}.[(5/8)^{1/2}\psi_{31}-(3/8)^{1/2}\psi_{3-3}]\,[(5/8)^{1/2}\psi_{3-1}-(3/8)^{1/2}\psi_{33}]$
G	A_{1g}	$(7/12)^{1/2}\psi_{40}+(5/24)^{1/2}(\psi_{44}+\psi_{4-4})$
	E_g	$(\psi_{42}+\psi_{4-2})/\sqrt{2},(5/12)^{1/2}\psi_{40}-(7/24)^{1/2}(\psi_{44}+\psi_{4-4})$
	T_{1g}	$(\psi_{44}-\psi_{4-4})/\sqrt{2},[(7/8)^{1/2}\psi_{41}-(1/8)^{1/2}\psi_{4-3}]\,[(7/8)^{1/2}\psi_{4-1}+(1/8)^{1/2}\psi_{43}]$
	T_{2g}	$(\psi_{42}+\psi_{4-2})/\sqrt{2},[(1/8)^{1/2}\psi_{41}-(7/8)^{1/2}\psi_{4-3}]\,[(1/8)^{1/2}\psi_{4-1}-(7/8)^{1/2}\psi_{43}]$
H	E_g	$(\psi_{52}-\psi_{5-2})/\sqrt{2},(\psi_{54}-\psi_{5-4})/\sqrt{2}$
	T_{1g}	$\psi_{50}[(126/256)^{1/2}\psi_{55}+(60/256)^{1/2}\psi_{51}+(70/256)^{1/2}\psi_{5-3}]$ $[(126/256)^{1/2}\psi_{5-5}+(60/256)^{1/2}\psi_{5-1}+(70/256)^{1/2}\psi_{53}]$
	T_{1g}	$(\psi_{54}+\psi_{5-4})/\sqrt{2},[(10/256)^{1/2}\psi_{55}+(84/256)^{1/2}\psi_{51}-(162/256)^{1/2}\psi_{5-3}]$ $[(10/256)^{1/2}\psi_{5-5}+(84/256)^{1/2}\psi_{5-1}-(162/256)^{1/2}\psi_{53}]$
	T_{2g}	$(\psi_{52}+\psi_{5-2})/\sqrt{2},[-(120/256)^{1/2}\psi_{55}+(112/256)^{1/2}\psi_{51}+(24/256)^{1/2}\psi_{5-3}]$ $[-(120/256)^{1/2}\psi_{5-5}+(112/256)^{1/2}\psi_{5-1}+(24/256)^{1/2}\psi_{53}]$

The compilation of entries in table T9.3 follows the requirement under R4. These are apparent for those split levels emanating from S P and D states. The character table and our knowledge of interconnection between real and imaginary forms of the wavefunctions reveal the information. Leaving aside the IR's of the levels, stemming from F state for which arguments have already been offered in Sec 9.2, a few remarks about the symmetry tags of levels originating from G state seem to be in order. That A_{1g} should be a combination of ψ_{40}, ψ_{44} and ψ_{4-4} becomes evident from our previous construction of totally symmetric $h_{crys(i)}(O_h)$ of Sec 7.3. Its orthogonal associate must, therefore, belong to either E_g IR or to one of the triply degenerate IR's viz T_{1g} or T_{2g}. Operations by C_4 and C_2' of the O-point group on this as well as $1/\sqrt{2}(\psi_{42}+\psi_{4-2})$ stamp these out to be of E symmetry. The remaining six functions, when operated upon by C_4 of the O-point group divide themselves up into two triplets T_1 and T_2 from their calculated character values of +1 and -1 respectively. This completes the assignment of symmetry tags. The symmetries of the wavefunctions of the levels emerging from the H-multiplet follow from some involved arguments which are not discussed here.

II. Energy Evaluations of Split Levels of ^1S, ^3P, ^1D, ^1G of d^2 systems in O_h field

We now address ourselves to the task of finding the energies of the split levels in O_h crystal field - For this, we shall employ, as will suit our convenience, both the methods, viz., using the full form of the wavefunction and the diagonal sum rule described in Sec 9.2. To the crystal field energies, so found, one has to add the respective coulomb perturbation energies of the corresponding multiplets of the Russell Saunders' states to get the total perturbation energies.

A. ^3P in O_h, field - d^2 configuration - energy calculation

Table $T_{9.3}$ indicates that the O_h crystal field energy of the ^3P state will be achieved readily once the det form of $\psi_{11}(^3T_{1g})$ and $\psi_{1-1}(^3T_{1g})$ i.e., of the kets $|1111>$ (cell H of table $T_{4.3}$) and $|11-11>$ (cell complementary to cell H) be known.

(i) $|1111> = \psi_{11}(^3P)$ can be obtained as the function orthogonal to $\psi_{31}(^3F)$ of cell H (table $T_{4.3}$). The latter is available through the L_- operation on cell E i.e. on $|3121>$. The wavefunction orthogonal to it is

$$|1111> = \psi_{11}(^3P) = \sqrt{2/5}(2^+ - 1^+) - \sqrt{3/5}(1^+ 0^+) \qquad \text{..........(9.21)}$$

The other wavefunction

$$|11-11> = \psi_{1-1}(^3P) = \sqrt{2/5}(-2^+ 1^+) - \sqrt{3/5}(-1^+ 0^+)$$

$$= -\sqrt{2/5}(1^+ - 2^+) + \sqrt{3/5}(0^+ - 1^+) \qquad \text{..........(9.22)}$$

wherein column exchange has been made to obtain the dets in keeping with standard convention.

$\varepsilon(^3T_{1g};{}^3P,d^2) = <1/\sqrt{2}(\psi_{11}+\psi_{1-1})\ |\ \mathcal{H}_{crys}(O_h)|\ 1/\sqrt{2}(\psi_{11}+\psi_{1-1})$ on using table $T_{9.3}$

Substituting (9.21) (9.22) in the RHS of the above expression, followed by expansion of the dets and simplification resulting from orthogonality, orthonormality and the effective number of integrals one gets, in terms of the surviving d-orbital integrals

$$\begin{aligned}\varepsilon({}^3T_{1g};{}^3P,d^2) &= 1/5 < d_2 \left|h_{crys}(O_h)\right| d_2 > +1/5 < d_{-1} \left|h_{crys}(O_h)\right| d_{-1} > \\ &\quad +3/10 < d_1 \left|h_{crys}(O_h)\right| d_1 > +3/10 < d_0 \left|h_{crys}(O_h)\right| d_0 > \\ &\quad -1/5 < d_1 \left|h_{crys}(O_h)\right| d_1 > -1/5 < d_{-2} \left|h_{crys}(Oh)\right| d_{-2} > \\ &\quad -3/10 < d_{-1} \left|h_{crys}(O_h)\right| d_{-1} > -3/10 < d_o \left|h_{crys}(O_h)\right| d_0 > \\ &= 0 \qquad \text{..........(9.23)}\end{aligned}$$

where $h_{crys}(O_h)$ is appropriately associated with the first or the second electron and the baricenter raising component $h_{crys}(O_h)$ has been omitted from $h_{crys}(O_h)$ rendering it to be identical with $h_{crys}{}^{//}(O_h)$ (cf Eq 7.24). It is thus seen that the ^3P state which has the symmetry of $^3T_{1g}$ in the O_h field remains unsplit.

B. ^{1}D in O_h field: (d^2 configuration): energy levels

The det forms of ψ_{22} and ψ_{2-2} (cf Table T9.3) are necessary to evaluate the energies of E_g and T_{2g} of ^{1}D state in O_h field.

(i) $\psi_{22}(^1D) = |2020> = a$ proper linear combination of the dets (2^+0^-), (2^-0^+) and (1^+1^-) [cf cell F, Table T4.3]

This cell contains a ^{1}G component $|4020>\psi_{42}$, available in det form from repeated (two times) operations of L_- starting with cell A of the table. The cell F also contains a component of ^{3}F, viz., $|3120>, \psi_{32}$, obtainable by a direct S_- operation on cell E of table T4.3. The two wavefunctions, so found, are

$$|4020> = \psi_{42}(^1G) = \sqrt{3/14}(2^+0^-) - \sqrt{3/14}(2^-0^+) + \sqrt{8/14}(1^+1^-) \qquad \text{.....(9.24)}$$

$$|3120> = \psi_{32}(^3F) = 1/\sqrt{2}[(2^+0^-) + (2^-0^+)] \qquad \text{.....(9.25)}$$

The wavefunction $\psi_{22}(^1D)$ which is orthogonal to both (9.24) and (9.25) may be readily found out. Thus

$$\psi_{22}(^1D) = |2020> = \sqrt{2/7}(2^+0^-) - \sqrt{2/7}(2^-0^+) - \sqrt{3/7}(1^+1^-) \qquad \text{.....(9.26)}$$

ii) $\psi_{2-2}(^1D) = |20-20>$, having a complementary det structure should be

$$\begin{aligned}\psi_{2-2}(^1D) = |20-20> &= \sqrt{2/7}(-2^+0^-) - \sqrt{2/7}(-2^-0^+) - \sqrt{3/7}(-1^+ -1^-) \\ &= -\sqrt{2/7}(0^- -2^+) + \sqrt{2/7}(0^+ -2^-) - \sqrt{3/7}(-1^+ -1^-) \qquad \text{.....(9.27)}\end{aligned}$$

$$\varepsilon(^1E_g : {}^1D, d^2) = <1/\sqrt{2}\{\psi_{22} + \psi_{2-2}\} \mid \mathscr{H}_{crys}(O_h) \mid 1/\sqrt{2}\{\psi_{22} + \psi_{2-2}\}>$$

On substituting (9.26) and (9.27) in the above energy expression, expanding the dets and simplifying after taking usual precautions and effecting changes in the columns of dets, where necessary , in the ket part of the bracket, the energy expressed in terms of surviving d-orbital matrix elements , is,

$$\varepsilon({}^1E_{1g};{}^1D,d^2) = 2/14\,[2<d_2|h_{crys}(O_h)|d_2>+2<d_{-2}|h_{crys}(O_h)|d_{-2}> +4<d_o|h_{crys}(O_h)|d_o>+2<d_2|h_{crys}(O_h)|d_{-2}> +2<d_{-2}|h_{crys}(O_h)|d_2>] +\{2<d_1|h_{crys}(O_h)|d_1>+2<d_{-1}|h_{crys}(O_h)|d_{-1}>\}3/14 \quad .....(9.28)$$

$$= 96/14\,Dq - 48/14\,Dq = 24/7\,Dq \quad ....(9.29)$$

In (9.28) the baricenter raising component $h_{crys}{}'(O_h)$ has been dropped from $h_{crys}(O_h)$ (cf Eq 7.24) which is appropriately associated with electron 1 or 2 occupying d-orbitals of the corresponding integrands. The necessary column changes mentioned above give rise to the last four matrix elements in the first parenthesis. Once again in arriving at (9.29), results of (7.54) have been utilised.

For the triply degenerate energy level we may write

$$\varepsilon({}^1T_{2g};{}^1D,d^2) = <1/\sqrt{2}\{\psi_{22}-\psi_{2-2}\}|\mathscr{H}_{crys}(O_h)|\,1/\sqrt{2}\{\psi_{22}-\psi_{2-2}\}> \quad ....(9.30)$$

Proceeding as in the previous case, with similar precautions and changes in column of dets we have finally

$$\varepsilon({}^1T_{2g};{}^1D,d^2) = 2/14\,[4Dq+4.6Dq-4.5Dq]+3/14[-4.4Dq]$$

$$= -16/7\ \mathrm{Dq} \quad ....(9.31)$$

C. ^{1}G state in O_h, crystal field (d^2 configuration): energy calculations

To calculate the crystal field energies of the four symmetry states into which the ^{1}G term splits in O_h environment we shall combine the methods of calculations, viz., (a) directly from the final wavefunction forms for $\varepsilon({}^1E_g;{}^1G,d^2)$ and $\varepsilon({}^1T_{1g};{}^1G,d^2)$, (b) use of diagonal sum rule for $\varepsilon({}^1T_{2g};{}^1G,d^2)$ and finally (c) using the principle of nonshift of baricenter of all the split levels of ^{1}G for the calculation of $\varepsilon({}^1A_{1g};{}^1G,d^2)$.

A look at Table $T_{9.3}$ suggests that the above calculations of energies are feasible if we have the det forms of $\psi_{44},\psi_{4-4},\psi_{4-2}$. These are available either directly from table $T_{4.3}$ or after some manipulations with shift operators on specific cells of that table. Thus

i) $\psi_{44}(^1G,d^2)=|4040>=(2^+2^-)$ [cell A, $T_{4.3}$] ....(9.32)

ii) $\psi_{4-4}(^1G,d^2)=|40-40>=(-2^+-2^-)$ [Complementary cell of cell A]..........(9.33)

iii) $\psi_{42}(^1G,d^2)$ = a proper linear combination of the dets

(2^+0^-), (2^-0^+), (1^+1^-) [cf cell table $T_{4.3}$]

$=\sqrt{3/14}(2^+0^-)-\sqrt{3/14}(2^-0^+)+\sqrt{8/14}(1^+1^-)$ as given by Eqn (9.24)

iv) $\psi_{4-2}(^1G,d^2)=\sqrt{3/14}(-2^+0^-)-\sqrt{14}(-2^-0^+)+\sqrt{8/14}(-1^+-1^-)$

$=-\sqrt{3/14}(0^-2^+)+\sqrt{3/14}(0^+-2^-)+\sqrt{8/14}(-1^+-1^-)$ ....(9.34)

a) Choosing the wavefunction $1/\sqrt{2}(\psi_{44}-\psi_{4-4})$ for $\psi(^1T_{1g})$ level from table $T_{9.3}$ and employing (9.32) and (9.33) we have

$\varepsilon(^1T_{1g};^1G,d^2)$ $=<1/\sqrt{2}\{(2^+2^-)-(-2^+-2^-)\}|\mathscr{H}_{crys}(O_h)|1/\sqrt{2}\{(2^+2^-)-(-2^+-2^-)\}>$

$=1/2[2<d_2|h_{crys}(O_h)|d_2>+2<d_{-2}|h_{crys}(O_h)|d_{-2}>$

= 2Dq on using (7.54) ....(9.35)

& on dropping the rise in energy of the baricenter.

Now $\psi(^1E_g:^1G,d^2)=1/\sqrt{2}(\psi_{42}-\psi_{4-2})$ and using (9.24) and (9.34) we find the diagonal matrix element of $\mathscr{H}_{crys}(O_h)$.

$\varepsilon(^1E_g;^1G,d^2)$ $=<1/\sqrt{2}\{\sqrt{3/14}(2^+0^-)-\sqrt{3/14}(2^-0^+)+\sqrt{8/14}(1^+1^-)$

$-\sqrt{3/14}(0^+-2^+)+\sqrt{3/14}(0^+-2^-)+\sqrt{8/14}(-1^+-1^-)\}$

$|\mathscr{H}_{crys}(O_h)|1/\sqrt{2}\{\sqrt{3/14}(2^+0^-)-\sqrt{3/14}(2^-0^+)+\sqrt{8/14}(1^+1^-)$

$-\sqrt{3/14}(0^--2^+)+\sqrt{3/14}(0^+-2^-)+\sqrt{8/14}(-1^+-1^-)\}>$

= 3/28 [4 Dq +4x6 Dq, +4×5 Dq]+8/28[4×(-4Dq)] ....(9.36)

= 4/7 Dq on dropping the baricenter raising contribution.

b) $\varepsilon(^1T_{2g};^1G,d^2)$: The octahedral perturbation matrix in the basis set ψ_{42} and ψ_{4-2} will involve an energy value equal to the sum of $\varepsilon(^1E_g)$ and $\varepsilon(^1T_{2g})$. Thus by "diagonal sum rule".

$\varepsilon(^1T_{2g};^1G,d^2)+\varepsilon(^1E_{2g};^1G,d^2)=\varepsilon(|42>+\varepsilon(|4-2>)$

Using the det forms of $|42>$ and $(|4-2>$ from (9.34), the foregoing expression reduces to

$$\varepsilon({}^1T_{2g};{}^1G,d^2)+\varepsilon({}^1E_g;{}^1G,d^2)= 3/14\ (4\ Dq+24Dq)\ -8/14\ (16Dq) = -\ 22/7Dq \quad ..(9.37)$$

Hence $\varepsilon({}^1T_{2g};{}^1G,d^2)=$ (-22/7-4/7) Dq = -26/7 Dq ....(9.38)

c) $\varepsilon({}^1A_{1g};{}^1G,d^2)$

Since the baricenter of ^{1}G state after its initial rise in the crystal field by the $h_{crys}{}'(O_h)$ part of $h_{crys}(O_h)$ remains unmoved on further splitting into ${}^1A_{1g}, {}^1E_g, {}^1T_{1g}$ and ${}^1T_{1g}$ levels we may write the following energy equation :

$\varepsilon({}^1A_{1g})+2\varepsilon({}^1E_g)+3\varepsilon({}^1T_{1g})+3\varepsilon({}^1T_{2g})=0$ for ^{1}G state in O_h field.

Making use of Eqs (9.35, 9.36, 9.38) we have

$$\varepsilon({}^1A_{1g};{}^1G,d^2)=4Dq \quad ....(9.39)$$

D. The ^{1}S state of d^2 configuration does not undergo splitting in O_h field.

To the octahedral crystal field energy values of the different split levels of the terms of d^2 configuration, appearing in the appropriate relations secs (9.2) and (9.3), are added the corresponding electrostatic perturbation energies of table T4.6 to yield total perturbation energies due to coulomb and crystal field energies. These are all put up in Table9.4 to have an overall and a comparative estimate.

Table $T_{9.4}$

Total perturbation energies (coulomb + crystal field) of d^2 system in O_h field F_0-$49F_4$ = A, F_2-$5F_4$ = B, $35F_4$=C, ε_b= baricenter raising energy

Russel Saunders' multiplets of d^2 configuration	O_h crystal field symmetry levels	Total perturbation energy	Energy gap between the highest and the lowest split levels $\Delta = 10\,Dq$
	$^3T_{1g}$	$A-8B-6Dq+\varepsilon_b$	
^3F	$^3T_{2g}$	$A-8B+2Dq+\varepsilon_b$	$18Dq=9/5\Delta$
	$^3A_{2g}$	$A-8B+12Dq+\varepsilon_b$	
^3P	$^3T_{1g}$	$A+7B+0Dq+\varepsilon_b$	0Dq
^1D	$^1T_{2g}$	$A-3B-16/7Dq+\varepsilon_b$	$40/7Dq=4/7\Delta$
	^1E_g	$A-3B+24/7Dq+\varepsilon_b$	
	$^1T_{2g}$	$A+4B+2C-26/7Dq+\varepsilon_b$	
^1G	^1E_g	$A+4B+2C+4/7Dq+\varepsilon_b$	$54/7Dq=27/35\Delta$
	$^1T_{1g}$	$A+4B+2C+2Dq+\varepsilon_b$	
	$^1A_{1g}$	$A+4B+2C+4Dq+\varepsilon_b$	
^1S	$^1A_{1g}$	$A+14B+7C+0Dq+\varepsilon_b$	$0\,Dq$

The relative energy levels are now diagrammatically shown in Fig. 9.1 choosing a scale where $A-8B+\varepsilon_b$, i.e., coulomb perturbation energy of ^3F plus the baricentre raising energy is set equal to zero.

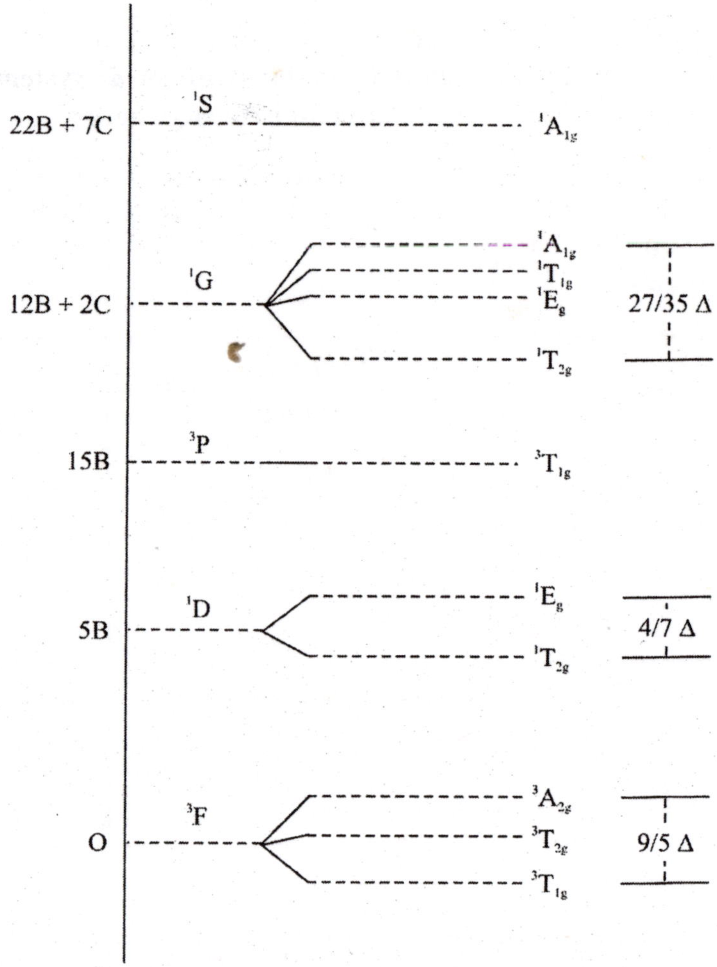

Fig. 9.1: Relative energy levels of the split terms of d^2 system in the electrostatic and O_h perturbation fields (weak field case).

9.4 Energies of Some Split Levels of Russel Saunders' Terms of d^3, d^4 and d^5 Configurations in Weak O_h Field.

A natural follow up of the method, described for d^2 system, can be applied to the higher d^n systems. Since the high spin states generally describe the low lying energy levels including the ground state, we shall take up the highest spin states of ^4F and ^4P of d^3 configuration and ^5D and ^6S states of d^4 and d^5 respectively and examine their crystal field splittings in the weak fields of octahedral environment.

I. d^3 Configuration

A. It may be seen from Table $T_{2.1}$ that the two highest spin multiplicity multiplets are ^4F and ^4P. In the O_h field the F multiplet splits into levels of A_{2g}, T_{2g} and T_{1g} symmetries. The final wavefunctions, which are symmetry adapted linear combinations of atomic kets $|LSM_LM_S>$ and noted briefly by ψ 's, can all be collected from the table T9.3 Let us choose, among the sets there, the specific wavefunctions

$$\left.\begin{array}{l}\psi\left({}^4T_{1g};{}^4F,d^3\right)=\psi_{30}\\ \psi\left({}^4T_{2g};{}^4F,d^3\right)=1/\sqrt{2}\left(\psi_{32}+\psi_{3-2}\right)\\ \psi\left({}^4A_{2g};{}^4F,d^3\right)=1/\sqrt{2}\left(\psi_{32}-\psi_{3-2}\right)\end{array}\right\} \quad \text{..........(9.40)}$$

Before deriving the energy values, we need to write down the det forms of the atomic kets $\psi_{30},\psi_{32},\psi_{3-2}$ occurring in Eq (9.40). Although we do not have appropriate "M_LM_S" microstate det table for d^3 system ready on hand to help us in this regard, we can nevertheless obtain few specific det forms needed here separately. Let us stick to the highest M_s spin value, viz., 3/2.

(i) $\psi_{30}=|3\,3/2\,0\,3/2>$: Starting with the ket $\psi_{33}=|3\,3/2\,3\,3/2>$ which is characterized by a single det, $(2^+\,1^+\,0^+)$ and working upon it by L. we have

$$\left.\begin{array}{l}L\_\psi_{33}=L\_1|3\,3/2\,3\,3/2>=\{l_{1-}+l_{2-}+l_{3-}\}(2^+1^+0^+)\\ \text{i.e., } \psi_{32}=|3\,3/2\,2\,3/2>=(2^+1^+-1^+)\end{array}\right\} \quad \text{..........(9.41)}$$

Repeating the operation with L. twice commencing from (9.41) one finally gets

$$\psi_{30}=|3\,3/2\,0\,3/2>=1/\sqrt{5}\left(1^+0^+-1^+\right)+2/\sqrt{5}\left(2^+0^+-2^+\right) \quad \text{..........(9.42)}$$

ii) $\psi_{32}=|3\,3/2\,2\,3/2>=\left(2^+0^+-1^+\right)$ as is already obtained in (9.41)

iii) $\psi_{3-2}=|3\;\;3/2-2\;\;3/2>=\left(1^+-1^+-2^+\right)$ ....(9.43)

which is evident from (9.41).

Energy Expressions:

$\varepsilon\left({}^4T_{1g};{}^4F,d^3\right)=<1/\sqrt{5}\left(1^+0^+-1^+\right)+2/\sqrt{5}\left(2^+0^+-2^+\right)\left|h_{crys(1)}(O_h)+h_{crys(2)}(O_h)+h_{crys(3)}(O_h)\right|$
$1/\sqrt{5}\left(1^+0^+-1^+\right)+2/\sqrt{5}\left(2^+0^+-2^+\right)>$ which on expansion of the dets and on separation of the integrals becomes

$$\varepsilon\left({}^4T_{1g};{}^4F,d^3\right)=1/5\{<1,1>+<-1,-1>,+5<0,0>+4<2,2>+4<-2,-2>\}=6Dq$$

..........(9.44)

wherein $<d_n\left|h_{crys}\right|d_n>$ is abbreviated as <n, n> and relations (7.54) have been used. The energy value (9.44) for the ${}^4T_{1g}$ level of d^3 is the same as $\varepsilon\left({}^3T_{1g}\right)$ of d^2 with the sign reversed.

On substituting ψ_{32}, ψ_{3-2} by their det forms from Eqs (9.41 and 9.43) in the appropriate equations of (9.40), we have

$$\varepsilon\left({}^4T_{2g};{}^4F,d^3\right)=1/2\{<2,2>+<1,1>,+<-1,-1>+<2,-2>+<1,1>+<-1,-1>$$
$$+<-2,-2>+<-2,2>\}=-2Dq \qquad \text{..........(9.45)}$$

and $\varepsilon\left({}^4T_{2g};{}^4F,d^3\right)=-12Dq$ ....(9.46)

where again the abbreviated notations for the matrix elements of $h_{crys},(O_h)$ between the complex d-functions have been used. It is thus seen that the crystal field effect on the (d^2) and the d^3 systems (multiplets ^3F and ^4F respectively) in their ground states is just an inversion of the signs of energy values of the same symmetry-type split levels.

B. ^4P multiplet of d^3 in O_h field

Table $T_{9.3}$ reveals that one of the final wavefunctions of the $^4T_{1g}$ levels of ^4P in octahedral field is ψ_{10}. If M_LM_S - microstate det table be drawn up for d^3, it will be found, ψ_{10}, having the atomic ket$|13/2\,03/2>$, would be orthogonal to a ^{4} F component ket $|3\,3/2\,0\,3/2>$ (cf Eq 9.42) for which the det form is already known. Therefore, its orthogonal function should have the det form

$$\psi_{10}\left({}^4T_{1g};{}^4P,d^3\right)=|13/2\,03/2>=2/\sqrt{5}\left(1^+0^+-1^+\right)-1/\sqrt{5}\left(2^+0^+-2^+\right) \qquad \text{..........(9.47)}$$

Thus from (9.47), we have, using the abbreviated notations and results of (7.54),

$$\varepsilon\left({}^4T_{1g};{}^4P,d^3\right)=1/5\,[4<1,1>+4<0,0>+4<-1,-1>+<2,2>+$$
$$<0,0>+<-2,-2>=0 \qquad \text{..........(9.48)}$$

Thus ^4P multiplets of d^3, like ^3P of d^2, remains unsplit in O_h field.

II. d^4 and d^5 multiplets in O_h field

The highest spin multiplets in d^4 and d^5 are ^5D and ^6S states respectively. Since the S term is symmetric spherically, it will not be affected by octahedral field (neglecting the raising of the baricenter)

As to the ^5D multiplets of d^4 , the crystal field wavefunctions in O_h field are of E_g and T_{2g} symmetries and their forms can readily be read off from the table $T_{9.3}$ Let us choose the following two wavefunctions from the foregoing table viz.,

$1/\sqrt{2}(\psi_{22}+\psi_{2-2})$ and $1/\sqrt{2}(\psi_{22}-\psi_{2-2})$ for the E_g and T_{2g} symmetries respectively. Sticking to $M_s=2, \psi_{22}$ and ψ_{2-2}, have the following representations respectively

$$\left.\begin{aligned}\psi_{22} &= |2222> = \left(2^{+}1^{+}0^{+} - 1^{+}\right) \\ \psi_{2-2} &= |22-22> = \left(1^{+}0^{+} - 1^{+} - 2^{+}\right)\end{aligned}\right] \qquad \text{..........(9.49)}$$

Employing these two functions in the final wavefunctions of ^5E_g and $^5T_{2g}$ symmetries mentioned before and proceeding as usual we finally get

$$\varepsilon\left({}^5E_g;{}^5D,d^4\right) = 1/2\{<2,2>+<-2,-2>,+2<0,0>+2<1,1>+$$

$$2<-1,-1>-<2,-2>-<-2,2>\}$$

$$= \text{-6Dq using relations (7.54)} \qquad \text{..........(9.50)}$$

And $\varepsilon\left({}^5E_{2g};{}^5D,d^4\right) = 4Dq$ ....(9.51)

To the crystal field energy values of Eqs (9.44,9.45, 9.46, 9.48, 9.50 and 9.51) are to be added the electrostatic perturbation energies of the corresponding multiplet terms of d^3, d^4 and d^5 configurations (Eqs 6.15, 6.16, 6.17) to obtain the total perturbation energies.

Table $T_{9.5}$

Total perturbation energies of coulomb and O_h field interaction matrix of some of the split states of d^3 .d^1 and d^5 configurations.

Multiplet terms	Split levels in O_h crystal field	Total perturbational energy (without the baricentre raising component)	Energy gap between the highest and the lowest split levels
	$^4A_{2g}$	$3A$-$15B$-$12Dq$	
$^4F(d^3)$	$^4T_{2g}$	$3A$-$15B$-$2Dq$	$18\ Dq = 9/5\ \Delta$
	$^4T_{1g}$	$3A$-$15B$+$6Dq$	
$^4P(d^3)$	$^4T_{1g}$	$3A$+$0Dq$	$0\ Dq$
$^5D(d^4)$	^5E_g	$6A$-$21B$-$6Dq$	$10\ Dq = \Delta$
	$^5T_{2g}$	$6A$-$21B$+$4Dq$	
$^6S(d^5)$	$^6A_{1g}$	$10A$-$35B$+$0Dq$	0Dq

9.5 Higher d^n configurations in O_h Field (Weak)

The hole formalism, described in chapter 6, shows that the multiplet terms of d^1 and d^9. d^2 and d^8, d^3 and d^7. d^4 and d^6 configurations are the same and that the relative energies in each set of the corresponding terms of the configurations forming the pair are identical . Though the relative energies are the same, the actual energies of the terms in the two sets will differ by a constant magnitude. The net order of tcrms will encrgywisc. however, remain unchanged for the pair of configurations.

The crystal field effect will also be similar within a pair of similar symmetry terms emanating from the pair of configurations with the only difference that the levels are inverted in the two cases energywise. This is readily understandable since the roles of n-d electrons ($n \leq 5$) in d^n configuration arc played by n positrons in d^{10-n} electron configuration.

We have already discussed the coulomb perturbation energies of some of the highest spin states (and also of some high spin states) of p^n and d^n configurations and tabulated their values (Table $T_{4.5}$, $T_{4.6}$ Eqs. 6.15, 6.16, 6.17). To these may be added the O_h crystal field energies which are already known for similar terms of lower d^n configuration ($n \leq 5$) after paying heed to the change in sign of Dq for systems with $n > 5$. This will give the total perturbation energies of the split levels due to coulomb and crystal field interactions.

Exercises

1. Verify the character table of the reducible representation of the O-pointgroup in the basis set of the seven F - functions (sec. 9.2). Find the symmetry types of the functions spanned by the basis set.
2. Using the normalised forms of the spherical harmonics (chap.1), write down the angular parts of the ^3F complex functions. How do you linearly combine these to form real f - orbitals, some of which are needed to compile the table $T_{9,1}$
3. Find the diagonal matrix elements of a weak $\mathscr{H}_{crys}(O_h)$ with respect to the wave function $\psi\{^1T_{2g};{}^3F,d_2\}=\sqrt{(5/8)}\psi_1-\sqrt{(3/8)}\psi_{-3}$. Show that the result is in agreement with (9.14).
4. Obtain the off-diagonal matrix element $<\psi_{A2g}|\mathscr{H}_{crys}(O_h)|\psi_{T2g}>$ of d^2 configuration after collecting the requisite expressions from appropriate equations and tables of sec. 9.2 and elsewhere. Can you predict the value directly? What theorem underlies your direct inference?
5. Justify the method of obtaining wave functions of a degenerate set by operation on a particular member by a symmetry operation (cf. Eqs. 9.3 and 9.4 of sec. 9.2). Can you apply such a technique to obtain the T_{1g} wave functions of P terms of d^n configuration as in table $T_{9.3}$?

Chapter Ten

Symmetries, Energies and Eigenfunctions of d^n Systems in Strong & Intermediate O_h Crystal Fields. Extension To Lower Symmetry Fields

General

In the previous chapter we have examined d^n systems in different environments where the dominating influence was due to the mutual coulomb repulsion with the crystal field sneaking in as a secondary perturbative influence. In the present chapter, the roles are reversed, the crystal field perturbation is much superior to and dominates over coulomb interactions. The spin orbit interaction, as a third perturbative source will be disregarded in the present chapter as was done in the previous chapter. In addition to the treatment for the strong crystal field case, there will be presented a few sample calculations for crystal fields which are neither too strong nor too weak compared to coulomb repulsions, $\sum_{k<l} e^2 / r_{kl}$.

The physical reality that should obtain in (hypothetically) infinitely strong crystal fields of d^n complexes is the complete and the total collapse of Russell Saunders' couplings of angular momenta of d electrons of the central atom or ion. This renders L and S completely ineffective and meaningless as good quantum numbers. In actual complexes which are departures from the above extreme situation, a two step build-up of a final situation may however be divined. At first the complete decoupling causes the electron to behave independently of their mutual electrostatic repulsions and the prevailing strong crystal field forces them to inhabit the single electron d orbitals energywise permitted by the symmetry of the surrounding crystal field . After this main site-symmetry effect, their primary energies are altered just a little bit by the weak perturbative electrostatic repulsions $\sum_{k<l} e^2 r_{kl}$, conferring now on L and S some meaning as worthwhile quantum numbers.

10.1 d^2 Systems in Strong Crystal Fields of O_h Environment - Energy Levels and Wavefunctions.

The ideas developed in previous paragraphs enable us to write for a d^n system

$$\mathscr{H} = \mathscr{H}_{\text{cent}} + \mathscr{H}_{\text{crys}}(O_h) + \sum_{k<l} e^2 / r_{kl} \qquad \ldots\ldots(10.1)$$

The principle of calculations in a strong field case is as follows :

For a d^n system we may make an initial selection of wavefunctions which are simultaneous eigenfunctions of $\mathscr{H}_{\text{cent}}$ and S^2 and hence of $\sum_{k<l} e^2 / r_{kl}$ (in a central field $\sum_{k<l} e^2 / r_{kl}$ commutes with $\mathscr{H}_{\text{cent}}$ and S^2)

Once this has been done, we may effect, if need be, suitable linear combinations of these initially selected wavefunctions to construct the final wavefunction set satisfying the symmetry needs demanded by environmental strong crystal field. The diagonal elements of $\sum_{k<l} e^2/r_{kl}$ between the final wavefunclions will yield the secondary perturbation energies. These coulombic interaction energies are then to be added to the crystal field energies, that are easily and separately calculable, to yield the final set of total perturbation energies.

d^2 System in Strong O_h Field

A. Role of Group Theory - Eigenfunctions:

In a strong O_h field the ground state configuration of d^2 is t_{2g}^2 - an effect of primary correlation on dismantling of the Russell Saunders' coupled state. It may be noticed that e_g and t_{2g} orbitals are just symmetry notated forms of the pure single electron d-orbital with different radial functions. The former in real forms are then to be identified with d_{z2}, d_{x2-y2}, d_{xz}, d_{yz}, and d_{xy}. The final symmetry states of t_{2g}^2 configuration, which incidentally also represent the symmetries of the final wavefunctions, are the resolved states of the dpr of T_{2g} of the O_h point group. Using Eqns (2.86 and 2.84) and the requisite character table it is found that

$$T_{2g} \times T_{2g} = A_{1g} + E_g + T_{1g} + T_{2g} \qquad \text{......(10.2)}$$

These are then the symmetry tags of the final would-be wavefunctions.

Let us first concern ourselves with the primary or initial selections mentioned before. The configuration t_{2g}^2 may have S= 0 or 1. (we bring in the spin factor here since the secondary correlation effect due to coulomb interaction depends on spin nature.) We indicate the initial t_{2g} orbitals by the real forms of d_{xz}, d_{yz}, and d_{xy} orbitals in O_h field. For S=1, the two-electron dets representing the wavefunctions of t_{2g}^2 configuration are the following Pauli permitted triplet states, a part of primary zero-order wavefunctions that are eigenfunctions[1.7] of $\mathscr{H}_{cent}$ and S^2.

$$\left.\begin{aligned} \psi_1 &= \left(d_{xz}^+ \quad d_{yz}^+\right) \\ \psi_2 &= \left(d_{xz}^+ \quad d_{xy}^+\right) \\ \psi_3 &= \left(d_{yz}^+ \quad d_{xy}^+\right) \end{aligned}\right\} \qquad \text{......(10.3)}$$

The other initially selected wavefunctions are the determinants with the singlet spin state, S= 0 for t_{2g}^2 configuration. These are

$$\left.\begin{aligned} \psi_4 &= 1/\sqrt{2}\left\{\left(d_{xz}^+ \quad d_{yz}^-\right) - \left(d_{xz}^- \quad d_{yz}^+\right)\right\} \\ \psi_5 &= 1/\sqrt{2}\left\{\left(d_{xz}^+ \quad d_{xy}^-\right) - \left(d_{xz}^- \quad d_{xy}^+\right)\right\} \\ \psi_6 &= 1/\sqrt{2}\left\{\left(d_{yz}^+ \quad d_{xy}^-\right) - \left(d_{yz}^- \quad d_{xy}^+\right)\right\} \end{aligned}\right\} \qquad \text{......(10.4)}$$

$$\left.\begin{aligned} \psi_7 &= \left(d_{xz}^{+} \quad d_{xz}^{-}\right) \\ \psi_8 &= \left(d_{yz}^{+} \quad d_{yz}^{-}\right) \\ \psi_9 &= \left(d_{xy}^{+} \quad d_{xy}^{-}\right) \end{aligned}\right\} \qquad \ldots\ldots(10.5)$$

We now try to compose the final symmetry adapted (cf 10.2) wavefunctions from the initial chosen sets (10.3, 10.4, 10.5) which are already eigenfunctions of $\mathscr{H}_{\text{cent}}$, S^2 and $\sum_{k<l} e^2/r_{kl}$ .We note that the triplet states do not mingle with the singlets to yield a pure symmetry state. The transformation properties of the triplet functions (10.3) and of singlet ones under the symmetry operators of the O-point group are, therefore, to be separately investigated. Taking the set (10.3) and remembering how x, y, z transform when the symmetry axis C_4^z is the quantizing axis (cf chap 7, Eq. 7.13), we derive the following table containing the set of characters:

O	E	C_4^z	C_2	$C_2^{/}$	C_3
ψ_1	ψ_1	ψ_1	ψ_1	$-\psi_1$	$-\psi_2$
ψ_2	ψ_2	$-\psi_3$	$-\psi_2$	$-\psi_3$	$-\psi_3$
ψ_3	ψ_3	$-\psi_2$	$-\psi_3$	$-\psi_2$	ψ_1
χ	3	1	-1	-1	0

……10.6

The character set is identified with that of IR T_1 of the O-point group. Hence the functions (10.3) are themselves symmetry adapted to ${}^3T_{1g}$. (cf 10.2) as demanded by O_h perturbation field and constitute three final wavefunctions.

Turning our attention now to the singlet set, we find that amongst them, ψ_4, ψ_5 and ψ_6 of (10.4) mix with each other under O-group symmetry operations and stay clear of ψ_7, ψ_8 and ψ_9 of (10.5). A transformation exercise like (10.6) with ψ_4, ψ_5 and ψ_6 as basis yields character values identical with those under T_2 of the O-point group. These three primarily selected wavefunctions, therefore, continue to be three final ${}^1T_{2g}$ symmetry-adapted wavefunctions (cf 10.2) under O_h crystal field perturbation.

The remaining three final wavefunctions of A_{1g} and E_g (cf 10.2) symmetries should therefore emerge from suitable linear combinations of ψ_7, ψ_8 and ψ_9 as none of these are either of pure A_{1g} or of pure E_g symmetry. We shall develop their symmetry adapted combinations by further use of group theory. As $C_2^{/}$ in O-group has different character values in A_1 and E, we utilise this operation to extract some useful information about possible combinations. Thus

$$C_2'\psi_7 = C_2'\left(d_{xz}^+ \quad d_{xz}^-\right) = \psi_8$$

$$C_2'\psi_8 = C_2'\left(d_{yz}^+ \quad d_{yz}^-\right) = \psi_7 \text{ hence } C_2' = \begin{bmatrix} 0 & 1 & 0 \\ 1 & 0 & 0 \\ 0 & 0 & 1 \end{bmatrix} \qquad \text{......(10.7)}$$

$$C_2'\psi_9 = C_2'\left(d_{xy}^+ \quad d_{xy}^-\right) = \psi_9$$

The eigenvalues are $\lambda = \pm 1$ from the (2×2) block and $\lambda = 1$ from the (1×1) block. The coefficients are thus $C_7 = C_8 = 1/\sqrt{2}$ and $C_9 = 1$. The eigenfunctions of C_2' are

$$\psi_9, 1/\sqrt{2}(\psi_7 + \psi_8) \text{ and } 1/\sqrt{2}(\psi_7 - \psi_8) \qquad \text{......(10.8)}$$

The characters of transformation under C_2' of ψ_9 and $1/\sqrt{2}(\psi_7 + \psi_8)$ are both + 1 like under A_1 of the point group O. That of $1/\sqrt{2}(\psi_7 - \psi_8)$ is -1. Since the representation space spanned by ψ_7, ψ_8 and ψ_9 comprises subspaces A_1 and E, we may tentatively suggest that the A_1 basis function consists of a linear combination of the first two functions of (10.8) and the third function is likely to belong to IR,E , as it is already orthogonal to and linearly independent of both ψ_9 and $1/\sqrt{2}(\psi_7 + \psi_8)$). Acting on this tentative proposal, we suggest the combination

$p\{1/\sqrt{2}(\psi_7 + \psi_8)\} + q\psi_9$ for A_1 symmetry. Applying C_3 on it

$$C_3\left[p\{1/\sqrt{2}(\psi_7 + \psi_8)\} + q\psi_9\right] = p\{1/\sqrt{2}(\psi_9 + \psi_7)\} + q\psi_8$$

and since $\chi(C_3)$ in A_1 is 1

$$p\{1/\sqrt{2}(\psi_7 + \psi_8)\} + q\psi_9 = p\{1/\sqrt{2}(\psi_9 + \psi_7)\} + q\psi_8 \qquad \text{......(10.9)}$$

Equating the coefficients of ψ_9 from both sides of (10.9), we have p $= \sqrt{2}$ and $q = \pm 1/\sqrt{3}$, the latter being the consequence of the normalization relation $p^2 + q^2 = 1$. Using the + signs for the coefficients, the wavefunction of A_{1g} symmetry

$$\psi_{A1g} = 1/\sqrt{3}(\psi_7 + \psi_8 + \psi_9) \qquad \text{......(10.10)}$$

That this is the real A_{1g} function may be confirmed by group operations under C_4 and C_2 upon (10.10) and comparing the characters. The second member of E_g symmetry must simultaneously be orthogonal to both $1/\sqrt{2}(\psi_7 - \psi_8)$ and (10.10). The two wavefunctions of E_g symmetry are

$$\left.\begin{aligned} \psi(E_g a) &= 1/\sqrt{2}(\psi_7 - \psi_8) \\ \psi(E_g b) &= \psi_7/\sqrt{6} + \psi_8/\sqrt{6} - 2\psi_9/\sqrt{6} \end{aligned}\right] \qquad \text{......(10.11)}$$

The second wavefunction of (10.11) has been obtained from three simultaneous equations satisfying the orthogonality and orthonormality conditions viz

$$\left.\begin{aligned} &1/\sqrt{3}a + 1/\sqrt{3}b + 1/\sqrt{3}c = 0 \\ &1/\sqrt{2}a - 1/\sqrt{2}b = 0 \\ \text{and}\quad &a^2+b^2+c^2 = 1 \\ \text{whence } &2a = 2b = -c = -(\pm 2/\sqrt{6}) \end{aligned}\right\} \quad \ldots\ldots(10.12)$$

B. Energy Eigenvalues of Split Levels

To find the energy levels relative to the baricenter as zero, we have to consider contributions by (i) the prominent perturbation by the crystal field in the first place and (ii) the comparatively weak correlative electrostatic perturbation arising from e^2/r_{12}

(i) For the d^2 system in the strong field, the resultant $(t_{2g})^2$ configuration would mean energy involvement of $2\times(-4Dq) = -8Dq$ relative to zero level in O_h field.

(ii) To compute the second contribution, it is just necessary to find the diagonal matrix elements of e^2/r_{12} with respect to the final wavefunctions (10.3, 10.4, 10.10, 10.11)

Thus using (10.3) the energy level will be

$$\begin{aligned} \varepsilon\left({}^3T_{1g};t_{2g}^2\right) &= -8Dq + <\left(d_{xz}^+\ d_{yz}^+\right)\left|e^2/r_{12}\right|\left(d_{xz}^+\ d_{yz}^+\right)> \\ &= -8Dq + J\left(d_{xz}\ \ d_{yz}\right) - K\left(d_{xz}\ \ d_{yz}\right) \end{aligned} \quad \ldots\ldots 10.13$$

To evaluate the last two integrals on RHS of (10.13) we write

$$J\left(d_{xz}\ \ d_{yz}\right)$$

$$= <\left\{1/\sqrt{2}(d_{-1}-d_1)i/\sqrt{2}(d_{-1}+d_1)\right\}\left|e^2/r_{12}\right|\left\{1/\sqrt{2}(d_{-1}-d_1)i/\sqrt{2}(d_{-1}+d_1)\right\}>$$

$$= 1/4 <\left\{d_{-1}d_{-1}+d_{-1}d_1-d_1d_{-1}-d_1d_1\right\}\left|e^2/r_{12}\right|\left\{d_{-1}d_{-1}+d_{-1}d_1-d_1d_{-1}-d_1d_1\right\}>$$

wherein the relationships between real and complex forms of d-functions [Appendix Chap 1] have been used and the i factor within the bra, part is equivalent to -i when shifted outside. Now remembering the vital conditions, viz., $m_{la} + m_{lb} = m_{lc} + m_{ld}$ for the survival of the integrals (Sec 2.8), we have

$$J\left(d_{xz}^{+}\ d_{yz}^{+}\right) = 1/4\{< d_{-1}d_{-1}|e^2/r_{12}|d_{-1}d_{-1} > + < d_{-1}d_1|e^2/r_{12}|d_{-1}d_1 > - < d_{-1}d_1|e^2 r_{12}|d_1 d_{-1} >$$

$$- < d_1 d_{-1}|e^2/r_{12}|d_{-1}d_1 > + < d_1 d_{-1}|e^2/r_{12}|d_1 d_{-1} > + < d_1 d_1|e^2/r_{12}|d_1 d_1 >$$

$$= 1/4\{J(-1,-1),+J(-1,1) - K(-1,1) - K(1,-1) + J(1,-1) + J(1,1)\}$$

$$= \text{A-2B+C on using table } T_{2.4} \qquad \text{......(10.14)}$$

The second integral on the RHS of (10.13) on similar treatment is

$$K\left(d_{xz}^{+}\ d_{yz}^{+}\right)$$

$$= 1/4 < \{d_{-1}d_{-1} + d_{-1}d_1 - d_1 d_{-1} - d_1 d_1\}|e^2/r_{12}|\{d_{-1}d_{-1} - d_{-1}d_1 + d_1 d_{-1} - d_1 d_1\}>$$

$$= 1/4\{J(-1,-1) - J(-1,1) + K(-1,1) + K(1,-1) - J(1,-1) + J(1,1)\}$$

$$= \text{3B+C on using table } T_{2.4} \qquad \text{......(10.15)}$$

Combining (10.14) and (10.15) and restoring this is (10.13) we get

$$\varepsilon\left({}^3T_{1g}; t_{2g}^2\right) = -8Dq + \{A - 5B\} \qquad \text{......(10.16)}$$

The energy of the ${}^1T_{2g}$ term t_{2g}^2 configuration is

$$\varepsilon\left({}^1T_{2g}; t_{2g}^2\right) = -8Dq + 1/2 < \{(d_{xz}^{+}\ d_{yz}^{-}) - (d_{xz}^{-}\ d_{yz}^{+})\}|e^2/r_{12}|\{(d_{xz}^{+}\ d_{yz}^{-}) - (d_{xz}^{-}\ d_{yz}^{+})\}> \qquad \text{......(10.17)}$$

Putting the integral part equal to I,

$$2I = j\left(d_{xz}, d_{yz}\right) + J\left(d_{xz}, d_{yz}\right) + K\left(d_{xz}, d_{yz}\right) + K\left(d_{xz}, d_{yz}\right)$$

The two K integrals come into being on changing the columns of the dets in the ket.

On changing over from real to complex forms of d-wavefunctions and using table $T_{2.4}$ to evaluate I, the required energy turns out to be

$$\varepsilon\left({}^1T_{2g}; t_{2g}^2\right) = -8Dq + \{A + B + 2C\} \qquad \text{.........(10.18)}$$

The other two energy levels, found out in similar way, are

$$\varepsilon\left({}^1E_{g}; t_{2g}^2\right) = -8Dq + \{A + B + 2C\} \qquad \text{.........(10.19)}$$

$$\varepsilon\left({}^1A_{1g}; t_{2g}^2\right) = -8Dq + \{A + B + 5C\} \qquad \text{.........(10.20)}$$

It may be noted that the terms ${}^1T_{2g}$ and ${}^1E_g$ arising from t_{2g}^2 configuration are accidentally degenerate.

As has been seen in the foregoing energy evaluations and as will turn up shortly in many cases, a knowledge of the values of J and K integrals involving real d-orbitals will be necessary[2]. It will, therefore, be advantageous to compile, in the manner shown before, a table of these J and K intergrals for a ready pick-up of such values as and when needed (See table $T_{10.1}$).

Table $T_{10.1}$

J and K integrals between real d-orbitals in Racah parameters
$J(d_{xz}\ d_{xz}) = J(d_{yz}, d_{yz}) = J(d_{xy}, d_{xy}) = J(d_{z2}, d_{z2}) = J(d_{x2-y2}, d_{x2-y2}) = A + 4B + 3C$
$J(d_{xz}\ d_{yz}) = J(d_{xz}, d_{xy}) = J(d_{yz}, d_{xy}) = A - 2B + C$
$J(d_{z2}, d_{xz}) = J(d_{z2}, d_{yz}) = A + 2B + C$
$J(d_{z2}\ d_{xy}) = J(d_{z2}, d_{x2-y2}) = A - 4B + C$
$J(d_{x2-y2}\ d_{xy}) = A + 4B + C$
$J(d_{x2-y2,}\ d_{xz}) = J(d_{x2-y2}, d_{yz}) = A - 2B + C$
$K(d_{xz}\ d_{yz}) = K(d_{xz}, d_{xy}) = K(d_{yz}, d_{xy}) = 3B + C$
$K(d_{z2}\ d_{xz}) = K(d_{z2}, d_{yz}) = B + C$
$K(d_{z2}\ d_{xy}) = K(d_{z2}, d_{x2-y2}) = 4B + C$
$K(d_{x2-y2}\ d_{xz}) = K(d_{x2-y2}, d_{yz}) = 3B + C$
$K(d_{x2-y2,}\ d_{xy}) = C$

One can conceive of other pairs of the real d-orbitals and as such the list of integrals in table $T_{10.1}$ can be enlarged. However, all such additional J and K integrals may be handled in a way as in the example leading to Eq.(10. 13).

10.2 Higher d^n Systems in Strong Octahedral Field - Ground Terms

We shall confine ourselves to finding the ground field terms and the energies of d^3 to d^9 systems in O_h environment. The ground state energies of d^n systems in strong field arise from the configuration t_2^m where $n = m \leq 6$ and $t_2^6 e^k$ when $k = n - 6$ with $n > 6$.

I d^3 System

The strong crystal field configuration is t_{2g}^3 in O_h environment. Using the real d-orbitals for the t_{2g} wavefunctions, the highest spin state configuration, conforming to the lowest energy of t_{2g}^3 configuration is given by the det, $\psi = (d_{xz}^+\ d_{yz}^+\ d_{xy}^+)$, a quartet spin state and an eigenfunction of $\mathscr{H}_{cent}$, S^2 and $\sum_{k<l} e^2/r_{kl}$. The transformation properties of the three-electron det under the group operators of O and

the character values are shown below.

O	E	$6C_4$	$3C_2$	$6C_2^{/}$	$8C_3$
$(d_{xz}\ d_{yz}\ d_{xy})$	$(d_{xz}\ d_{yz}\ d_{xy})$	$(d_{yz}\ -d_{xz}\ -d_{xy})$ $=-(d_{xz}\ d_{yz}\ d_{xy})$	$(-d_{xz}\ -d_{yz}\ d_{xy})$ $=(d_{xz}\ d_{yz}\ d_{xy})$	$(-d_{yz}\ -d_{xz}\ d_{xy})$ $=-(d_{xz}\ d_{yz}\ d_{xy})$	$(d_{xy}\ d_{xz}\ d_{yz})$ $(d_{xz}\ d_{yz}\ d_{xy})$
χ	1	-1	1	-1	1

The set of characters belongs to A_2 of O-group and hence the det $(d_{xz}^{+}d_{yz}^{+}d_{xy}^{+})$ is symmetry adapted to ${}^4A_{2g}$ under O_h point group.

To find the energy, relative to the baricenter level equal to zero, we have to add the coulombic perturbation energy to the initial strong field correlative energy, viz., $3\times(-4Dq)=-12Dq$

Thus,

$$\varepsilon({}^4A_{2g};t_{2g}^3) = -12Dq + <(d_{xz}^{+}d_{yz}^{+}d_{xy}^{+})|\Sigma_{k<l}\, e^2/r_{kl}|(d_{xz}^{+}d_{yz}^{+}d_{xy}^{+})>$$

$$= -12Dq + J(d_{xz},d_{yz}) - K(_{xz},d_{yz}) + J(d_{xz},d_{xy}) - K(d_{xz},d_{xy}) + J(d_{yz},d_{xy}) - K(d_{yz},d_{xy})$$

$$= -12Dq + \{3A - 15B\} \qquad \text{......(10.21)}$$

on using table $T_{10.1}$ and remembering that r_{kl} takes on the possible values r_{12}, r_{13} and r_{23}.

II. d^4 System in Strong O_h Field

In t_{2g}^4 crystal field configuration, the strong O_h, field causes the distribution of electrons in any of the three equivalent triplet det forms which correspond to the initial set of eigenfunctions of $\mathscr{H}_{cent}$,S^2 and $\Sigma_{k<l}\, e^2/r_{kl}$ of lowest energy. The dets are

$$\left.\begin{aligned} \psi_1 &= (d_{xz}^{+}d_{xz}^{-}d_{yz}^{+}d_{xy}^{+}) \\ \psi_2 &= (d_{xz}^{+}d_{yz}^{+}d_{yz}^{-}d_{xy}^{+}) \\ \psi_3 &= (d_{xz}^{+}d_{yz}^{+}d_{xy}^{+}d_{xy}^{-}) \end{aligned}\right\} \qquad \text{......(10.22)}$$

Using these as base functions of O-group symmetry transformations, results in the following table

O	E	6C$_4$	3C$_2$	$6C_2^{/}$	8C$_3$
ψ_1	ψ_1	ψ_2	$-\psi_1$	$-\psi_2$	ψ_3
ψ_2	ψ_2	$-\psi_1$	$-\psi_2$	$-\psi_1$	$-\psi_1$
ψ_3	ψ_3	ψ_3	ψ_3	$-\psi_3$	$-\psi_2$
χ	3	1	-1	-1	0

The characters indicate that the three initially chosen functions are symmetry –adapted to $^3T_{1g}$ and constitute a degenerate set of three final wavefunctions in O$_h$ field

The energy value of this triply degenerate state is

$$\varepsilon\left({}^3T_{1g};t_{2g}^4\right) = <\left(d_{zx}^+d_{xz}^-d_{yz}^+d_{xy}^+\right)\left|\Sigma_{k<l}\,e^2/r_{kl}\right|\left(d_{xz}^+d_{xz}^-d_{yz}^+d_{xy}^+\right)> -16Dq$$

$$= -16Dq + J\left(d_{xz}^+d_{xz}^-\right) + J\left(d_{xz}^+d_{yz}^+\right) - K\left(d_{xz}^+d_{yz}^+\right) + J\left(d_{xz}^+,d_{xy}^+\right) - K\left(d_{xz}^+.d_{xy}^+\right)$$

$$+ J\left(d_{xz}^-,d_{yz}^+\right) + J\left(d_{xz}^-,d_{xy}^+\right) + J\left(d_{yz}^+,d_{xy}^+\right) - K\left(d_{yz}^+,d_{xy}^+\right)$$

$$= -16Dq + \{6A - 15B + 5C\} \qquad \text{......(10.23)}$$

to arrive at which table T$_{10.1}$ has been used.

III. d^5 system in strong O$_h$ field

Following in the footsteps of d^4-system, it can be shown that for the strong field t_{2g}^5 configuration, the highest spin state wavefunctions of lowest energy are triply degenerate and belong to the IR, $^2T_{2g}$ of the O$_h$, point group. The wavefunclions are

$$\left.\begin{aligned} \psi_1 &= \left(d_{xz}^+d_{xz}^-d_{yz}^+d_{yz}^-d_{xy}^+\right) \\ \psi_2 &= \left(d_{xz}^+d_{xz}^-d_{yz}^+d_{xy}^+d_{xy}^-\right) \\ \psi_3 &= \left(d_{xz}^+d_{yz}^+d_{yz}^-d_{xy}^+d_{xy}^-\right) \end{aligned}\right\} \qquad \text{......(10.24)}$$

The octahedral stabilisation energy is $5\times(-4D_q) = -20Dq$ ......(10.25)

The coulombic perturbation energy is readily found out by evaluating the diagonal matrix element of $\Sigma_{k<l}\,e^2/r_{kl}$ from any one of the three wavefunctions of (10.24).

$$\varepsilon\left(elec.{}^2T_{2g};t_{2g}^5\right) \quad =\{10A-20B+10C\} \qquad \text{......(10.26)}$$

$$\therefore \varepsilon\left({}^2T_{2g};T_{2g}^5\right) \quad =-20Dq+\{10A-20B+10C\} \qquad \text{......(10.27)}$$

IV. d^6 System in strong O_h field.

In a strong field all the levels of t_{2g} will be filled by electrons with paired spins. Without going into details, it can be easily ascertained that the symmetry of the final wavefunction in the octahedral crystal field as well as in the weaker coulombic field will be ${}^1A_{1g}$ and its det form is $\left(d_{xz}^+d_{xz}^-d_{yz}^+d_{yz}^-d_{xy}^+d_{xy}^-\right)$

The octahedral field stabilisation energy is $6\times(-4D_q) = -24D_q$ and the electrostatic field energy is

$$\begin{aligned}
\varepsilon\,(\text{elec. } {}^1A_{1g}) = {} & J\left(d_{xz}^+,d_{xz}^-\right)+J\left(d_{yz}^+,d_{yz}^-\right)+J\left(d_{xy}^+,d_{xy}^-\right)+\\
& J\left(d_{xz}^+,d_{yz}^+\right)-K\left(d_{xz}^+,d_{yz}^+\right)+J\left(d_{xz}^+,d_{xy}^+\right)-K\left(d_{xz}^+,d_{xy}^+\right)+\\
& J\left(d_{xz}^+,d_{yz}^-\right)+J\left(d_{xz}^+,d_{xy}^-\right)+J\left(d_{xz}^-,d_{yz}^+\right)+J\left(d_{xz}^-,d_{xy}^+\right)+\\
& J\left(d_{xz}^-,d_{yz}^-\right)-K\left(d_{xz}^-,d_{yz}^-\right)+J\left(d_{xz}^-,d_{xy}^-\right)-K\left(d_{xz}^-,d_{xy}^-\right)+\\
& J\left(d_{yz}^+,d_{xy}^+\right)-K\left(d_{yz}^+,d_{xy}^+\right)+J\left(d_{yz}^+,d_{xy}^-\right)+J\left(d_{yz}^-,d_{xy}^+\right)+\\
& J\left(d_{yz}^-,d_{xy}^-\right)-K\left(d_{yz}^-,d_{xy}^-\right)\\
= {} & 15A-30B+15C
\end{aligned}$$

To arrive at which table $T_{10.1}$ has been used

$$\text{Hence} \quad \varepsilon\left({}^1A_{1g};t_{2g}^6\right)=-24Dq+\{15A-30B+15C\} \qquad \text{......(10.28)}$$

The ground state energies of d^7, d^8, d^9 and d^{10} in strong O_h field can be obtained in a manner already exhibited for lower d^n cases. The procedure involves tackling of larger number of electrons and of J and K integrals. An alternative method was suggested by Griffith[3]. Just as we find the energies of $t_{2g}^m\,(m=1\ to\ 6)$ in strong fields, we may, likewise, find energies of the symmetry adapted terms emerging from the configuration e_g^k (k = 1 to 4). We now compile a table showing the electrostatic energies of t_{2g}^m and e_g^k configurations in strong O_h field. The relevance of the table will be apparent later on.

Table $T_{10\cdot2}$: Electrostatic energies of terms of t_{2g}^m and e_g^k configurations in strong O_h field

Strong field config.	Sym of terms and wavefunctions	Energy (Electrostatic)	Vide Eq.
e_g^1	^2E_g	0	
e_g^2	$^3A_{2g}$ ^1E_g $^1A_{1g}$	A-8B A+2C A+8B+4C	
e_g^3	^2E_g	3A-8B+4C	
e_g^4	$^1A_{1g}$	6A-16B+8C	
t_{2g}^1	$^2T_{2g}$	0	
t_{2g}^2	$^3T_{1g}$ $^1A_{1g}$ ^1E_g $^1T_{2g}$	A-5B A+10B+5C A+B+2C A+B+2C	(10.16) (10.20) (10.19) (10.18)
t_{2g}^3	$^4A_{2g}$ ^2E_g $^2T_{1g}$ $^2T_{2g}$	3A-15B 3A-6B+3C 3A-6B+3C 3A+5C	(10.21)

For configurations, where the electrostatic energies have not been previously derived in this chapter, these can easily be found out and may be checked against the data recorded in table $T_{10\cdot2}$.

For d^n systems with mixed configurations, like $t_{2g}^m\ e_g^4$ (m= n-4), $t_{2g}^6\ e_g^k$ (k= n-6), where either the e_g level is completely filled or the t_{2g} level is so, the electrostatic matrix is altered by an additive constant common to all the states of e_g^k when the closed shell t_{2g}^6 is adjoined to it in the total configuration . The same is true of t_{2g}^m with : a t_{2g}^m adjoined to a completely filled e_g^4.

The electrostatic energies for such mixed configurations were formulated by Griffith:[4]

$$\varepsilon^{elec}\left(t_{2g}{}^{m}e_{g}{}^{4}\right)=\varepsilon^{elec}\left(t_{2g}^{m}\right)+(4m+6)A+(2m+8)(-2B+C) \qquad \ldots\ldots(10.29)$$

$$\varepsilon^{elec}\left(t_{2g}{}^{m}e_{g}{}^{k}\right)=\varepsilon^{elec}\left(e_{g}^{k}\right)+(6k+15)A+(3k+15)(-2B+C) \qquad ......(10.30)$$

V. d^7 System in Strong O_h Field

In a strong O_h field, the configuration, with lowest energy is $t_{2g}^{6}\,e_{g}^{1}$. Since t_{2g}^{6} has the symmetry ${}^{1}A_{1g}, t_{2g}^{6}, e_{g}^{1}$ has the overall symmetry ${}^{2}E_{g}$ like e_{g}^{1} of table T10.2 Using Eq (10.30) with k=1

$$\varepsilon^{elec}\left({}^{2}E_{g};t_{2g}^{6}\,e_{g}^{1}\right)=\varepsilon^{elec}\left({}^{2}E_{g};e_{g}^{1}\right)+21A-36B+18C=0+21A-36B+18C$$

$$\therefore \varepsilon\left({}^{2}E_{g};t_{2g}^{6}\,e_{g}^{1}\right)=-18Dq+21A-36B+18C \qquad ......(10.31)$$

since O_h field stabilization energy = 1×6Dq + 6×(-4Dq) = -18 Dq

VI. d^8 System in Strong O_h Field

The lowest energy configuration in strong O_h field is $t_{2g}^{6}\,e_{g}^{2}$ and it possesses wavefunctions of symmetries similar to those e_{g}^{2} of configuration of table T10.2. Utilising the table and using Eq (10.30) and remembering that the octahedral field stabilization energy is -12 Dq for the eight electrons involved, the d^8 energies are

$$\varepsilon\left({}^{3}A_{2g};t_{2g}^{6}\,e_{g}^{2}\right)=-12Dq+(A-8B)+27A+21(-2B+C)$$

$$=-12Dq+(28A-50B+21C) \qquad ......(10.32)$$

$$\varepsilon\left({}^{1}E_{g};t_{2g}^{6}\,e_{g}^{2}\right)=-12Dq+(A+2C)+27A-42B+21C$$

$$=-12Dq+(28A-42B+23C) \qquad ......(10.33)$$

$$\varepsilon\left({}^{1}A_{1g};t_{2g}^{6}\,e_{g}^{2}\right)=-12Dq+(A+8B+4C)+27A-42B+21C$$

$$=-12Dq+(28A-34B+25C) \qquad ......(10.34)$$

VII. d^9 System in Strong O_h Field

In strong field the configuration is $t_{2g}^{6}\,e_{g}^{3}$ and its overall symmetry is ${}^{2}E_{g}$. Since -6Dq is the octahedral field stabilization energy

$$\varepsilon\left({}^{2}E_{g};t_{2g}^{6}\,e_{g}^{3}\right)=-6Dq+(36A-56B+28C) \qquad ......(10.35)$$

VIII. d^{10} system is strong O_h field:

Here the configuration is $t_{2g}^{6}\,e_{g}^{4}$ and the symmetry of the term ${}^{1}A_{1g}$. The octahedral stabilisation energy being nil, we have from Eq (10.30)

$$\varepsilon\left({}^{1}A_{1g};t_{2g}^{6}\,e_{g}^{4}\right)=0+(45A-70B+35C) \qquad ......(10.35A)$$

Table $T_{10.3}$

Ground term energies of d^n systems in weak and strong octahedral fields

Configuration	Free ion ground term	Weak field configuration	Weak field ground term	Weak field ground term energy
d^1	^2D	t_{2g}^1	$^2T_{2g}$	$-4Dq+0$
d^2	^3F	t_{2g}^2	$^3T_{1g}$	$-6Dq+(A-8B)$
d^3	^4F	t_{2g}^3	$^4A_{2g}$	$-12Dq+(3A-15B)$
d^4	^5D	$t_{2g}^3\ e_g^1$	^5E_g	$-6Dq+(6A-21B)$
d^5	^6S	$t_{2g}^3\ e_g^2$	$^6A_{1g}$	$0Dq+(10A-35B)$
d^6	^5D	$t_{2g}^4\ e_g^2$	$^5T_{2g}$	$-4Dq+(15A-35B+7C)$
d^7	^4F	$t_{2g}^5\ e_g^2$	$^4T_{1g}$	$-6Dq+(21A-40B+14C)$
d^8	^3F	$t_{2g}^6\ e_g^2$	$^3A_{2g}$	$-12Dq+(28A-15B+21C)$
d^9	^2D	$t_{2g}^6\ e_g^3$	^2E_g	$-6Dq+(36A-56B+28C)$
d^{10}	^1S	$t_{2g}^6\ e_g^4$	$^1A_{1g}$	$0Dq+(45A-70B+35C)$

Configuration	Free ion ground term	strong field configuration	Strong field ground term	Strong field ground term energy
d^1	^2D	t_{2g}^1	$^2T_{2g}$	$-4Dq+0$
d^2	^3F	t_{2g}^2	$^3T_{1g}$	$-8Dq+(A-5B)$
d^3	^4F	t_{2g}^3	$^4A_{2g}$	$-12Dq+(3A-15B)$
d^4	^5D	t_{2g}^4	$^3T_{1g}$	$-16Dq+(6A-15B+5C)$
d^5	^6S	t_{2g}^5	$^2T_{2g}$	$-20Dq+(10A-20B+10C)$
d^6	^5D	t_{2g}^6	$^1A_{1g}$	$-24Dq+(15A-30B+15C)$
d^7	^4F	$t_{2g}^6\ e_g^1$	^2E_g	$-18Dq+(21A-36B+18C)$
d^8	^3F	$t_{2g}^6\ e_g^2$	$^3A_{2g}$	$-12Dq+(28A-50B+21C)$
d^9	^2D	$t_{2g}^6\ e_g^3$	^2E_g	$-6Dq+(36A-56B+28C)$
d^{10}	^1S	$t_{2g}^6\ e_g^4$	$^1A_{1g}$	$0Dq+(45A-70B+35C)$

For each configuration, the energy due to its own H_{cent} (Eq. 2.25) and the baricentre-raising energy component due to $\sum h_{crys}(O_h)$ of Eq. (7.20a) have been clubbed together as the zero of the energy scale of the corresponding configuration. The energy values are thus relative to each such separately conceived zero levels.

It is seen from table $T_{10.3}$, that the spin multiplicities and the symmetry of ground terms of d^1 d^2, d^3, d^8, d^9, d^{10} remain unaltered in both weak and the strong O_h fields. However, for d^4, d^5, d^6 and d^7 these two characteristics change; the spin multiplicities decrease on going over to the high field side yielding ground states of lower spins.

As to the energy, both the weak and the strong field values are the same for d^1, d^3, d^8, $d^9$and d^{10}, d^2 incidentally spinning away from this group. But for each of d^4, d^5, d^6 and d^7, the energy values markedly differ in weak and high fields. Of the two components of the net energy, the site symmetry (octahedral stabilization) energy expressed in Dq units decreases on passage from weak to strong fields for d^4 - d^7 systems. On the other hand, the coulombic perturbation energy for each of the latter d^n members (d^4 - d^7) increases on the strong field side. This is on the whole due to larger number of forced pairing of electron spins, generating pairing repulsion energy, to enable accommodation of more electrons in the lower t_{2g} orbitals in the strong octahedral field.

Besides, higher number of parallel spins of d^4, d^5, d^6 and d^7 hint that the complexes of such systems in weak O_h, field will tend to be highly paramagnetic. Paramagnetism is greatly reduced or even diamagnetic behaviour may ensue on the strong field side of (d^4 - d^7) complexes due to reduction of spin multiplicity.

10.3 Crystal Field Terms and Energies at Intermediate Field Strengths of O_h Environment

The energies of different crystal field terms at intermediate field strengths are not necessarily derivable from the corresponding wavefunction of just a pure symmetry term within a particular symmetry classified terms of the parent weak field (w.f) or strong field (s.f) limit terms, such as ${}^3F$ or t_{2g}^2 etc. of d^2 configuration interaction. One has to consider configuration interaction[5], i.e. the possibility of mixing of two similar symmetry notated terms lying in close proximity from the viewpoint of energy at the prevailing field strength. To illustrate the two ${}^3T_{1g}$ terms originating from two parent w.f..limit terms viz., ${}^3F$ and ${}^3P$of d^2configuration, if near enough in energy at intermediate field strength, will interact .Or starting from s.f limit terms of t_{2g}^2 and (t_{2g}^1, e_{2g}^1), the two ${}^3t_{2g}$ term's stemming from them, if near enough in energy at the existing field, will undergo configuration interaction. By similar symmetry terms we mean here terms with the same IR and the same spin multiplicity. Fig 10.1 strives to depict such configuration interaction of the two ${}^3t_{2g}$ terms when $e^2 / r_{kl} \approx \mathscr{H}_{crys}(O_h)$ in their respective perturbational effect.

Fig 10.1 would tend, to make us believe that the ${}^1T_{2g}$ terms originating from the configurations and (t_{2g}^2) and $\left(t_{2g}^1 - e_g^1\right)$ on s.f side are similarly to undergo configuration interaction. We can, however, neither predict nor reject this feasibility. The actual phenomenon will depend on whether their mutual energy gap becomes narrow enough at intermediate field strength to induce mixing or not.

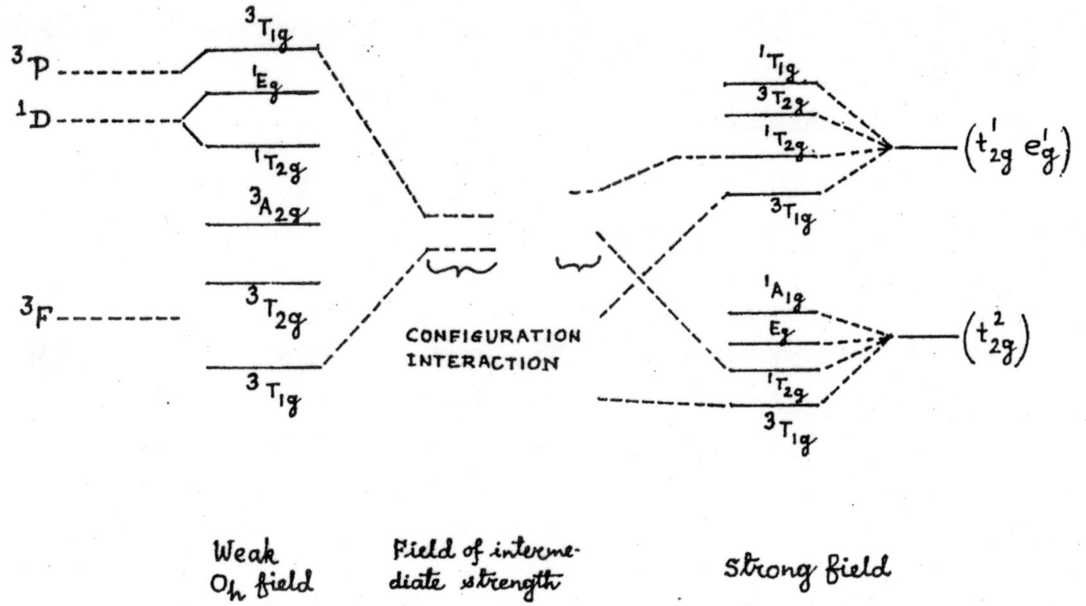

Fig. 10.1: Some weak and strong field terms of d^2 in O_h environment – Configuration Interaction.

Before we turn to a specific illustration of the calculation procedure, it would be advisable to note the following remarks.

1. The basis sets chosen for perturbational calculations, say for d^n, are n-electron dets or their linear combinations on the w.f side and some different dets or their linear combinations on the s.f side . Since the same determinantal space, comprising d-electron kets. are spanned by the different basis sets, there should exist a unitary transformation connecting the basis sets.

2. Each symmetry term of a given spin multiplicity on the w.f side should merge into a term of identical symmetry and spin multiplicity on the s.f side. But the relative energy gap between two such terms on the w.f side is in general different from the one between the corresponding merging terms on the s.f side.

3. A subtle group theoretical orthogonality theorem[6] demands that the integral of the product of two members of two different sets of degenerate wavefunctions of the same symmetry but originating from two different atomic/crystal field configurations will vanish unless the chosen components correspond. The correspondence is determined by identical values of M_L, M_s of the functions or by

similarity of transformation behaviour accompanying a descent in symmetry.

This orthogonality theorem also holds for the matrix element of the hamiltonian or any other operator commuting with it between the chosen wavefunctions from the two sets.

The actual calculation of configuration interaction at intermediate field strengths may be made by approaching either from the w.f side or from the s.f side. Let us consider the energy of $^3T_{1g}$ state i.e. $\varepsilon(^3T_{1g};d^2)$ at intermediate field values in an O_h environment.

(i) Approach from weak field side

If configurational mixing takes place for the $^3T_{1g}$ state, it ought to mix the wavefunctions $\psi(^3T_{1g};^3F,d^2)$ and $\psi(^3T_{1g};^3P,d^2)$ (cf. fig. 10.1). We, therefore, construct the total perturbational energy matrix, comprising both the coulombic repulsion and the symmetry field effects, in the basis set of the above two wavefunctions. The components chosen here are ψ_{30} and ψ_{10} of Table $T_{9.3}$, The det forms of these components, (Eqs 9.5b, 9.8) and Table $T_{4.3}$ are ,

$$\psi_{30} = \psi(^3T_{1g};^3F,d^2) = 1/\sqrt{5}(2^+ - 2^+) + 2/\sqrt{5}(1^+ - 1^+) \text{ where } M_L = 0 \text{ and } M_s = 1$$

$$\psi_{10} = \psi(^3T_{1g};^3P,d^2) = 2/\sqrt{5}(2^+ - 2^+) - 1/\sqrt{5}(1^+ - 1^+) \text{ where } M_L = 0 \text{ and } M_s = 1$$

The required matrix is

$$\begin{bmatrix} <\psi_{30} \mid \mathscr{H}_{crys}(O_h) + e^2/r_{12} \mid \psi_{30}> & <\psi_{30} \mid \mathscr{H}_{crys}(O_h) + e^2/r_{12} \mid \psi_{10}> \\ <\psi_{10} \mid \mathscr{H}_{crys}(O_h) + e^2/r_{12} \mid \psi_{30}> & <\psi_{10} \mid \mathscr{H}_{crys}(O_h) + e^2/r_{12} \mid \psi_{10}> \end{bmatrix} \quad \ldots(10.36)$$

The energy values, associated with the diagonal matrix elements, are scripted from Eqs. (9.16, 9.23) and Table T.9.4, leaving out the baricenter raising energy component. Thus the diagonal matrix elements of (10.36) are

$$\left.\begin{aligned} a_{11} &= A - 8B - 6Dq \\ a_{22} &= A + 7B - 0Dq \end{aligned}\right\} \quad \ldots\ldots(10.37)$$

The off-diagonal matrix elements

$$\left.\begin{aligned} a_{12} = a_{21} &= <\psi_{30} \mid \mathscr{H}_{crys}(O_h) + e^2/r_{12} \mid \psi_{10}> \\ &=< \{1/\sqrt{5}(2^+ - 2^+) + 2/\sqrt{5}(1^+ - 1^+)\} \mid \mathscr{H}_{crys}(O_h) + e^2/r_{12}) \mid \\ &\qquad \{2/\sqrt{5}(2^+ - 2^+) - 1/\sqrt{5}(1^+ - 1^+)\}> \end{aligned}\right\} \quad \ldots\ldots(10.38)$$

The coulombic contribution

$$=\frac{2}{5}\{J(2,-2)-k(2,-2)\}-\frac{2}{5}\{J(1,-1)-K(1,-1)\}-\frac{1}{5}\left\{\sum_k C^k(2,1)C^k(-1-2)F^k\right.$$

$$\left.-\sum_k C^k(2-1)C^k(1-2)F^k\right\}+\frac{4}{5}\left\{\sum_k C^k(13)C^k(-2-1)F^k-\sum_k C^k(1-2)C^k(2-1)F^k\right\}$$

= 0, on using Table $T_{2.1}$ and $T_{2.2}$

Similarly, the octahedral field contribution, using the abbreviated notations (cf. Eq. 9.44)

$$=\frac{2}{5}\langle 2,2\rangle+\frac{2}{5}\langle -2,-2\rangle-\frac{2}{5}\langle -1,-1\rangle-\frac{2}{5}\langle -1,-1\rangle$$

= 4 Dq. on using Eq. (7.54)

The energy determinant of the matrix, written it terms of available results of (10.37) and (10.38), is equated to zero for finding the given values.

$$\begin{vmatrix} A-8B-6Dq-E' & 4Dq \\ 4Dq & A+7B-E' \end{vmatrix}=0 \qquad \text{......(10.39)}$$

If, however, we wish to express the energy roots ε referred to a new zero level of energy which is (A - 8B) energy units away from the former zero level, then this new $\varepsilon = E' - (A-8B)$ [This is equivalent to setting the electrostatic energy level of ^3F of d^2 system as zero]. The det there assumes the form

$$\begin{vmatrix} -6Dq-\varepsilon & 4Dq \\ 4Dq & 15B-\varepsilon \end{vmatrix}=0 \qquad \text{......(10.40)}$$

Numerical values of ε are obtainable only when suitable values of B and of increasing values of Dq for the d^2 metal ion concerned are inserted in (10.40) enabling us to assess the effect of configuration interaction.

(ii). Approach from strong field side:

One can make a start from the extremely strong crystal field limit term. With increasing electrostatic interaction a situation will be encountered where an erstwhile pure wavefunction of certain symmetry starts mixing with wavefunction of a higher term of the same symmetry. The example we intend to deal with here is the interaction of $^3T_{1g}$ term of T_{2g}^2 configuration with a higher $^3T_{1g}$ term originating from T_{2g}^1, e_{2g}^1 configuration (Fig. 10.1).

The three dets of $^3T_{1g}$ originating from $\left(t_{2g}^2\right)$ are (cf. Eq. 10.3)

a) $\psi_1 = \left(d_{xz}^+ d_{yz}^+\right)$, b) $\psi_2 = \left(d_{xz}^+ d_{xy}^+\right)$, c) $\psi_3 = \left(d_{yz}^+ d_{xy}^+\right)$, ...(10.41)

and those emanating from $t_{2g}^1.e_g^1$ are

d) $\phi_1 = \left(d_{xy}^+ d_{x^2-y^2}^+\right)$, e) $\phi_2 = \left(d_{xz}^+ d_{x^2-y^2}^+\right)$, f) $\phi_3 = \left(d_{yz}^+ d_{x^2-y^2}^+\right)$, ...(10.42)

as may be verified by considering their transformation properties.

Any text on symmetry groups[(7) (8)] describes the resolution or branching of an IR of a higher symmetry group into a number of IR's when there occurs a descent in symmetry due to some application of perturbation. Thus on lowering the symmetry group from O_h to D_{4h}, there occurs the following resolution.

Group $O_h \rightarrow D_{4h}$

IR $^3T_{1g} \rightarrow {}^3A_{2g} + {}^3E_g$

If we subject the basis functions of (10.41) and of (10.42) separately to the transformations of D_{4h} the following functions are found to belong to the IR $^3A_{2g}$ each $\psi_1 = \left(d_{xz}^+ d_{yz}^+\right)$ and $\phi_1 = \left(d_{xy}^+ d_{x^2-y^2}^+\right)$. As in the foregoing example of the w.f. case, we are now to form the perturbation matrix in the basis set $\psi\left({}^3T_{1g}; t_{2g}^2\right)$and $\phi\left({}^3T_{1g}; t_{2g}^1\, e_g^1\right)$.

The strong crystal field contributions to the above two states are already known, viz., -8Dq and 2Dq respectively.

Having regard to the observation (3) on correspondence made in the beginning of the present section, the wavefunctions, one each from $\psi\ \&\ \Phi$, with which we need to construct the perturbation (configuration interaction) matrix, are ψ_1 and ϕ_1 respectively each being of $^3A_{2g}$ symmetry in D_{4h}. The matrix is

$$\begin{pmatrix} \left\langle \psi_1 \left| \frac{e^2}{r^{12}} \right| \psi_1 \right\rangle - 8Dq & \left\langle \psi_1 \left| \left(H_{crys}(O_h) + \frac{e^2}{r^{12}} \right) \right| \phi_1 \right\rangle \\ \left\langle \phi_1 \left| \left(H_{crys}(O_h) + \frac{e^2}{r_{12}} \right) \right| \psi_1 \right\rangle & \left\langle \phi_1 \left| \frac{e^2}{r_{12}} \right| \phi_1 \right\rangle + 2Dq \end{pmatrix} \quad ......(10.43)$$

Now $\varepsilon^{elec}\left({}^3T_{1g}; t_{2g}^2\right) = \left\langle \psi_1 \left| \frac{e^2}{r_{12}} \right| \psi_1 \right\rangle = A - 5B$ from Eq (10.16) and $\varepsilon^{elec}\left({}^3T_{1g};t_{2g}^1\,e_g^1\right)=\left\langle \phi_1\left|\frac{e^2}{r_{12}}\right|\phi_1\right\rangle=\left\langle\left(d_{xy}^+d_{x^2-y^2}^+\right)\left|\frac{e^2}{r_{12}}\right|\left(d_{xy}^+d_{x^2-y^2}^+\right)\right\rangle$

$$= J\left(d_{xy}^+,d_{x^2-y^2}^+\right)-K\left(d_{xy}^+d_{x^2-y^2}^+\right)$$

$$= A + 4B \text{ from Table } T_{10.1}$$

The off-diagonal matrix element, viz.,

$$\left\langle \psi_1\left|\left(H_{crys}(O_h)+\frac{e^2}{r_{12}}\right)\right|\phi_1\right\rangle = ODq+\left\langle\left(d_{xz}^+\,d_{yz}^+\right)\left|\frac{e^2}{r_{12}}\right|\left(d_{xy}^+d_{x^2-y^2}^+\right)\right\rangle$$

$$= -6B$$

which is obtained from the available standard table[(2)]. This value may also be obtained independently by expressing the real d orbitals in their respective equivalents of linear combination of the complex d functions $(d_{\pm2},d_{\pm1},d_o\ etc)$ and using the c^k table. The secular determinant of the matrix (10.43) is set equal to zero.

$$\begin{vmatrix} A-5B-8Dq-E' & -6B \\ -6B & A+4B+2Dq-E' \end{vmatrix}=0 \qquad \text{.........(10.44)}$$

We may raise our reference zero level energy by A - 5B making the roots E' change to ε, viz., $\varepsilon = E'-(A-5B)$ and in this way simplify the look of the determinant

$$\begin{vmatrix} -8Dq-\varepsilon & -6B \\ -6B & 9B+2Dq-\varepsilon \end{vmatrix}=0 \qquad \text{.........(10.44a)}$$

Once again if numerical values of solutions are demanded, suitable values of B accompanied with increasing values of Dq are to be introduced in (10.44a) to have an idea of the extent of configuration interaction at intermediate field strengths.

10.4. d^n Systems in Symmetry Environments Lower than O_h's:

In Chapter 7, we have already developed the forms of crystal field hamiltonians of T_d, D_{4h}, D_3, and other symmetry environments. Further in a number of cases, it has been shown how the hamiltonians are connected with the $H_{crys}(O_h)$ We may use these different hamiltonians, independently of $H_{crys}(O_h)$, to obtain results in the new situations. Alternatively, we can exploit their connectivities with the octahedral hamiltonian and treat these as perturbation of an ideal octahedral system of which the wavefunctions and energies are either already known to us or can be easily obtained (cf. Sec. 8.4). This latter type of approach has already been illustrated for single d-electron complex (Chap. 8) and in finding matrix elements in Sec. 7.9 (D,

E, G). When we are using a d^n system (n >1), no new principle pops up except that we have to use now polyelectronic wavefunctions like the ones used earlier in this chapter.

A. Tetrahedral Systems:

For any pure d^n case (i.e., where pd mixing is not considered the results of T_d symmetry are readily obtainable from those of O_h symmetry (cf. Sec 8.3). However, one should remember that the e levels lie lower than t_2's in T_d symmetry. Also, unlike the O_h environment, spin multiplicities of the ground terms in both the strong and weak fields remain unaltered for d^4, d^5, d^6 and d^7 systems (cf. concluding remarks Sec. 10.2). There is thus no cross-over point for high spin complexes into low spin ones with increasing crystal field in T_d environment.

B. d^n Systems in D_{4h} (tetrag.) Symmetry: Distortion of Octahedral Symmetry. Eigenfunctions and Energy levels

Let us consider a d^3 system in D_{4h} surrounding as in a complex, MA_2X_4 where the two A ligands are in trans positions along the vertical axis. We regard this as a perturbation of an ideal octahedral system, MX_6, and shall, therefore, use the perturbation hamiltonian (Eq. 7.50, 7.51) part scooped from

$$H_{crys}(D_{4h} - tetrag) = H_{crys}(O_h) + Ar^2Y_{20} + Br^4Y_{40} + \left(-2ze^2/a + 2ze^2/b\right).3$$

The part $\left(Ar^2Y_{20} + Br^4Y_{40}\right)$ acts as distortion hamiltonian. The brace of terms, $3.\left(-\frac{2ze^2}{a} + \frac{2ze^2}{b}\right)$, being constants, serve only to alter the position of the baricentre of the octahedral molecule.

It we make an s.f. approach, the high field O_h ground term, accruing from t_{2g}^3 configuration of the d^3 system, is ${}^4A_{2g}$. We shall, however, deal with the more complex situation, viz., the excited stage $t_{2g}^2\ e_g^1$ configuration with all electrons having parallel spins. Amongst the states are the two quarters, ${}^4T_{2g}$ and ${}^4T_{1g}$ in O_h symmetry. The wavefunctions corresponding to these two states are either distributed over the following dets or should be some linear combinations of these. These dets, noted by f_1 through f_6 for brevity, are,

$$\left\{\begin{matrix} f_1 = \left(d_{xz}^+ d_{yz}^+ d_{x^2-y^2}^+\right), f_2 = \left(d_{xz}^+ d_{xy}^+ d_{x^2-y^2}^+\right), f_3 = \left(d_{yz}^+ d_{xy}^+ d_{x^2-y^2}^+\right) \\ f_4 = \left(d_{xz}^+ d_{yz}^+ d_{z^2}^+\right), f_5 = \left(d_{xz}^+ d_{xy}^+ d_{z^2}^+\right), f_6 = \left(d_{yz}^+ d_{xy}^+ d_{z^2}^+\right) \end{matrix}\right\} \quad \ldots\ldots(10.45)$$

Now when distortion of symmetry occurs (as two X ligands of MX_6 are substituted by two A's) there ensues a descent in symmetry viz.

$O_h \rightarrow D_{4h}$, and the IR's of O_h branches into IR's D_{4h}

$$\left.\begin{array}{l} {}^4T_{2g}(O_h) \rightarrow {}^4B_{2g}(D_{4h}) + {}^4E_g(D_{4h}) \\ {}^4T_{1g}(O_h) \rightarrow {}^4A_{2g}(D_{4h}) + {}^4E_g(D_{4h}) \end{array}\right\} \quad \text{and} \qquad \ldots\ldots(10.46)$$

By carrying out trial transformations D_{4h} (or even D_4) on the functions f_1, to f_6, it is found that the function f_1, satisfies the criteria of a base function of ${}^4B_{2g}$. Thus

$$\psi_1 = f_1 = \left(d^+_{xz} d^+_{yz} d^+_{x^2-y^2}\right) \quad = \psi\left({}^4T_{2g}; t^2_{2g}, e^1_g\right) \qquad \ldots\ldots(10.47)$$

$$\equiv \psi\left({}^4B_{2g}; t^2_{2g}, e^1_g; D_{4h}\right) \text{ on reduction}$$

Similarly the function f_4 during the above trial transformation, is singled out to belong to ${}^4A_{2g}$ of D_{4h}. We, therefore, have

$$\psi_4 = f_4 = \left(d^+_{xz}, d^+_{yz}, d^+_{z^2}\right) \quad = \psi\left({}^4T_{1g}; t^2_{2g}, e^1_g\right) \qquad \ldots\ldots(10.48)$$

$$\equiv \psi\left({}^4A_{2g}; t^2_{2g}, e^1_g; D_{4h}\right) \text{ on reduction}$$

Having found the two wavefunctions (10.47, 10.48) in O_h environment (and incidentally also of D_{4h}) of symmetries ${}^4T_{2g}$ and ${}^4T_{1g}$ of d^3 in s.f of O_h, we now try to find out the remaining pair of wavefunctions of ${}^4T_{2g}$ and also of ${}^4T_{1g}$ of O_h. We recall the group theoretical principle, viz., if ψ_1 be an eigenfunction of $H_{crys}(O_h)$, then $R\psi_1$ is also so, where R is a relevant group operator. To find the remaining two wavefunctions of ${}^4T_{2g}$, we apply C_3^2, and C_3 operators on ψ_1(10.47) and also on ψ_4(10.48) to find the pair of ${}^4T_{1g}$. To effect these transformations readily, the following transformation table containing results of group operators upon the real forms of d-orbitals turns out to be very useful[(9)(10)].

	E	C_4^z	C_2	C_2''	C_3
d_{z^2}	d_{z^2}	d_{z^2}	d_{z^2}	d_{z^2}	$-\frac{1}{2}d_{z^2} + \frac{\sqrt{3}}{2}d_{(x^2-y^2)}$
$d_{x^2-y^2}$	$d_{(x^2-y^2)}$	$-d_{(x^2-y^2)}$	$d_{(x^2-y^2)}$	$-d_{(x^2-y^2)}$	$-\frac{1}{2}d_{(x^2-y^2)} - \frac{\sqrt{3}}{2}d_{z^2}$
d_{xy}	d_{xy}	$-d_{xy}$	d_{xy}	d_{xy}	d_{yz}
d_{xz}	d_{xz}	d_{yz}	$-d_{xz}$	$-d_{yz}$	d_{xy}
d_{yz}	d_{yz}	$-d_{xz}$	$-d_{yz}$	$-d_{xz}$	d_{xz}

Using this table

$$\psi_2 = C_3^2\psi_1 = C_3^2 f_1 = C_3^2\left(d_{xz}^+ d_{yz}^+ d_{x^2-y^2}^+\right)$$

$$= \left(d_{yz}^+ d_{xy}^+ \left\{-\frac{1}{2}d_{x^2-y^2} + \frac{\sqrt{3}}{2}d_{z^2}\right\}\right)$$

$$= -\frac{1}{2}\left(d_{yz}^+ d_{xy}^+ d_{x^2-y^2}^+\right) + \frac{\sqrt{3}}{2}\left(d_{yz}^+ d_{xy}^+ d_{z2}^+\right) \qquad \text{.........(10.49)}$$

(N.B. Recall the principle of multiplying a det by a constant)

$$\psi_3 = C_3\,\psi_1 = C_3\,f_1 = C_3\,\left(d_{xz}^+\,d_{yz}^+\,d_{x2-y2}^+\right)$$

$$= -\frac{1}{2}\left(d_{xy}^+\,d_{xz}^+\,d_{x^2-y^2}^+\right) - \frac{\sqrt{3}}{2}\left(d_{xy}^+ d_{xz}^+ d_{z2}^+\right) \qquad \text{......(10.50)}$$

Hence the three wavefunctions of ${}^4T_{2g}$ state in strong O_h field are given by (10.47, 10.49 and 10.50). The last two transform as the IR, E_g in the lower symmetry group D_{4h}.

These C_3 and C_3^2 operations on ψ_4, i.e. f_4 generate in a similar manner the two remaining wavefunctions, ψ_5 and ψ_6 of the ${}^4T_{1g}$ symmetry in O_h environment. The expressions, on using the foregoing transformation table properly, are

$$\psi_5 = C_3^2\psi_4 = C_3^2 f_4 = C_3^2 \quad = C_3^2\left(d_{xz}^+ d_{yz}^+ d_{z^2}^+\right)$$

$$= -\frac{1}{2}\left(d_{yz}^+\,d_{xy}^+\,d_{z2}^+\right) - \frac{\sqrt{3}}{2}\left(d_{yz}^+\,d_{xy}^+\,d_{x^2-y^2}^+\right) \qquad \text{......(10.51)}$$

$$\psi_6 = C_3\psi_4 = C_3^2 f_4 \qquad = -\frac{1}{2}\left(d_{xy}^+\,d_{xz}^+\,d_{z^2}^+\right) + \frac{\sqrt{3}}{2}\left(d_{xy}^+\,d_{xz}^+\,d_{x^2-y^2}^+\right) \qquad \text{......(10.52)}$$

These two (10.51, 10.52) together with (10.48) complete the members of the ${}^4T_{1g}$ wavefunction. The wavefunctions (10.51 and 10.52) also behave as bases of the IR, E_g, in the lower group D_{4h}.

Energies of ψ (D_{4h}) staes of T_{2g}^2, e_g^1 configuration (strong field)

We now try find out the perturbation energy in ψ (D_{4h}) states by using the O_h field wavefunctions ${}^4T_{2g}$ and ${}^4T_{1g}$ as the initial wavefunctions and the distortion hamiltonion for perturbation calculations. The initial energies in the unperturbed state

$$\varepsilon_0\left({}^4T_{2g},O_h\right) = -2Dq+\left\langle {}^4T_{2g}\left|\sum\frac{e^2}{r_{kl}}\right|{}^4T_{2g}\right\rangle \qquad \text{......(10.53)}$$

$$\varepsilon_0\left({}^4T_{1g},O_h\right) = -2Dq+\left\langle {}^4T_{1g}\left|\sum\frac{e^2}{r_{kl}}\right|{}^4T_{1g}\right\rangle \qquad \text{......(10.54)}$$

Thus from (10.53)

$$\left.\begin{aligned} &\varepsilon_{D_{4h}}^{Total}\left({}^4B_{2g};t_{2g}^2\,e_g^1,d^3\right)=\varepsilon_0\left({}^4T_{2g},O_h\right)+n\varepsilon_D^{/} \\ \text{and}\quad &\varepsilon_{D_{4h}}^{Total}\left({}^4E_g;t_{2g}^2\,e_g^1,d^3\right)=\varepsilon_0\left({}^4T_{2g},O_h\right)+(n-3)\varepsilon_D^{//}\end{aligned}\right\} \qquad \text{......(10.55)}$$

where $\varepsilon_D^{/}$ and $\varepsilon_D^{//}$ are the perturbational level energies obtained on splitting ${}^4T_{2g}$ level by lowering the symmetry from O_h to D_{4h} and n represents the number of electrons occupying $\varepsilon_D^{/}$ level. Instead of banking on ${}^4T_{2g}$ functions of O_h, we can as well evaluate $\varepsilon_D^{/}$ and $\varepsilon_D^{//}$ separately and directly from ${}^4B_{2g}$ function and from ${}^4E_g$ function of (10.47) and (10.50) respectively, both belonging to D_{4h}. Thus using the wavefunction (10.47)

$$\begin{aligned}\varepsilon_D^{/} &= \left\langle\left(d_{xz}^+d_{yz}^+d_{x^2-y^2}^+\right)\middle|\Sigma\left(Ar^2Y_{20}+Br^4Y_{40}\right)\middle|\left(d_{xz}^+d_{yz}^+d_{x^2-y^2}^+\right)\right\rangle \\ &= \left\langle d_{yz}^+\middle|\Sigma\left(Ar_1^2Y_{20}+Br_1^4Y_{40}\right)\middle|d_{xz}^+\right\rangle \\ &\qquad +\left\langle d_{yz}^+\middle|\Sigma\left(Ar_2^2Y_{20}+Br_2^4Y_{40}\right)\middle|d_{yz}^+\right\rangle \\ &\qquad +\left\langle d_{x^2-y^2}^+\middle|\Sigma\left(Ar_3^2Y_{20}+Br_3^4Y_{40}\right)\middle|d_{x^2-y^2}^+\right\rangle \end{aligned} \qquad \text{......(10.56)}$$

= 7 Dt, on using perturbational energy parts only of Eqs. (7.66 and 7.67)

Similarly using the E_g wavefunction (10.50)

$$\begin{aligned}\varepsilon_D^{//} &= \left\langle d_{xy}^+\middle|\left(Ar_1^2Y_{20}+Br_1^4Y_{40}\right)\middle|d_{xy}^+\right\rangle \\ &\qquad +\left\langle d_{xz}^+\middle|\left(Ar_2^2Y_{20}+Br_2^4Y_{40}\right)\middle|d_{xz}^+\right\rangle \\ &\qquad +\frac{1}{4}\left\langle d_{x^2-y^2}^+\middle|\left(Ar_3^2Y_{20}+Br_3^4Y_{40}\right)\middle|d_{x^2-y^2}^+\right\rangle \\ &\qquad +\frac{3}{4}\left\langle d_{z^2}^+\middle|\left(Ar_3^2Y_{20}+Br_3^4Y_{40}\right)\middle|d_{z^2}^+\right\rangle \\ &= -\frac{7}{4}Dt \end{aligned} \qquad \text{......(10.57)}$$

In the foregoing expression for $\varepsilon_{D''}$, many component integrals vanish due to either orthogonality or for symmetry reasons such as the one $\left\langle d^{+}_{x^2-y^2}\middle|\left(Ar^2Y_{20}+Br^4Y_{40}\right)\middle|d^{+}_{z^2}\right\rangle$.

Furthermore, in arriving at Eqs. (10.56) and (10.57) r_1, r_2, r_3 refer to the radius vectors of the electrons 1,2 and 3 respectively. The levels $\varepsilon_{D'}$ and $\varepsilon_{D''}$ represent the departures from the strong field total unperturbed energy value of the d^3 system in O$_h$ surroundings.

The energy interval between $^4B_{2g}$ and ^4E_g levels of D$_{4h}$ is

$$\Delta\varepsilon = 7Dt - \left(-\frac{7}{4}Dt\right) = \frac{35}{4}Dt$$

The perturbational energy levels $\varepsilon_{D'}$ and $\varepsilon_{D''}$ can be evaluated independently, as in the preceding example, by using the wavefunctions ψ_4 and ψ_6 of Eqs. (10.48 and 10.52) respectively and finding the diagonal matrix elements of the distortion hamiltonian. The results are

$$\varepsilon_{d'} = -4Ds + 2Dt \qquad \text{......(10.59)}$$

and $$\varepsilon_{D''} = 2Ds + \frac{3}{4}Dt \qquad \text{......(10.60)}$$

relative to the strong field total energy value of the unperturbed d^3 complex of $t_{2g}^2\, e_g^1$ configuration in O$_h$ environment.

The energy interval between ^4E_g and $^4A_{2g}$ levels in actual D$_{4h}$ complex

$$\Delta\varepsilon = 6Ds - \frac{5}{4}Dt \qquad \text{.........(10.61)}$$

The ^4E_g level is energetically higher.

The crystal field theory, however, suits best the results for *O$_h$* and somewhat less so for T$_d$ environments. But the results of lower symmetries, either obtained independently or from perturbation of *O$_h$* environment, incorporate a number of energy parameters due to site symmetry alone (such as *Ds, Dt, Dσ, Dτ*). It is not always possible to determine unambiguously the magnitudes of all these parameters for all such systems. Crystal field theory is thus less successful for low symmetry environments of d^n systems.

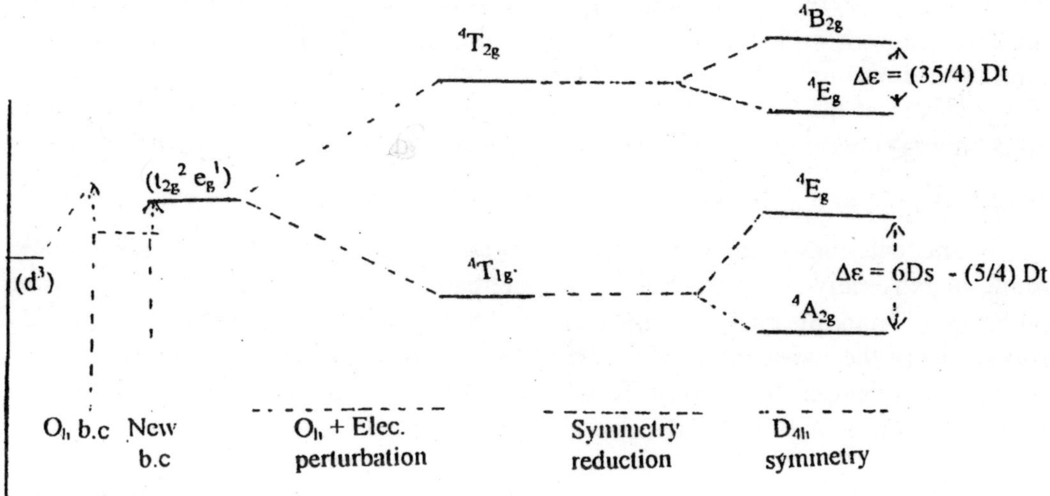

10.5 Term Correlation Diagrams; Energy Diagrams

Now that we know (chaps 9 and 10) how to find the specific terms and their energy values of different configurations in both the weak and strong field limits of any particular symmetry environment of a d^n atom/ion, we can correlate each term of specific symmetry and spin multiplicity on the weak field side with a strong field term on the basis of one-to-one correspondence .The complete picture of such one-to-one correspondence of terms of like symmetry and spin multiplicity is called term correlation diagram.

The technique of construction depends on the group theoretical method of descent in symmetry[11]. We skirt this method , but instead, take advantage of our knowledge of energy values in arranging the terms in a sequential manner on the two sides of a diagram for a d^n system in the given symmetry environment. In connecting a specific weak field term with a strong field term, the following principles are obeyed:

1. Conservation of symmetry type and spin: the symmetry of a term and its spin multiplicity do not change with the change in the field strength of the environment.

2. One-to-one correspondence: there are as many terms of different symmetries and spin multiplicities on the weak field side as there are on the strong field end. And one term on the left merges with just one term on the right of the same symmetry and spin.

3. Any two correlating lines emanating from two weak field terms of the same symmetry and spin multiplicity do not cross in their passage to the corresponding merging terms on the strong field side.

All these three principles can be theoretically established, although we desist from doing so here. Thus arranging the terms vertically on the two sides of an intended diagram and acting according to the three broad rules stated above, the correlation diagrams are constructed without even a rough or precise knowledge of energy values at the intermediate field strengths[12]. The respective diagrams for d^2 d^3 and d^8 atoms (ions) in O_h, environment are presented[13] (figs. 10.2, 10.4 and 10.3).

Energy Diagrams: d^n Systems in O_h, environment

The calculation of term energies over the entire stretch of field, weak to strong, is extremely complex. Such complete calculations have been made by Tanabe and Sugano[14] making an approach from the strong field side. Orgel[15] approached the problem from the weak field side. Owing to its very complex and lengthy nature, we just point out several features of Tanabe and Sugano's method with no pretension at all for a complete or even at effective sum up.

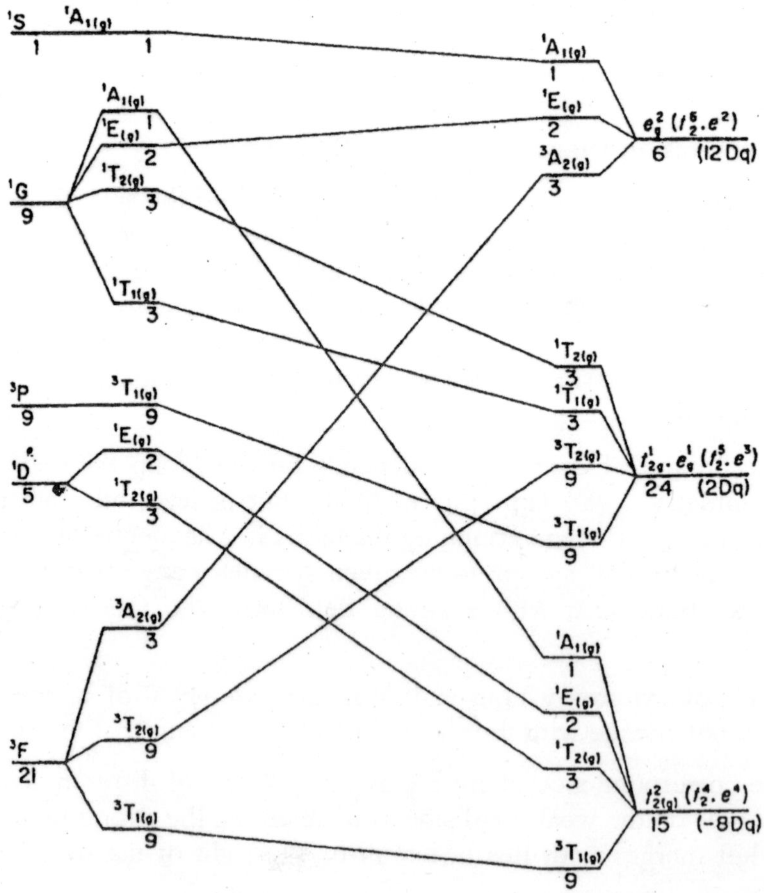

Fig. 10.2: Correlation diagram for free ion terms → strong-field configuration for: d^2 in O_h (V^{3+}, oct) [Numbers below energy levels indicate total degeneracies.] and d^8 in T_d (Ni^{2+}, tet) [For T_d symmetry, subscript g is to be dropped.]

(a) Atomic kets for free atoms are expressed as $|LSM_LM_S$ in one type of angular momentum coupling scheme. Likewise. a ket for an atom in a symmetry environment arising from a strong field configuration is expressed in the form. $|S\Gamma M_S M_\Gamma$ where Γ is the 1R of the term. S and Ms carry their usual significance and M_Γ a standard basis component of Γ and plays the role akin to M_L's in the ket $|LSM_L M_S>$. For example, the $^3T_{1g}$ term $^2t_{2g}$ of d^2 in strong field O_h environment bears the ket notation $|1T_{1g}1f>$ or more succinctly by $|^3T_{1g}1f>$, where f is one standard base function of T_{1g} of O_h symmetry group.

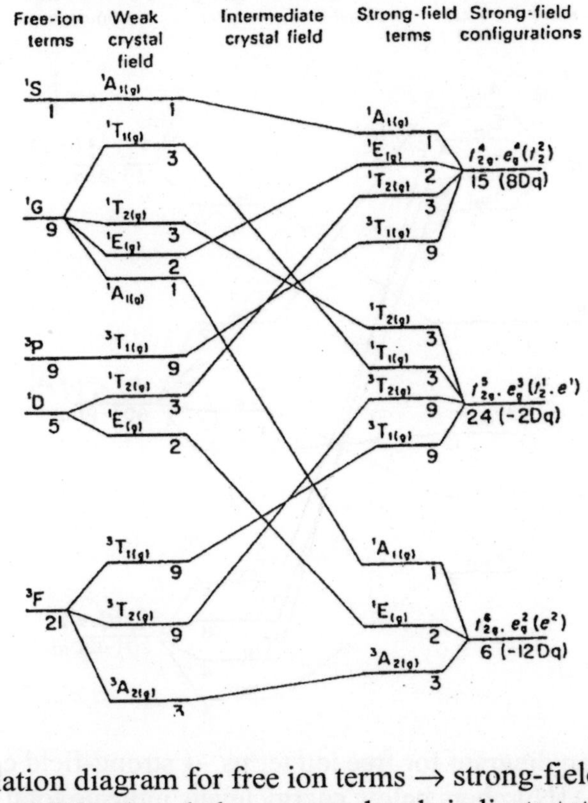

Fig. 10.3: Correlation diagram for free ion terms → strong-field configuration for: d^2 in T_d (V^{3+}, tet) [Numbers below energy levels indicate total degeneracies.] and d^8 in O_h (Ni^{2+}, oct) [For T_d symmetry, subscript g is to be dropped.]

(b) If the term emanating from the strong field configuration be such that it belongs to one of the resolved IR's of the dpr of two coupled kets $|S\Gamma' M_{S'} M_{\Gamma'}>$ and $|S''\Gamma'' M_{S''} M_{\Gamma''}>$ of the configuration, the coupled ket is written as $|\Gamma'\Gamma''\Gamma c>$. The latter can be expressed as a linear combination of the product of the component kets, briefly written as $|\Gamma'\Gamma'' M_{\Gamma'} M_{\Gamma''}>$ i.e. $|\Gamma'\Gamma'' ab>$ where a,b refer to standard bases $M_{\Gamma'}$, $M_{\Gamma''}$ of Γ' and Γ'' representations respectively.

$$\left|\Gamma'\Gamma''\Gamma c\right> \;=\; \Sigma_{a,b}\left|\Gamma'\Gamma''ab\right>\left<\Gamma'\Gamma''ab\right|\left.\Gamma'\Gamma''\Gamma c\right> \quad \ldots\ldots10.62$$

The coefficients represented by the integral may be looked upon as the analogues of Wigner's coefficients[16] in the expansion of the angular momentum coupled kets of atoms. The methods of evaluation of such coefficients have been worked out by Tanabe and Sugano and also independently by Griffith[17]

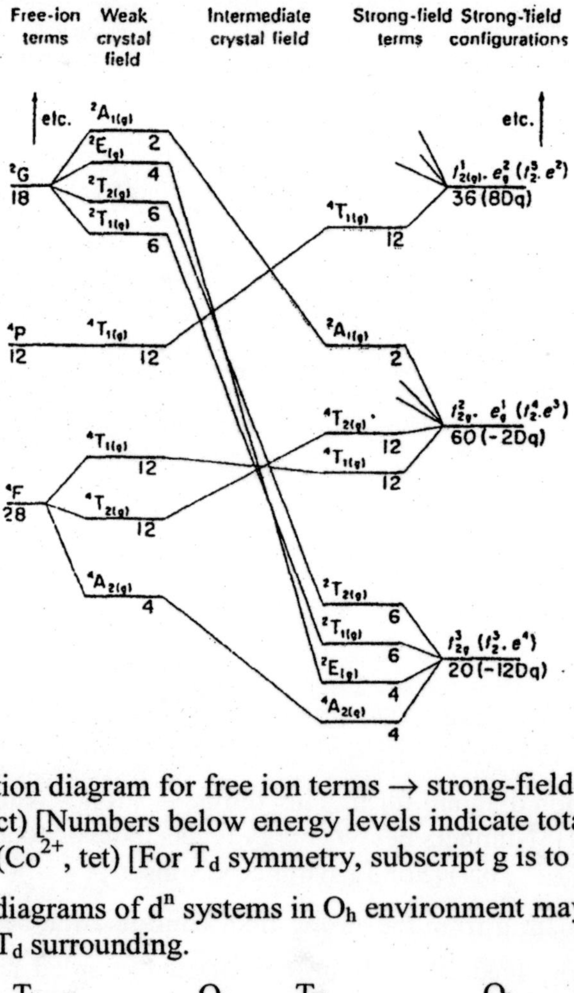

Fig. 10.4: Correlation diagram for free ion terms → strong-field configuration for: d^3 in O_h (Cr^{3+}, oct) [Numbers below energy levels indicate total degeneracies.] and d^7 in T_d (Co^{2+}, tet) [For T_d symmetry, subscript g is to be dropped.]

The correlation diagrams of d^n systems in O_h environment may also be used for d^{10-n} systems in T_d surrounding.

O_h	T_d	O_h	T_d	O_h	T_d
$d^1 \equiv$	d^9	$d^4 \equiv$	d^6	$d^7 \equiv$	d^3
$d^2 \equiv$	d^8	$d^5 \equiv$	d^5	$d^8 \equiv$	d^2
$d^3 \equiv$	d^7	$d^6 \equiv$	d^4	$d^9 \equiv$	d^1

(c) The concept of fractional parentage has been used by Tanabe and Sugano. The idea of fractional parentage (cf. Chisholm[18]) is roughly like this. If we have a configuration

$\left(t_{2g}^{p+q}, e_g^{r+s}\right)$, the terms arising out of this may be considered as 'offsprings or children' of their parents $t_{2g}^{p}\ e_g^{r}$ and $t_{2g}^{q}\ e_g^{s}$

The states of the children may be expressed as the combination of the states of their parents. This idea of fractional parentage leads to expressing the energies of matrix elements of n-electron systems in terms of (n-1) electron systems.

(d) The complete set of matrix elements of d^n system is determined with the help of an induction formula.

The energy diagrams, constructed by Tanabe and Sugano for d^2 through d^8 systems in octahedral symmetry, appear in Figs (10.5 - 10.11) with appropriate legends below the diagrams. The following remarks may be noted while reading and analyzing the diagrams.

1. These diagrams are possible to construct only for certain specific values of the parameters B and C/B remaining fixed over the trek from the weak to the strong field. However, in order to make each diagram amenable to utilization for a few rather than for just one specific ion of d^n system in O_h symmetry, energy values relative to that of the lowest term are plotted on the ordinate in units of the repulsion parameter B (i.e E/B) of the central ion . The plots are thus of E/B against increasing values of Dq./B along the abscissa. It turns out that a few transition metal ions of a particular d^n system have nearly the same E/B values calculated for such systems in field-free state. The energy diagram for this d^n system will thus be valid for all these specific ions instead of just one. Obviously, however, the energy diagram cannot be valid for all ions imaginable for this d^n system.
2. The ions and their respective values of B and C/B are indicated in the legends below the figures. Many of the diagrams are incomplete in the sense that many of the excited State energy levels are missing. This does not constitute any great disadvantage because, in their applications to electronic transitions, we are generally concerned with the lower states including the ground states.
3. The energy of the lowest term of a d^n configuration in the existing field is set equal to zero. The other energy values plotted along the ordinate are relative to this zero. In the diagrams of d^2. d^3 and d^8, the highest spin multiplicity terms with the lowest energy in field-free state continues to be so further on for any value of Dq/B along the abscissa up to the extreme strong field limit.
4. In the diagrams of d^4 and d^5 and d^6 and d^7 in O_h, field, however, while the highest spin multiplicity term forms the lowest energy level initially over a stretch of Dq/B values, at and beyond a critical value of the latter, another spin multiplicity term crosses the former energy curve to dip lower in energy. This constitutes a

new ground term on the strong field side. Its energy value beyond the cross-over point on the strong field side up to the extreme limit is set equal to zero. Hence in such diagrams (Figs 10.7 to 10.10) we have two different terms having zero energy defined on the left and right ordinates. The energy diagrams, therefore, suffer breaks at the crossover points in their total plot from the weak to the strong field side. The break is indicated by a vertical line in each of these four diagrams. In spite of the difference in the absolute magnitudes of the two zero level energies, the energy gap between any two given ordinate points on either the left or the right of the vertical line through the critical cross-over point remains the same.

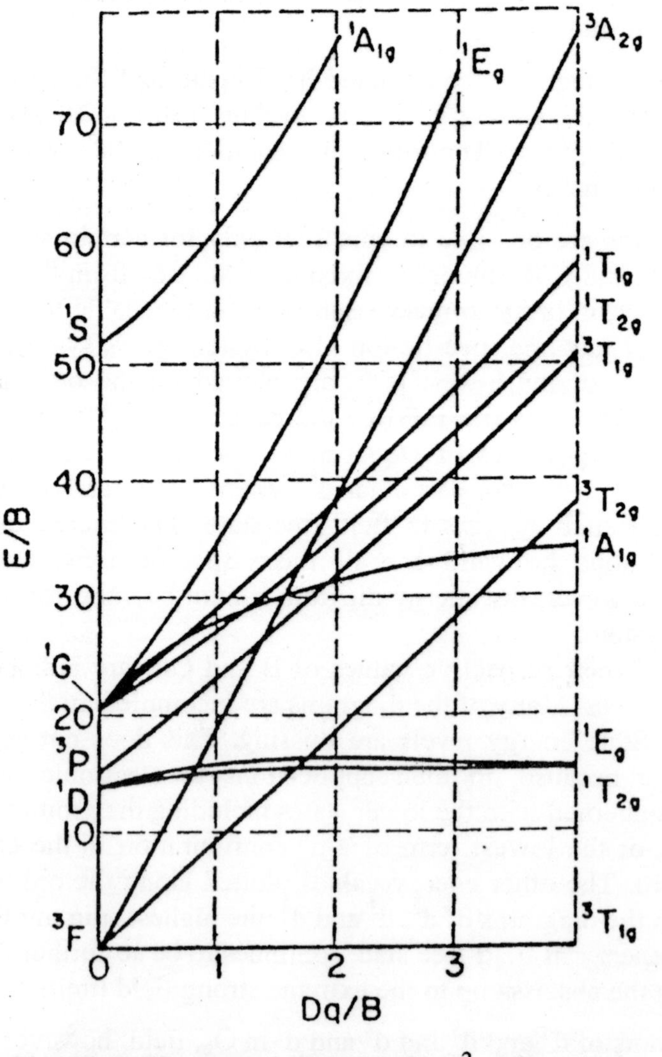

Fig. 10.5: The energy diagram for d^2. C/B = 4.42.

[V^{3+}, B ~ 861 cm^{-1}]

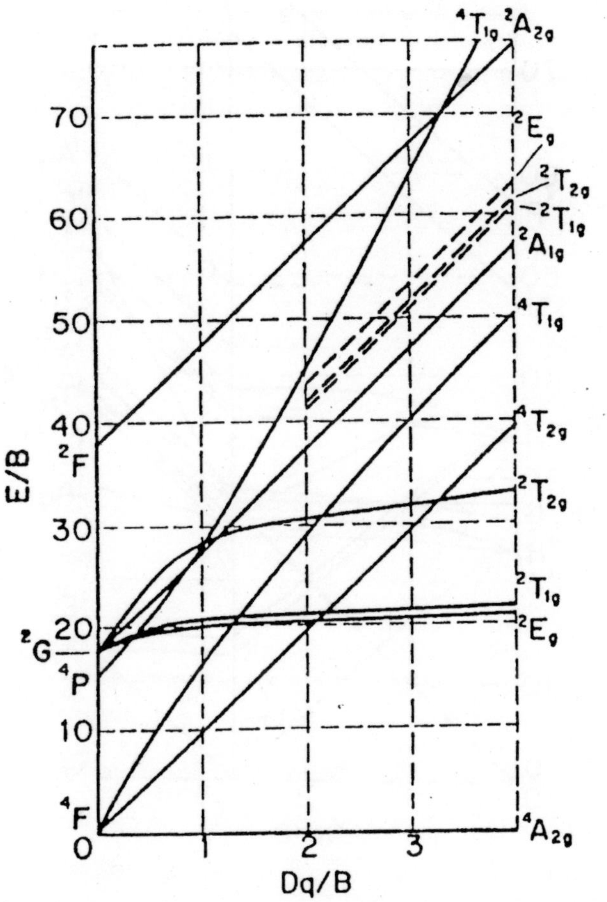

Fig. 10.6: The energy diagram for d^3. C/B = 4.50.
[Cr^{3+}, B ~ 1030 cm^{-1}]

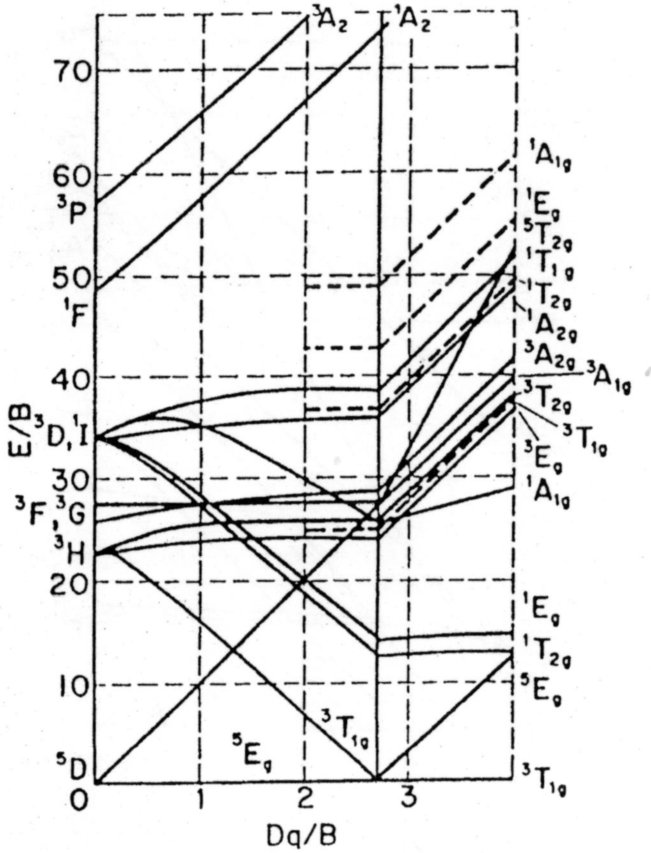

Fig. 10.7: The energy diagram for d^4. C/B = 4.6.
[Cr^{3+}, Mn^{3+}, B ~ 900 cm^{-1}]

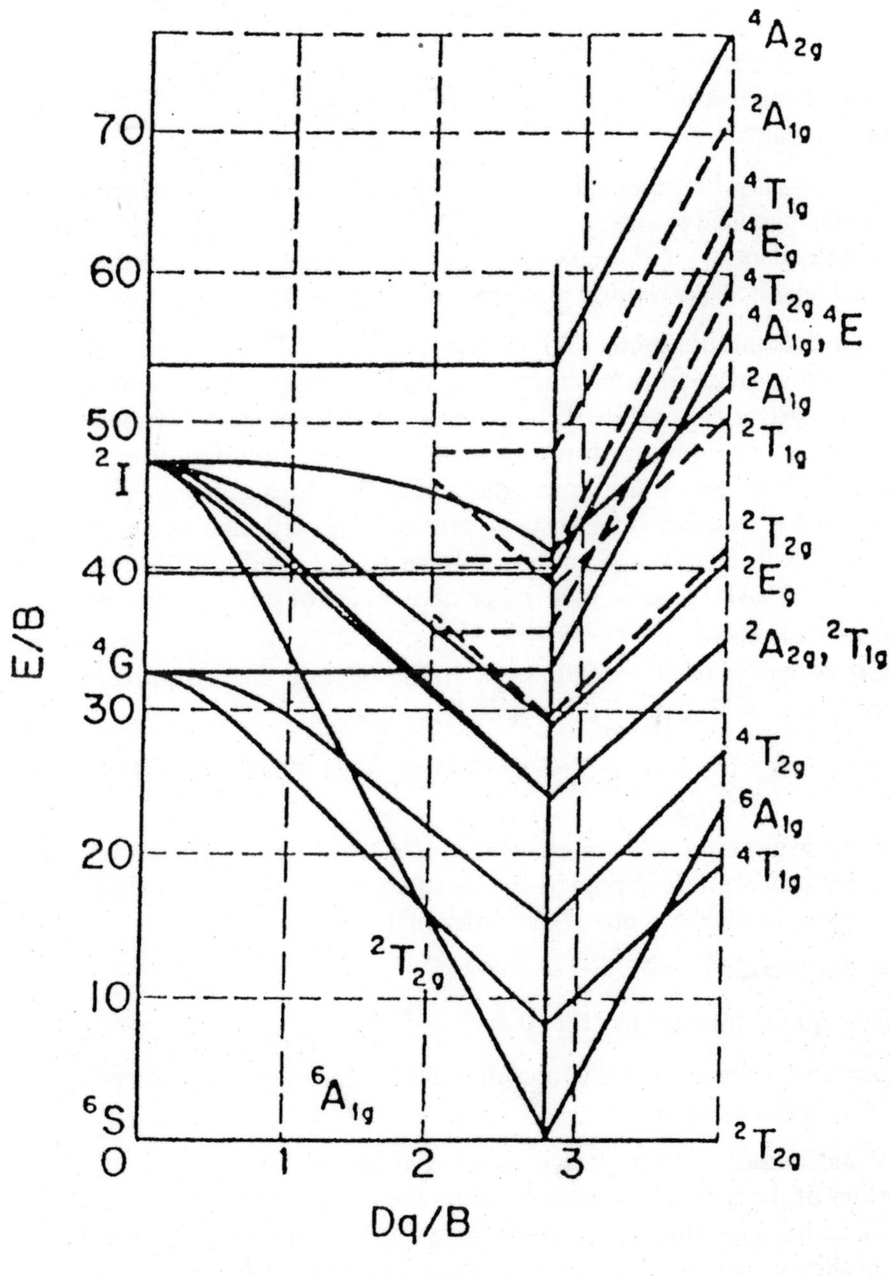

Fig. 10.8: The energy diagram for d^5. C/B = 4.48.
[Mn^{2+}, Fe^{3+}, B ~ 950 cm^{-1}]

5. If anyone of a number of d-d absorption bands of a given metal ion complex (B value known) be assigned to a certain specific transition, then its Dq value may be obtained- This is done by equating experimental energy of transition with the theoretical transition energy expressed in terms of Dq and B. Once this is done, other approximate transition energies may be read off from the Tanabe Sugano diagrams.

6. The process may be reversed when we desire to assign the observed transitions between the specific symmetry states. This is done by trial fitting of and matching the theoretical ratio of the absorption energies expressed in terms of Dq and B with the Tanabe Sugano diagram corresponding to the specific value of Dq/B.

7. It is seen that the diagrams (7.7 - 7.10) for d^4, d^5, d^6 and d^7 ions exhibit critical points of cross-over from higher spin ground states to lower spin ground states. Such a cross-over is possible if the pairing repulsion energy is more than offset by decrease in energy through forced pairing and accommodation in the lower level in a strong field. Since the points lie at Dq/B values equal to 2,2.2,2.7,2.8 for d^6,d^7, d^4 and d^5 respectively, ease of spin pairing will also occur in that order in their complexes. This is, of course, subject to the condition that the B values of the concerned metal ions in their respective states of ionization remain practically the same. Complexes of Cr^{3+}, Mn^{2+}, Co^{2+}, Co^{3+}, besides some others, provide instances of spin paired ground states at and beyond the cross-over points, [cf. B values of metal ions Table $T_{6.1}$ in field free states.]

8. The following types of transitions are feasible in octahedral complexes of metal ions.

(i) vibronic transitions which are electric dipole transitions coupled with vibrational contribution. Pure electric dipole d-d transition is symmetry-forbidden. Oscillator strengths for such transitions are of the order of 10^{-4}.

(ii) electric quadrupole transitions ($f \sim 10^{-7}$)

(iii) magnetic dipole transitions ($f \sim 10^{-6}$)

These transitions, called intrasystem combinations, occur between crystal field states of the same spin multiplicity.

There may additionally occur intersystem transitions between states of different spin contamination of the orbital symmetry states. Such transitions can lake place between levels of the same electronic configuration with different spin multiplicities and are possible only for d^5, d^6 and d^7 systems at field strengths beyond the cross-over points ($Dq/B > 2B$).

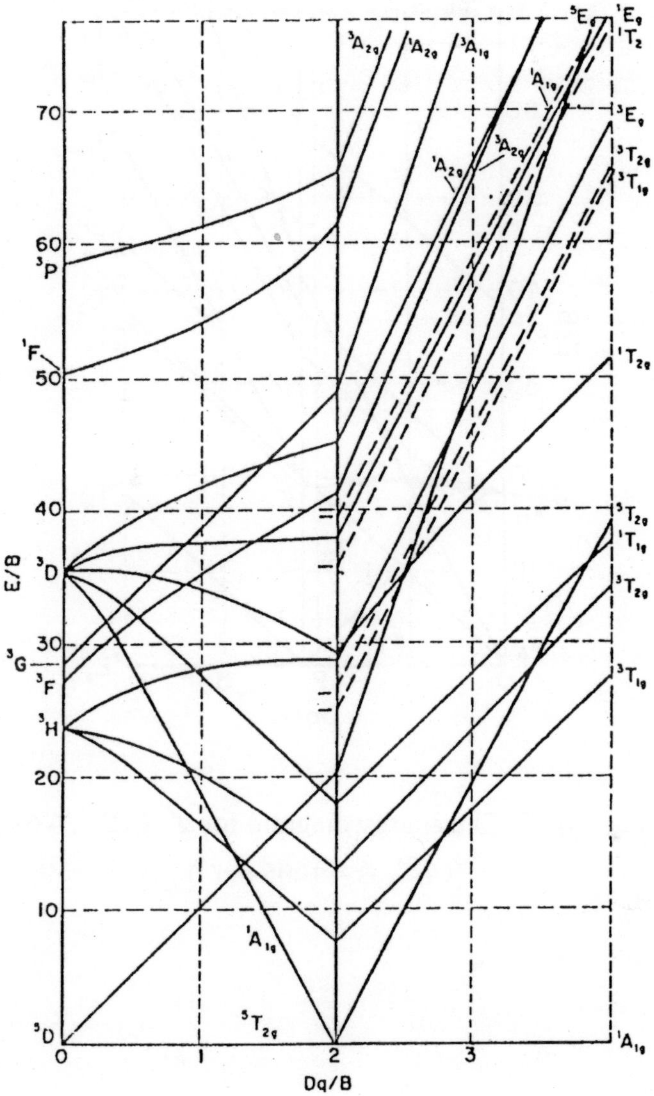

Fig. 10.9: The energy diagram for d^6. C/B = 4.81.
[Fe^{3+}, Co^{3+}, B ~ 1050 cm^{-1}]

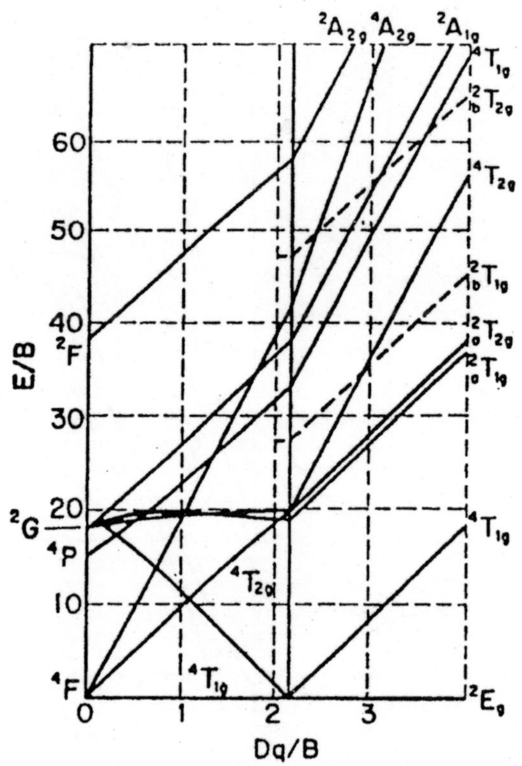

Fig. 10.10: The energy diagram for d^7. C/B = 4.63.
[Co^{2+}, B ~ 1005 cm^{-1}]

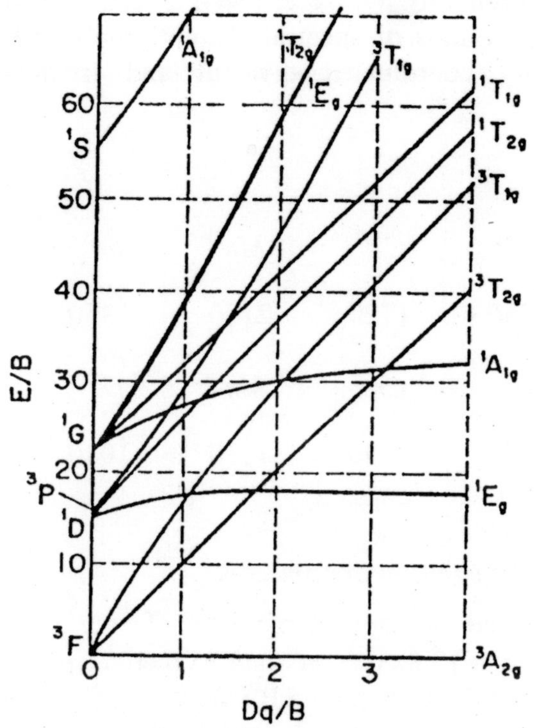

Fig. 10.11: The energy diagram for d^8. C/B = 4.71.
[Ni^{3+}, B ~ 1084 cm^{-1}]

10.6 Worked out Example

Tanabe and Sugano in their original paper listed a series of Dq values, ascertained from experimental observations of spectral absorptions by other researchers, for a number of octahedrally coordinated divalent and trivalent metal ions, viz. $[M(H_2O)_6]^{2+/3+}$

	Ti^{3+}	V^{3+}	C^{3+}	Mn^{3+}	Fe^{3+}	Co^{3+}	
Dq(cm^{-1})	2030	1860	1720	2100	1350	1920	
			Cr^{2+}	Mn^{2+}	Fe^{2+}	Co^{2+}	Ni^{2+}
Dq(cm^{-1})			1390	1230	1030	840	820

These values are useful for solution of problems.

Example 1. The observed absorption maxima of two d-d transition bands of $CsV(SO_4)_2$, 12 H_2O occur at 17.8×10^3 cm^{-1} and 25.6×10^3 cm^{-1} respectively. Using the requisite Dq value as given before and the B value from the energy graph (Fig 10.5),)(i) how far are these energy values supported by the energy curves when the first transition is assigned to the absorption $^3T_{2g} \leftarrow {}^3T_{1g}$.

(ii) How is the second transition to be assigned?

(i) Dq/B = 1860 cm^{-1} /861 cm^{-1} =2.16

The transition energy of $^3T_{2g} \leftarrow {}^3T_{1g}$ corresponding to Dq/B = 2.16 is found to be 19.5B (from Fig 10.5) i.e 16789.5 cm^{-1} which is in fair agreement with the experimental value, considering the condensed scale of the energy curve.

(ii) Some of the nearby theoretical absorptions which can be read off from the diagram corresponding to the same Dq/B = 2.16 are

a) $^3T_{1g}(^3P) \leftarrow {}^3T_{1g}(^3F)$ transition energy 32B = 2721 cm^{-1}

b) $^1A_{1g} \leftarrow {}^3T_{1g}(^3F)$ transition energy 31B = 25691 cm^{-1}

But the second transition (intersystem one) is ruled out for d^2 system. Hence the actual transition must be of type (a) though the graphical energy value is a little bit higher.

Example 2:

The maxima for two transitions viz. ${}^1T_{1g} \leftarrow {}^1A_{1g}$ and ${}^1T_{2g} \leftarrow {}^1A_{1g}$ of a Co^{3+} octahedral complex arc 16.0×10^3 em^{-1} and 25.0×10^3 cm^{-1} respectively. Utilise the requisite Tanabe Sugano diagram to obtain an approximate Dq value of the ion. '

The Co^{3+} ion is a d^6 system. Thus we are to refer to Fig 10.9 and the concerned B value =1050 cm^{-1}.

The ratio of the two absorption maxima = 25/16 = 1.56

We are to scan the graph to find which two ordinates (E/B) ending on the energy curves of ${}^1T_{2g}$ and ${}^1T_{1g}$ corresponding to a common value of Dq/B, bear a ratio 1.56. It is seen from the graphs that at the cross-over point, where Dq/B = 2. the ratio of the two ordinates is 1.6 which is near enough to 1.56.

$$\therefore Dq\left(Co^{3+}\right) = 2\times1050cm^{-1} = 2100cm^{-1}$$

Exercises

1. "d_{xz} is an eigen function of e^2/r_{kl} in a central field". How would you confirm or disprove the statement ?
2. Find the diagonal matrix element of the electrostatic operator e^2/r_{kl} with respect to the function $\left(d_{z2}^{+}\ d_{xz}^{+}\right)$ and check your result by referring to table $T_{10.1}$
3. Find the wave functions of the terms, ${}^3A_{2g}$ and ${}^1A_{1g}$ stemming from strong crystal- field configuration, e_g^2 due to weak coulombic perturbation of the strong field terms What are these perturbational energies ?
4. How can you show independently that, ψ_9 must mix with $(\psi_2+\psi_8)/\sqrt{2}$ to form a basis of A_1 representation of the O- point group as demanded by Eq. (10.10) ?
5. Can you theoretically establish the principle that any two correlating lines emanating from two weak-field terms of the same symmetry and spin multiplicity do not cross in their passage to the corresponding merging terms on the strong field side ?

References

1,2,5,10 C.J. Ballhausen : *Introduction to Ligand Field Theory* (McGraw-Hill)

3,4,17 J. S. Griffith : *Theory of Transition Metal Jons* (Cambridge Univ. Press)

6 M. Tinkham : *Group Theory and Quantum Mechanics* (Tata Mc+Graw-Hill)

7.9 S. C. Rakshit : *Molecular Symmetry Groups and Chemistry* (New Book Stall)

8 D. M. Bishop : *Group Theory and Chemistry* (Clarendron Press)

11 F. A. Cotton : *Chemical Applications of Group Theory* (Interscience).

12. A. P. B. Lever : *Inorganic Electronic Spectroscopy* (Elsevicr).

13. B. N. Figgis : *Introduction to Ligand Fields* (Wiley Eastern).

14. Y. Tanabe and S. Sugano (J. *Phys. Soc. Japan, 9, 753, 766, 1954).*

15. L. E. Orgel : *(J. Chem Phys, 23, 1819, 1955).*

16. E. Wigner : *Group Theory, its Applications to the Theory of Atomic Spectra.* (Acad.Press).

18. C. D. H. Chisholm : *Group Theoretical Techniques in Quantum Chemistry,* (Acad. Press).

Chapter Eleven

Further Perturbations in Cubic System: Spin Orbit Interaction and External Magnetic Field. Some Magnetic Properties.

We have already solved in Chapters 8-10 for the eigenfunctions and the energy eigenvalues of the perturbation matrices of coulomb repulsion and crystal field when d^n ion is placed in a symmetry environment both from weak and strong field viewpoints. Additionally, in Chapter 8 dealing with d^1 systems, the effect of spin orbit interaction on crystal field terms and the reversed consequence have also been treated. In some cases, where the final perturbed wavefunctions could be formulated in advance, the energy values were obtained directly without any need for the preconstruction of the perturbation matrices.

In the present chapter two additional perturbations - spin orbit and external magnetic field - will be imposed on the strong crystal field terms of d^1 and d^2 ions in O_h environment. The treatment will be based on some isomorphism property and, in this sense, will differ from the one meted out to d^1 of chapter 8.

The present chapter will be divided into two parts, part 1 dealing with the spin orbit interaction on strong crystal field terms and part II with the magnetic field effect on the crystal field terms either already mauled or still untarnished by spin orbit interaction. In fact part II deals with the basics of calculations of zeeman levels, magnetic susceptibilities and magnetic moments of d^1 and d^2 ions in O_h complexes.

Part I

To deal with spin orbit interaction in complexes, it will be helpful to have some knowledge of p^n isomorphism, transformation behaviour of spin under operators of double groups and the resolution of direct products of IR's of double groups (cf. Sec 8.5).

11.1 A. $(t_2d)^n$, p^n Isomorphism:

It was pointed out by Abragam and Pryce[1] and later by many other workers [2.3] that the transformation behaviour of the p-orbitals, viz., p_1, p_0 and p_{-1} under l.s or angular momentum operators or the matrix elements of these operators between the p-functions are similar to those of the t_{2g}/t_2 d-orbital functions provided we agree to set up a correspondence depicted by (I 1.1) below. Moreover, the magnetic susceptibilities of a ^2P state of a p-electron of a free ion and ^2D of d^1 in O_h environment are the same. The similarities in results mentioned above lead to the following correspondences.

p-electron ket	d-electron ket	notation of correspondence	
$\|p_1 >\sim$	$\|-1>$ i.e., d_{-1}	(t_2d1)	
$\|p_0 >\sim$	$\frac{1}{\sqrt{2}}[\|2>-\|-2>]$ i.e., $\frac{1}{\sqrt{2}}[d_2 - d_{-2}]$	$(t_2\zeta_1 0)$ or $(t_2 d0)$	.......(11.1)
$\|p_{-1} >\sim$	$-\|1\rangle$ i.e., $-d_1$	$(t_2d\text{-}1)$	

These correspondences spring from the similarities of results, viz.,

$$\left.\begin{aligned} &< p_m \mid l_z \mid p_n > = n\hbar\delta_{mn} \qquad < t_2 d_m \mid l_z \mid t_2 d_n > = -n\hbar\delta_{mn} \\ &< p_m \mid l_k \mid p_n > = < t_2 dm \mid l_k \mid t_2 dn > \text{for k=x, y, z components} \\ &< p_m \mid -l.s \mid p_n > = < t_2 dm \mid l.s \mid t_2 dn > \end{aligned}\right\} \quad ....(11.2)$$

The last relation can be utilized more effectively in the following way. The spin orbit matrix element $< t_2dm|\xi(r)l.s|t_2dm >$, after integration of its radial part, can be rewritten as

$$\xi'_{3d}\frac{\hbar^2}{\hbar^2} < t_2dm|l.s|t_2dm > \text{ where } \xi'_{3d} = \int R_{3d}^2\xi(r)r^2 \, dr$$ and the kets now stand only for the angular parts of the d-functions.

$$= \frac{\zeta_{3d}}{\hbar^2}\langle p_m|-l.s|p_m\rangle$$ on using the relation (11.2)

$$= \frac{\xi_{3d}}{2}\{j\,(j+1) - l\,(l+1) - s\,(s+1)\} \qquad ....(11.3)$$

where j, l, and s all refer to p-electron quantum numbers. This relation will come in handy for later use.

Such similarities also cover more than one d-electron configuration. For instance, if we have a (t_{2g}^n) of d-electrons having one of its wavefunctions, say, Ψ_{d^n}, we can correspondingly visualize a (p^n) configuration of which the component one-electron kets of ψ_{p^n} may replace the component equivalent d-electron kets in ψ_{d^n} for the purpose of investigating the behaviour of the latter under the operation of the angular momentum. As a concrete example, let us take

$\psi_{t_2^2} = (t_2d1^+, t_2d0^+) = (-1^+\zeta_1^+) \equiv (p_1^+ p_0^+) = \psi_{p^2}$ of ^{3}p state where ( ) represents the determinantal wavefunctions. It is because of such similarities in the behaviour of p and t_2d functions, that the term "$(t_2d)^n$, p^n" isomorphism has been coined.

B. The t_{2g} orbitals and the spin wavefunctions. Direct product of the IR's in the double group.

We are already familiar with the real d-orbitals forming the basis for the IR's t_{2g} and E_g in O_h environment. In dealing with spin orbit interaction it becomes easier to handle the treatment if one uses the complex d-functions as basis as far as possible. Transformation properties of d_2, d_1, d_0, d_{-1}, d_{-2} under O-group operators reveal the following as the basis of T_{2g} and E_g in O_h symmetry

$$T_{2g}\text{ symmetry}\begin{cases} t_{2g}^{+} = -d_1 = -|1> \\ t_{2g}^{-} = d_{-1} = |-1> \\ t_{2g}^{0} = \left|\frac{1}{\sqrt{2}}(d_2 - d_{-2})\right\rangle = |\zeta_1> \end{cases}$$

$$E_g\text{ symmetry}\begin{cases} e_g^a = d_z^2 \\ e_g^b = d_{x^2-y^2} \end{cases}$$

When the spin functions α and β are introduced, the basis sets are

$$\left.\begin{array}{l} 1)\ t_{2g}^{+}\alpha = -|1^{+}>,\quad t_{2g}^{+}\beta = -|1^{-}>,\quad t_{2g}^{-}\alpha = |-1^{+}>,\quad t_{2g}^{-}\beta = |-1^{-}>, \\ \quad t_{2g}^{0}\alpha = |\xi_1^{+}\rangle,\quad t_{2g}^{0}\beta = |\zeta_1^{-}\rangle \end{array}\right\} \qquad \text{..........(11.4)}$$

$$2)\ e_g^a\alpha,\ e_g^a\beta,\ e_g^b\alpha,\ e_g^b\beta \qquad \text{..........(11.5)}$$

The IR of α and β in the double group O' is known [4] to be E_2', i.e., Γ_6'. The IR of the space orbitals t_2^{+} and t_2^{-} in the group O' is T_2' (i.e., Γ_5'). Hence, the IR's of the spin orbitals (products of space orbitals and spin wavefunctions) of the set 1 (Relation 11.4) will be given by the resolution of the dpr of

$$E_2' \times T_2' \text{ (i.e., } \Gamma_6' \times \Gamma_5' \text{ of the O}' \text{ group)} = E_3' + G'$$

$$\text{(i.e., } \Gamma_7' + \Gamma_8') \qquad \text{..........11.4a}$$

Similarly, the resolution of the dpr, in O' group, of the set (2) (relation 11.5)

$$E_1' \times E_2' \text{ (i.e., } \Gamma_6' \times \Gamma_5') = G' \text{ (i.e., } \Gamma_8') \qquad \text{..........11.5a}$$

11.2 spin Orbit Interaction in d^n complexes of Cubic Symmetry:

General discussion – We shall confine ourselves to the ground and low crystal field terms of d^n (actually d^1 and d^2 only) while investigating the effect of spin orbit perturbation. These terms, especially, the ground ones, mainly determine the magnetic susceptibilities of the complexes. For d^n complexes of cubic symmetry, T_2/T_{2g}, T_1/T_{1g}, E/E_g, A_1/A_{1g}, A_2/A_{2g} are generally the low lying crystal field terms including the ground terms. Since the term separation due to crystal field is much larger than that due to spin orbit effect, we shall investigate the latter perturbation effect within a particular crystal field multiplet, say within a $^2T_{2g}$, ^2E_g or $^3T_{1g}$ etc. We shall not presently consider the mixing of high field terms of different symmetry with a low field one under this perturbation.

The calculation of energies and the final eigenfunctions is based on perturbations of the zero order (i.e., unperturbed) states that are taken to be the respective crystal field states. To start with, we can arrive at some general conclusions regarding the spin orbit effect on the ground terms of d^n systems in O_h or T_d symmetries.

(a) Ground term $^{2S+1}E_g$ or ^{2S+1}E

The initial wavefunctions, i.e., the crystal field terms including spins may be factored clearly into an orbital part and a spin part. The matrix element of spin orbit interaction will be nonvanishing if,

$\left\langle \psi_{\Gamma_8'} \middle| \Sigma\xi(r_i) l_i . s_i \middle| \psi_{\Gamma_8'} \right\rangle$ be nonzero where Γ_8' is the IR of the E_g wavefunction including spin in O_h' double group.

It is thus necessary that subject to any or all of the spin matrix parts not integration to zero,

$\Sigma\xi(r_i)\left\{\left\langle \psi_{Eg} \middle| l_i \middle| \psi_{Eg} \right\rangle\right.$. spin matrix of s_{xi}+$\left\langle \psi_{Eg} \middle| l_i \middle| \psi_{Eg} \right\rangle$. spin matrix of s_{yi}

$+\left\langle \psi_{Eg} \middle| l_i \middle| \psi_{Eg} \right\rangle$ spin matrix of s_{zi} be non zero ....(11.6)

Since l_x, l_y and l_z components of l_i have the same transformation properties as those of the rotational vectors R_x, R_y, and R_z in O_h group the latter set as well the former set form the basis of the IR; T_{1g}, of that point group.

The survival or disappearance of (11.6) will thus depend on whether or not an A_{1g} IR exists in the dpr of ($E_g \times T_{1g} \times E_g$) in O_h group. This dpr does not yield, on resolution, any IR of A_{1g} symmetry [cf. Table $T_{2.3}$].

$$E_g \times T_{1g} \times E_g = (A_{1g} + A_{2g} + E_g)T_{1g} = T_{1g} + T_{2g} + T_{1g} + T_{2g}$$

This means spin orbit interaction causes no splitting of either the E_g levels of O_h symmetry or, as can be similarly shown of the E levels of T_d symmetry complexes.

b) Ground terms $^{2S+1}A_{1g}$ (or A_1), $^{2S+1}A_{2g}$ (or A_2)

Arguing in the same vein as in the previous case of $^{2S+1}E_g$ (or E), it can also be shown that these crystal field levels are not affected by spin orbit interaction.

c) Ground or low lying terms ${}^{S2s+1}T_{2g}$ (or T_2), ${}^{2S+1}T_{1g}$ (or T_1)

Here the crystal field terms are generally split by spin orbit perturbation. Each specific case has to be individually treated to find the nature of splitting.

11.3. Spin Orbit Interaction in d^1 complexes of O_h/T_d symmetry

The d^1 system in octahedral environment gives rise to the ground ${}^2T_{2g}$ crystal field states. As is the custom in the perturbation method, we take the six orbitals (Eq. 11.4) of this orbitally triply degenerate state to serve as base functions of the perturbation matrix of the spin-orbit hamiltonian, viz.

$$\frac{\zeta_{3d}}{\hbar^2}(l.s),\ \text{i.e.,}\ \frac{\zeta_{3d}}{\hbar^2}\left[l_z s_z + \frac{1}{2}(l_+ s_- + l_- s_+)\right] \quad \text{(cf. 2.28, 2.29)}$$

As an aid to forming the matrix elements, the results of operation $\left(\zeta_{3d}/\hbar^2\right)$ on the spin orbitals (11.4) are first collected. Remembering the operational effects of the raising and the lowering operators (chaps 1 and 2) we have, writing ζ for ζ_{3d},

$$\text{a)}\ \frac{\zeta}{\hbar^2}\left\{l_z s_z + \frac{1}{2}(l_+ s_- + l_- s_+)\right\}\left|-1^+\right> = \zeta\left\{-\frac{1}{2}\left|-1^+\right>\right\} + \frac{\zeta}{\hbar^2}\left\{\frac{\sqrt{6}}{2}\hbar l_+\left|-1^-\right>\right\}$$

$$= \zeta\left\{-\frac{1}{2}\left|-1^+\right> + \frac{\sqrt{6}}{2}\left|0^-\right>\right\}$$

$$\text{b)}\ \frac{\zeta}{\hbar^2}\left\{l_z s_z + \frac{1}{2}(l_+ s_- + l_- s_+)\right\}\left|-1^-\right> = \zeta\left\{\frac{1}{2}\left|-1^-\right>\right\} + \frac{\zeta}{\hbar^2}\frac{\hbar}{2\sqrt{2}}l_-\left|-1^+\right>$$

$$= \zeta\left\{\frac{1}{2}\left|-1^-\right> + \left|-2^+\right>\right\}$$

$$\text{c)}\ \frac{\zeta}{\hbar^2}\left\{l_z s_z + \frac{1}{2}(l_+ s_- + l_- s_+)\right\}\left|\zeta_1^+\right> = \zeta\frac{1}{\sqrt{2}}\left\{\left|2^+\right> + \left|-2^+\right> - \left|-1^-\right>\right\}$$

$$\text{d)}\ \frac{\zeta}{\hbar^2}\left\{l_z s_z + \frac{1}{2}(l_+ s_- + l_- s_+)\right\}\left|\zeta_1^-\right> = \zeta\frac{1}{\sqrt{2}}\left\{-\left|2^-\right> - \left|-2^-\right> + \left|1^+\right>\right\} \quad \text{...(11.7)}$$

$$\text{e)}\ \frac{\zeta}{\hbar^2}\left\{l_z s_z + \frac{1}{2}(l_+ s_- + l_- s_+)\right\}\left|1^+\right> = \zeta\frac{1}{2}\left\{\left|1^+\right> + 2^-\right>\right\}$$

$$\text{f)}\ \frac{\zeta}{\hbar^2}\left\{l_z s_z + \frac{1}{2}(l_+ s_- + l_- s_+)\right\}\left|1^-\right> = \zeta\left\{-\frac{1}{2}\left|1^-\right> + \frac{\sqrt{6}}{2^\circ}\left|0^+\right>\right\}$$

The diagonal matrix elements of the perturbation operator $\frac{\zeta}{\hbar^2}(l.s)$ are readily calculated from the results collected in (11.7). Further a little scrutiny of and reflection on the results in (11.7) reveal that while the functions in (a) and (f) do not mix with any other nor between themselves under the effect of $\frac{\zeta}{\hbar^2}(l.s)$, those in (b) and (c) will mix giving offdiagonal matrix elements. So also do the functions of (d) and (e). Using the sequence of basis functions (11.7) in the order as they appear in (a) to (f), the total spin-orbit-perturbation matrix will be in block diagonal form, two submatrices of (1×I) dimension each and two others each of which is of (2×2). The total spin orbit matrix for d^1 in O_h symmetry is

$$\begin{array}{cccccc} |-1^+\rangle & |-1^-\rangle & |\zeta_1^+\rangle & |\zeta_1^-\rangle & |1^+\rangle & |1^-\rangle \end{array}$$

$$\begin{bmatrix} -\frac{\zeta}{2} & & & & & \\ & \frac{\zeta}{2} & -\frac{\zeta}{\sqrt{2}} & & & \\ & -\frac{\zeta}{\sqrt{2}} & 0 & & & \\ & & & 0 & \frac{\zeta}{\sqrt{2}} & \\ & & & \frac{\zeta}{\sqrt{2}} & \frac{\zeta}{2} & \\ & & & & & -\frac{\zeta}{2} \end{bmatrix}$$

Energy eigenvalues

$$\left.\begin{array}{l} \text{(a) } -\frac{\zeta}{2} \text{ from the 1st (1×1) matrix} \\ \text{(b \& c) } -\frac{\zeta}{2} \text{ and } \zeta \text{ from the 1st (2×2) matrix} \\ \text{(d\&c) } -\frac{\zeta}{2} \text{ and } \zeta \text{ from the 2nd (2×2) matrix} \\ \text{(f) } -\frac{\zeta}{2} \text{ from the 2nd (1×1) matrix} \end{array}\right\} \qquad \text{...........(11.8)}$$

where ζ is the spin orbit coupling constant for a 3d electron.

We can now readily solve for the eigenvectors and find the eigenfunctions. The final set appears below with their respective eigenvalues.

Eigenfunctions	Eigenvalues	
a) $\left\|-1^+\right>$		
b) $\frac{1}{\sqrt{3}}\left\|-1^-\right> + \sqrt{\frac{2}{3}}\left\|\zeta_1^+\right>$	$-\frac{\zeta}{2}$	..........(11.9a)
c) $\sqrt{\frac{2}{3}}\left\|\zeta_1^-\right> - \frac{1}{\sqrt{3}}\left\|1^+\right>$		
d) $\left\|1^-\right>$		
e) $\sqrt{\frac{2}{3}}\left\|-1^-\right> - \frac{1}{\sqrt{3}}\left\|\zeta_1^+\right>$		
f) $\sqrt{\frac{1}{3}}\left\|\zeta_1^-\right> + \sqrt{\frac{2}{3}}\left\|1^+\right>$	ζ	..........(11.9b)

It is thus seen that the six spin orbitals (11.4) forming the zero order crystal field states in O_h' field divide themselves into a fourfold degenerate linear combination set of Γ_8' symmetry and two-fold degenerate linear combination set of Γ_7' symmetry (cf. results of Sec. 8.7). The symmetries are confirmed by actual transformation properties of the eigenfunctions of (11.9a) and of (11.9b) under the group operators of O_h' double group.

Instead of the above matrix based approach, one could have first formed Γ_8' and Γ_7' - symmetry adapted wavefunctions[5] comprising linear combinations of zero-order spin orbitals (11.4). The energy eigenvalues could then be obtained by evaluating just two diagonal matrix elements of $\frac{\zeta}{\hbar^2}(l.s)$ of two wavefunctions selected, one each, from the Γ_8' and Γ_7' symmetry sets respectively.

It may be remarked that the results, i.e., the eigenvalues and eigenfunctions of d^1 in O_h environment are just replica of the results (3.12 a, b; 3.13 a,b; 3.13 c,d) for a p-electron in a free atom when the latter results are recast in the light of $(t_2d)^n - p^n$ isomorphism (Sec. 11.1). One ought to remember that in such a recasting ζ_d is to be read as ζ_p and that the operator $l \equiv -l_p$. We have not used the p^n isomorphism in our present treatment of d^1 complex because such a process would not have lessened our labour to any appreciable extent. But the labour is lightened considerably by resorting to isomorphism principle, as will be shown in the next section (11.4), when we deal with the energy levels of spin-orbit effect on d^2 octahedral complexes.

The treatment of d^1 complex of T_d symmetry is not anything different. The ground crystal field state of E symmetry is unaffected by spin orbit-interaction (Sec. 11.2). The higher t_2 orbitals will split, in a manner as does the lower T_{2g} level in O_h symmetry, by spin-orbit perturbation.

11.4. Spin-Orbit Effect in d^2 Octahedral Complexes:

The d^2 system in strong O_h field gives rise to the configurations t_{2g}^2 or the excited ones $t_{2g}^1 e_g^1$ and e_g^2. Our interest lies in the spin-orbit effect on the ground state. The crystal field terms arising from t_{2g}^2 are ${}^3T_{1g}$, ${}^1T_{2g}$, ${}^1E_g$ and ${}^1A_{2g}$, of which the last two are insensitive to spin-orbit perturbation. Energywise, again, ${}^1T_{2g}$ and ${}^1E_g$ are accidentally degenerate (Eqs. 10.19 & 10.20). Using the isomorphism of $(t_2d)^n$ with p^n, wavefunctions of t_{2g}^2, i.e., $\psi({}^3T_{1g})$may be correspondingly substituted (cf. 11.1) by the wavefunctions of p^2 configurations of free ions, i.e., by $\psi({}^3P_j)'s$.

Similarly, $\psi({}^1T_{2g})$, $\psi({}^1E_g)$are to be replaced by $\psi({}^1D_2)$ and $\psi({}^1A_{2g})$by $\psi({}^1S_0)$.

Now the spin-orbit levels of ${}^3T_{1g}$ state of t_{2g}^2 are

$$\varepsilon^{sp.or}({}^3T_{1g}) = \left\langle \psi({}^3T_{1g}) \left| \frac{\lambda}{\hbar^2} L.S \right| \psi({}^3T_{1g}) \right\rangle$$

λ = multielectron sp in-orbit coupling constant for D^2 configuration $= \frac{\zeta_{3d}}{2} = \lambda({}^3P)^{6.7}$

$$= \left\langle \psi({}^3T_{1g}) \left| \frac{\zeta_{3d}}{2\hbar^2} L.S \right| \psi({}^3T_{1g}) \right\rangle$$

$$= -\left\langle \psi({}^3P_J) \left| \frac{\lambda}{\hbar^2}({}^3P) L_p.s \right| \psi({}^3P_J) \right\rangle$$

on using the above equality and the isomorphism remembering $L= -L_p$

$$= -\frac{\lambda}{2}({}^3P)\,[J(J+1)-L(L+1)-S(S+1)] \qquad \text{..........(11.10)}$$

$$= -\varepsilon^{sp.or}({}^3P_J)$$

J = 2, 1, 0 and L and S refer to L and S values of ${}^3P$ states of p^2

On using L= 1, S= 1 and the permissible values of J in (11.10) we have

$$\left.\begin{aligned}\varepsilon_1^{sp.or}\left({}^3T_{lg}\right) &= -\lambda\left({}^3P_0\right) = -\frac{1}{2}\zeta_{3d}\\ \varepsilon_2^{sp.or}\left({}^3T_{lg}\right) &= \lambda\left({}^3P\right) = \frac{1}{2}\zeta_{3d}\\ \varepsilon_3^{sp.or}\left({}^3T_{lg}\right) &= 2\lambda\left({}^3P\right) = \zeta_{3d}\end{aligned}\right\} \quad \text{..........(11.11)}$$

The spin-orbit effect will be nil for ${}^1T_{2g}$ and ${}^1E_g$ as may be confirmed by putting in the J, L and S values in (11.10) of the isomorphic state of ${}^1D_2$. Similarly ${}^1A_{2g}$ will also remain unaffected.

External Magnetic Field Perturbations in Cubic Systems. Magnetic Quantities and Properties.

11.5. Micro- and Macro-Magnetic Quantities

We shall at first consider a few of such terms and quantities as are relevant to the discussion of magnetic properties of complexes. Some of these terms concern microsystems, such as atoms, ions and molecules and will be called here micromagnetic quantities. As opposed to this, there occur magnetic terms which pertain to macroamounts of materials of which magnetic properties are studied. These quantities are described as macromagnetic quantities. Of the micromagnetic quantities, the following will claim our attention: magnetic dipole moment μ, perturbation hamiltonian in a magnetic field $H_m^{/}$, energy levels in magnetic field $\varepsilon(B/H)$, magnetic polarisability of an atom/molecule α_m in presence of a field, first and second order Zeeman coefficients $W^{(1)}$, $W^{(2)}$.

As to the macroscopic magnetic quantities that will creep in are the magnetic polarisation M, the magnetic susceptibility X_m, the magnetic permeability in vacuum $\mu_{vac,}$, the magnetic field H, the magnetic flux density or inductions B (also called B field), the paramagnetic and the diamagnetic properties of materials. Of course, the last four H, B, paramagnetism and diamagnetism are also used in connection with microsystems, viz., atoms, molecules and ions.

A. Magnetic moment of a free atom (or ion):

The magnetic moment of an atom or ion is made up of contributions coming from orbital and spin moment terms.

(i) The orbital magnetic moment μ_l of an electron is related to its orbital angular momentum 1^* as

$\mu_l = g_l \frac{e}{2mc} 1^* = \left|\frac{e}{2mc}\right| \sqrt{l(l+1)}\hbar$ where $g_l = 1$, the absolute value of electronic charge e is to be considered when considering the magnitude of μ_l. The sign of charge is to be taken into account for ascertaining the vector direction of μ_l relative to that of 1*.

(ii) In a similar manner, spin moment s* is related to the spin magnetic moment μ_s through $\mu_s = g_s \frac{e}{2mc} s^* = 2\left|\frac{e}{2mc}\right| \sqrt{s(s+1)}\hbar$ where $g_s = 2$ and the sign of electron charge is to be considered when treating the vectorial property of μ_5 and s*.

(iii) The total magnetic moment μ_j is the vector sum of μ_l, and μ_5 and is related to the total angular (mechanical) momentum, j*, in much the same way as the former two

$$\mu_j = (\mu_1 + \mu_s) = g_j \left|\frac{e}{2mc}\right| j^* = \left\{1 + \frac{j(j+1) - l(l+1) + s(s+1)}{2j(j+1)}\right\} \left|\frac{e}{2mc}\right| \sqrt{j(j+1)}\hbar$$

where g_j called the Lande' g factor (cf. Sec. 3.3), is given by the expression within the parenthesis,

(iv) If a multielectron free atom or ion is involved, such that its wavefunctions can be expressed in (LSM_LM_S) or ($LSJM_J$) schemes, then all the foregoing relations will be valid with the only modification of substituting the quantum numbers I, s, j by capital quantum numbers L, S and J respectively relevant for multielectron atom itself and not the individual electrons.

$$\mu_J = g_J \left|\frac{e}{2mc}\right| J^* = g_J \left|\frac{e}{2mc}\right| \sqrt{J(J+1)}\,\hbar = g_J \sqrt{J(J+1)}\beta \qquad \text{...........(11.12)}$$

where $g_J = \left\{1 + \frac{J(J+1) - L(L+1) + S(S+1)}{2J(J+1)}\right\}$ and β = Bohr magneton

(v) If, however, the atom or ion be present in an external magnetic field H or a B field, the total magnetic moment does not become a measurable quantity, but its component in the field direction becomes the effective observable magnetic moment, μ_{eff}. Its magnitude is $\mu_{eff} = \mu_J \cos\theta = g_J M_J \beta$,, where θ represents one of the angles of space quantisations between the J^* vector and the field direction. M_J is one of the (2J+1) number of quantum numbers associated with different space quantisations of the J^* vector.

The relations given under (iii), (iv) and (v) are all valid only when the particular primary Russell Saunders' term undergoes a fairly strong LS coupling causing the width of the multiplet, dependent on J value, large compared to the thermal energy kT.

(vi) If the LS coupling of the primary Russell Saunders' term is weak rendering the width of the multiplet samll compared to kT, then J ceases to be a good quantum number and the magnetic moment becomes

$$\mu_{L.S} = \sqrt{L(L+1)+4S(S+1)}\beta \qquad \text{..........(11.13)}$$

B. Magnetic moment of a molecule

For this purpose we may visualise the molecular type as either

(i) a complex, such that its magnetic moment is effectively determined by that of a central ion/atom, which unlike a free ion/atom, is now present in a symmetry environment

or (ii) a species, such that its magnetic moment is made up of contributions coming, in general, from all the atoms present and not just from one atom.

Coming to B (i), since the central atom is in a crystal field, the orbital contribution to the magnetic moment may be partially or fully quenched. The orbital contribution will depend on the extent of LS coupling (Sec. 11.2) in a particular crystal field.

a) In A_1, A_2, E states in O_h/T_d symmetry, the orbital contribution to magnetic moment is nil, the total moment being determined by the "spin-only" value. This can also be easily established quantamechanically.[8]

b) In T_1 and T_2 states there is, in general, intrinsic LS interaction (Sec.11.3). This results in some contribution from the orbital motions, in addition to their spin contribution, to the total magnetic moment value.

c) For the sake of higher accuracy, if one takes into account the mixing of a higher T_1 / T_2 states with A_2 /A_1 or E states under the coupling LS operator in O_h/T_d symmetry, then there occurs a small orbital contribution to the magnetic moment calculated for the ground state of the complex.

Quantitative formulation of magnetic moments in complexes pertaining to foregoing situations (a), (b) will be given in Sec 11.8.

C. The other micromagnetic quantities, viz., magnetic perturbation hamiltonian (cf. chap. 2, Eq. 2.37), magnetic polarisability, first and second order Zeeman coefficients will make their debut in appropriate places when we deal with crystal field complexes in external magnetic field.

Macromagnetic quantities

(a) Magnetic polarisation or magnetisation, M.

It is defined as the average magnetic dipole moment per unit volume of the ma'erial. If μ be the magnetic moment of a molecule and n, the number of molecules per unit vollume then $M = n\bar{\mu}$, $\bar{\mu}$ =average magnetic moment of a molecule

(b) Magnetic susceptibility χ_v, χ_g, χ_M

The magnetic polarisation of a medium is proportional to the magnetic field H to which it is subjected. Thus for unit volume of the subatance M= χ_v H, where χ_v called the volume susceptibility, is the proportinality constant. Of course, of more importance than χ_v is the specific susceptibility, $\chi_g = \chi_v / p$ where p = density. For one gm. mole of a substance of molecular weight, M', there is defined another susceptibility term, viz., the molar susceptibility

$\chi_M = M' \chi_g = M' \chi_v / p = V_M \chi_v$ where V_M is the molar volume.

(c) Magnetic permeability of a medium and Flux density

When a H field is switched on, it generates due to polarisation of the medium an induction or flux density B at any point within the medium. This flux density is proportional to H in general and related to it as

B= $\mu_{med} H$, where μ_{med} is the magnetic permeadibility of the medium. For vacuum B = $\mu_{vac} H$. If the medium be such that it develops a magnetic polarization M and χ_v is its susceptibility

$$B = H + 4\pi M = (1 + 4\pi \chi_v) H = \mu_{med} H$$

where it is seen $\mu_{med} = 1 + 4\pi \chi_v$

If we consider a molecule, the space within it will mostly be vacumn. So that a B field or magnetic flux within a molecule due to magnetic field H is

B $\approx \mu_{vac} H$, where the permeability within the molecule is the same as it is outside, viz. μ_{vac}

(d) Paramagnetism and Diamagnetism:

In general, materials in all phases - solid, liquid or gas - behave in either a magnetophobic or magnetophilic way when exposed to an external magnetic field H. Some materials, on exposure to such a field, try to draw themselves away from the stronger parts of the field to its weaker regions. Thus these materials intrinsically reduce as much as possible the ingress of the lines of force through their bulk and help to decrease the effective magnetic field at any point within the bulk. These are called diamagnetic materials. Examples are sand, water, marble and others where the electrons in their molecules or crystals are all spin-paired.

As opposed to these, are there substances, which on exposure to external magnetic field, try to lodge themselves in positions where the magnetic field is strong. These, called paramagnetic materials, permit a larger flux density within them increasing the effective magnetic field at any point inside. Illustrations are nitric oxide, oxygen and many coordination compounds. The paramagnetic compounds are characterised by molecules or complex ions having one or more unpaired electrons.

The diamagnetic or paramagnetic properties, defined above qualitatively, may be regarded as macro-magnetic behaviour. But such behaviour is ultimately attributable to their diamagnetic or paramagnetic response at micro-level as well, viz., the presence or absence of all spin paired electrons in the molecules.

It can be shown, by employing the principles of electromagnetic'induction at the elementary electronic circuit level, that the motions [orbital and spin (?)] of both the electrons of a spin-paired unit will be accelerated in directions setting up new magnetic fields opposing any externally applied field. Even if we discard the naive idea of spin as a motion, then there would still be the negative contribution from the orbital motions with the coupled spin playing no such direct roles. Thus one would expect an intrinsic diamagnetic character of materials, the molecules of which contain spin-paired duplets of electrons.

But the orbital magnetic field and spin magnetic moment of an unpaired electron in each molecule of a compound undergo preferential orientation in the presence of an external magnetic field. This augments the effective magnetic field at the molecule. Of course, toeing this line of argument one can imagine plenty of diamagnetic contributions by plethora of spin-paired electrons already present in the molecule of which the spin-unpaired electron or electrons only donate to paramegnetism. It is thus a net effect of one or **a** few versus many. But the paramagnetic contribution by a single unpaired electron is several folds of power of ten times the diamagnetic contribution by a spin-paired unit. Hence is the resultant paramagnetic character of the molecules containing one or a few unpaired electrons each.

Thus a study of magnetic susceptibilities is essentially a study of paramagnetic and diamagnetic susceptibilities. Paramagnetic susceptibility is temperature dependent (Sec. 11.6) whereas diamagnetic susceptibility is independent of temperature.

11.6. General Forms of Perturbational Energies in External Magnetic Field:

Consider a molecule with a nondegenerate wavefunction ψ_m^0 and energy ε_m^0 in the unperturbed state. When an external magnetic field H is applied causing a magnetic flux B within the molecule, the same as in vacuum, the perturbation operator (cf. Eq. 2.37).

$H_m^{/} = \frac{\beta}{\hbar}(L + 2S).B = \frac{\beta}{\hbar}(L_z + 2S_z)B$ when the B field is along the Z-direction.

The expression for the total energy of the nondegenerate system is

$$\varepsilon_m = \varepsilon_m^0 + \langle \psi_m^0 | H_m^{/} | \psi_m^0 \rangle + \sum_n^{/} \frac{\langle \psi_m^0 | H_m^{/} | \psi_n^0 \rangle \langle \psi_n^0 | H_m^{/} | \psi_m^0 \rangle}{\varepsilon_m^0 - \varepsilon_n^0} \quad \text{..........(11.14)}$$

But with $H_m^{/} = -\hat{\mu}.B = \frac{\beta}{\hbar}(L_z + 2S_z)B$ .with $\hat{\mu} = -\frac{\beta}{\hbar}(L_z + 2S_z)$

$$\varepsilon_m = \varepsilon_m^0 - \langle m | \hat{\mu} | m \rangle B + \sum_n^{/} \frac{\langle m | \hat{\mu} | n \rangle \langle n | \hat{\mu} | m \rangle B^2}{\varepsilon_m^0 - \varepsilon_n^0} \quad \text{..........(11.15)}$$

where the kets, $|m>, |n>$ represent the unperturbed wavefunction ψ_m^0, ψ_n^0 and the B field is parallel to Z-axis.

For the magnetic perturbation of a degenerate state, we have to build up the perturbation matrix of the operator $-\mu.B$ with the degenerate basis functions.

Applying Taylor's expansion to ε_m as a function of B first and then using Hellmann-Feynman[9] theorem for the variation of ε_m with the qth (q=x, y, z coordinates) component of the B field in the form

$$\frac{\partial \varepsilon_m}{\partial B_q} = \left\langle \frac{\partial H}{\partial B_q} \right\rangle = \left\langle \frac{\partial}{\partial B_q}\left(H^0 - \mu_m.B\right) \right\rangle = -\langle \mu_{mq} \rangle \quad \text{..........(11.16)}$$

we can finally end up in two general definitions (11.17 & 11.18) and two general relations (11.19 & 11.20), viz.

μ_{mq}^0 = qth component of magnetic moment in the state $|m>$, i.e.

$$\text{the unperturbed state at zero field} = -\left\{\frac{\partial \varepsilon_m}{\partial B_q}\right\}_0 \quad \text{..........(11.17)}$$

$\alpha_{mqq'} = (qq')$ tensor component of magnetic polarisability α_m

$$\text{In the } |m> \text{ state} = \left\{\frac{\partial^2 \varepsilon m}{\partial B_q . \partial B_{q'}}\right\}_0 \quad \text{..........(11.18)}$$

$$\mu_m = \mu_m^0 + \alpha_m.B \quad \text{..........(11.19)}$$

$$\varepsilon_m = \varepsilon_m^0 - \mu_m^0.B - \frac{1}{2}\alpha_m : BB = \varepsilon_m^0 + \varepsilon^{(m)} \quad \text{..........(11.20)}$$

The distinctions in the definitions and significances of $\langle \mu_{mq} \rangle$ and μ_{mq}^0 of Eqs. (11.16 and 11.17) are to be noted.

Comparing Eqs. (11.15 and 11.20), we can write, when B is parallel to Z axis.

$$\mu_m^0 = \langle m|\hat{\mu}|m\rangle = -\frac{\beta}{\hbar}\langle m|(L_z + 2S_z)|m\rangle = -W_m^{(1)} \text{ (say)} \qquad \text{..........(11.21)}$$

$$-\alpha_m = 2\sum_n' \frac{\langle m|\hat{\mu}|n\rangle\langle n|\hat{\mu}|m\rangle}{\varepsilon_m^0 - \varepsilon_n^0} = 2W_m^{(2)} \text{ (say)} \qquad \text{..........(11.22)}$$

where $W_m^{(1)}$ and $W_m^{(2)}$ are the first and second order Zeeman coefficients in the $|m>$ state. The perturbation energy is, therefore, given by

$$\varepsilon^{(m)} = W_m^{(1)}B + W_m^{(2)}B^2 \qquad \text{..........(11.23)}$$

and that μ_m of Eq. (11.19) can be written as, since $\varepsilon_m = \varepsilon_m^0 + \varepsilon^{(m)}$,

$$\mu_m = -W_m^{(1)} - 2W_m^{(2)}B = -\frac{\partial \varepsilon^m}{\partial B} \qquad \text{..........(11.24)}$$

In the case of degenerate system, the contributions to second order Zeeman energies come from the off-diagonal terms. The net second order correction can usually be read off from the form of the final perturbation energy.

11.7. Magnetic Field Effect along with Spin Orbit Perturbation in d^1 and d^2 Systems in O_h Crystal Field:

Magnetic susceptibility, which is related to average magnetic moment $\overline{\mu}$ of an atom, ion or molecule (e.g., central ion in a complexes), is always measured in the presence of an external magnetic field. Due to Boltzmann population of the energy levels, the magnetic moment, which is calculated from χ_M, is always an average representing contributions from the mth level and the adjacent populated levels. Since μ_m, in a general sense, is related to the variation of energy with B (Eq. 11.16), we need to consider the total magnetic interaction energy, comprising both the spin-orbit and Zeeman energy, of any crystal field state for the calculation of μ_m and thus finally of $\overline{\mu}$ and of χ_m

Now, of the different ground levels possible for various d^n systems in O_h field, A_1, A_2 and E do not have any spin-orbit perturbation effect as has been stated and proved earlier [Sec. 11.2 (a)].

For the present calculation, the unperturbed state of the central ion of the complex is taken to be the coulomb field interaction and crystal field interaction eigenfunction. This can be expressed either in (LSM_LM_S) or (LSJM,) schemes. If expressed in (LSM_LM_S) scheme, the energy of the unperturbed state will be independent of M_L and M_s. Moreover, since the L_z part of the magnetic field hamiltonian H_m will have no matrix element for any of the $^{2S+1}A_1$,$^{2S+1}A_1$ or $^{2S+1}A_2$ ^{2S+1}E symmetry states of d^n complexes [Sec.11.5 B(a)], the kets $|LSM_LM_s>$ with different values of M_s for any given L, S and M_L should be chosen as the basis eigenfunctions for the calculation of the magnetic field effect of the above mentioned symmetry states. In such a situation the matrix of the magnetic field operator will be diagonal for

$$\left\langle LSM_LM_s^{/}\left|\frac{\beta}{\hbar}2S_z\right|LSM_LM_S\right\rangle B = 2\beta M_S B\delta_{M_S,M_s^{/}} \qquad \text{..........(11.25)}$$

Hence the Zeeman energy levels are $2\beta M_s B$'s, relative to the initial (unperturbed) level equal to zero energy, with M_s taking its permissible values $-S \leq Ms \leq S$.

We now turn to the total magnetic perturbation (including spin orbit) of $^{2s+1}T_{2g}$ and $^{2s+1}T_{1g}$ octahedral crystal field states which constitute the ground states of d^1 and of d^2 (also d^3) complexes respectively.

(a) d^1-complex with octahedral ground states $^{2s+1}T_{2g}$ (i.e., $^2T_{2g}$)

The spin-orbit eigenfunctions (11.9a, 11.9b) of d^1 octahedral complex in $^2T_{2g}$ states are also eigenfunctions of the unperturbed initial state hamiltonian H_o + $h_{crys}(O_h)$. These functions may thus constitute a new basis set for constructing the perturbation matrix of both $\xi(r)$l.s and $\frac{\beta}{\hbar}(l_z+2s_z)$B that are present in the total magnetic hamiltonian part. While it is apparent that $\xi(r)l.s$ will be diagonal in this basis set, the second operator part is not so. A little trial exercise shows that $\frac{\beta}{\hbar}(l_z+2s_z)B$ will have off diagonal matrix elements between the pair (b) and (e) and between the functions (c) and (f) of the relations (11.9a and 11.9b). The functions (a) and (d) will remain separate from one another and also from the rest.

We thus rearrange the basis functions in the order (a), (d), (b), (e), (c) and (f) representing the ordinal serial order 1 to 6 respectively in constructing the matrix. The matrix which is formed easily and in which the eigenvalues of (11.9a and 11.9b) are also incorporated is block-diagonal comprising two (1×1) and two (2×2) submatrices.

$$\begin{bmatrix} -\frac{\zeta}{2} & & & & & \\ & -\frac{\zeta}{2} & & & & \\ & & -\frac{\zeta}{2} & -\sqrt{2}\beta B & & \\ & & -\sqrt{2}\beta B & \zeta-\beta B & & \\ & & & & -\frac{\zeta}{2} & -\sqrt{2}\beta B \\ & & & & -\sqrt{2}\beta B & \zeta+\beta B \end{bmatrix} \quad \ldots\ldots\ldots\ldots(11.26)$$

The general solutions for wavefunctions[10,11] formed by mixing a pair or more of component functions and their eigenvalues can be easily written down if the matrix elements be known. Thus relative to the $^2T_{2g}$ level as the reference zero of energy the energy eigenvalues obtainable from the matrix (11.26) are

$$\left.\begin{array}{ll} \text{i)}\ -\frac{\zeta}{2} & \\ \text{ii)}\ -\frac{\zeta}{2} & \text{iv)}\ \zeta-\beta B+\frac{4}{3}\frac{\beta^2B^2}{\zeta} \\ \text{iii)}\ -\frac{\zeta}{2}-\frac{4}{3}\frac{\beta^2B^2}{\zeta} & \text{v)}\ \zeta+\beta B+\frac{4}{3}\beta^2B^2 \\ \text{vi)}\ -\frac{\zeta}{2}+\frac{4}{3}\frac{\beta^2B^2}{\zeta} & \end{array}\right\} \quad \ldots\ldots\ldots\ldots(11.27)$$

The values are such that energy corrections upto the first and the second orders are evident. The energy values, collected here, will be used in connection with the calculation of magnetic susceptibility of a material consisting of octahedral d^1 complex molecules.

(b) d^2-Octahedral Complex with ground $^{2s+1}T_{1g}$ states

The calculations for d^2-O_h complex under the joint perturbation of $\frac{\lambda}{\hbar^2}$L .S and $\frac{\beta}{\hbar}(L_z+2S_z)B$ are made by having recourse to $(t_2d)^n$-p^n isomorphism of Secs (11.1 & 11.4). We have already seen that $^3T_{1g}$, $^1T_{2g}$, ^1E_g and $^1A_{2g}$ are amenable to substitutions by the terms $^3P_{2,1,0}$, $^1D_2$ and $^1S_0$ of the p^2 free ion configuration for such calculations.

The perturbation energies due to $\left(\frac{\lambda}{\hbar^2}L.S\right)$ Eq.(11.11) being already available, the additional contributions due to $\frac{\beta}{\hbar}(L_z + 2S_z)$ can be gathered from the matrix elements noted below wherein the ψ functions are those of (p^2) configurations corresponding to the triplet and the singlet states

$$\left\langle \psi\left({}^3P_1\right) \middle| \left\{\frac{\beta}{\hbar}(L_z + 2S_z)B\right\} \middle| \psi\left({}^3P_J\right)\right\rangle J = 2,\ 1,\ 0$$

$$\left\langle \psi\left({}^1D_2\right) \middle| \left\{\frac{\beta}{\hbar}(L_z + 2S_z)B\right\} \middle| \psi\left({}^1D_2\right)\right\rangle J = 2$$

$$\left\langle \psi\left({}^1S_0\right) \middle| \left\{\frac{\beta}{\hbar}(L_z + 2S_z)B\right\} \middle| \psi\left({}^1S_0\right)\right\rangle J = 0$$

Let us first confine our attention to the ^3P_J, states involved in the foregoing matrices. This state in O_h crystal field remains unsplit. Hence, the electrostatic cum spin-orbit perturbed eigenfunctions (nine in all) recorded in Table $T_{4.7}$ for the $\psi\left({}^3P_J\right)$'s of (p^2) will also be eigenfunctions in the O_h crystal field. Using three basis sets separately, comprising five $\psi\left({}^3P_2\right)$'s, three $\psi\left({}^3P_1\right)$'s, and one $\psi\left({}^3P_0\right)$ respectively of Table $T_{4.7}$, all expressed in two-electron dets of single-electron microket elements in (lsm_lm_s) scheme, one can evaluate all the matrix elements of $\frac{\beta}{\hbar}(L_z + 2S_z)$B recalling that $L_z = \sum_i l_{zi}$ and $S_z = \sum_i s_{zi}$. The results compiled in Sec. 4.5 thus become relevant.

Similar results can be worked out for the $^1D_2$ and ^1S_o terms. Of course, it should be remembered, while calculating the results for $^1D_2$, that the latter splits in the octahedral field. It is, therefore, necessary to linearly combine the five basis functions into two groups of two and three functions respectively having the transformation properties of the 1R, E_g, and the IR, T_{2g}. These two sets of basis functions are then separately used before evaluating the perturbation effect of the external magnetic field operator.

11.8. Magnetic Susceptibility and Magnetic Moments in Crystal Fields (O_h, T_d):

Experimentally we determine magnetic susceptibility χ_M in an applied magnetic field H. Theoretically, therefore, we need to find macroscopic magnetic polarisation, M, in terms of ensemble average of the magnetic moments of molecules taking into account the Boltzmann population of the other possible neighbouring states besides the mainly contributing ground state. The expression for χ_M follows immediately. Our interest in magnetic susceptibility and magnetic moment originates from the roles these quantities play in the structural aspects of metal ion complexes and in their paramagnetic and diamagnetic behaviour.

The molar susceptibility,

$$\begin{aligned}\chi_M &= M^{/}\chi_g = M^{/}\chi_v/\rho \\ &= \chi_v V_m = \frac{M}{H}V_m \\ &= \frac{1}{H}(n_1\mu_1 + n_2\mu_2 + \cdots) \\ &= \frac{N\bar{\mu}}{H} \qquad \text{..........(11.28)}\end{aligned}$$

where N=Avogadro number = $(n_1 + n_2 + n_3 \ldots)$i.e., the respective number of molecules with magnetic moments $\mu_1,\ \mu_2,\ \mu_3$ etc. that are present within the molar volume V_m of the material with molecular weight $M^{/}$

Now,

$$N\bar{\mu} = N\frac{\sum_m \mu_m e^{-\varepsilon_m/kT}}{\sum_m e^{-\varepsilon_m/kT}} = N.\frac{\sum_m \mu_m e^{-\varepsilon_m^0/kT}e^{-\varepsilon^{(m)}/kT}}{\sum_m e^{-\varepsilon_m^0/kT}e^{-\varepsilon^{(m)}/kT}}$$

Using Eqs. (11.19 & 11.20) and the fact that kT is large compared to the width of the magnetically split levels, viz,. $\varepsilon^{(m)} << kT$, we have, after expansion and suitable neglect of terms,

$$N\bar{\mu} = N\frac{\sum_m\left(\mu_m^0 + \alpha_m.B + \mu_m^0.\frac{B}{kT+\cdots}\right)e^{-\varepsilon_m^0/kT}}{\sum_m\left(1+\mu_m^0.\frac{B}{kT}\right)e^{-\varepsilon_m^0/kT}} \qquad \text{..........(11.29)}$$

Since, $-\varepsilon^{(m)} = \mu_m^0.B + \frac{1}{2}\alpha_m : BB$ by (11.20)

The molecular partition function $Z = \sum_m e^{-\varepsilon_m^0/kT}$ (in field-free-state), occurs in the denominator of (11.29). To simplify Eq.(11.29), we examine both the denominator and the numerator separately.

$$\text{Denominator} = Z + \frac{Z\sum_m\left(e^{-\varepsilon_m^0/kT}\mu_m^0\right).B}{Z}$$

The last part is ZB times the ensemble average of μ_m^0 in the absence of the field. Since in a fluid, the molecules tumble over in all directions with equal probability in the absence of any imposed field, the ensemble average is nil. This leaves only Z in the denominator.

Numerator: Since the partition function Z occurs in the denominator the first term in the numerator vanishes for reasons as stated above. The second and the third terms, however, remain. If the applied field coincides with the Z-direction, the Eq. (11.29) will assume an effective form

$$N\overline{\mu} = \frac{NB\sum_m \alpha_{mzz} e^{-\varepsilon_m^0/kT}}{z} + \frac{NB}{kT}\frac{\sum_m \mu_{mz}^0\ \mu_{mz}^0 e^{-\varepsilon_m^0/kT}}{z} \quad \text{..........(11.30)}$$

where α_{mzz} is the (ZZ)th component of the magnetic polarisability tensor α_m in the mth state i.e., $|m>$of the molecule. The averages of the quantities α_{mZZ} and $\mu_{mZ}^0\ \mu_{mZ}^0$ for freely tumbling molecules are[12]

$$\overline{\alpha}_{mzz} = \frac{1}{3}\left(\alpha_{mxx} + \alpha_{myy} + \alpha_{mzz}\right) = \alpha_m$$

and $\quad \mu_{mz}^0\ \mu_{mz}^0 = \left(\mu_{mz}^0\right)^2 = \frac{1}{3}\mu_m^{0^2} = \frac{1}{3}\mu^2$

Thus Eq. (11.30) can be in a form with μ representing magnetic moment in the absence of any field.

$N\overline{\mu} = N\left(\alpha_m + \dfrac{\mu^2}{3kT}\right)B,$ Av. magnetic moment in the $|m>$ state in the presence of field B applied along the z-direction ....(11.31)

This gives the expression for molar magnetic moment. The molar magnetic suspceptibility

$$\chi_M = \frac{N\overline{\mu}}{H} = N\left(\alpha_m + \frac{\mu^2}{3kT}\right)\mu_{vac,} \text{ since } B = \mu_{vac}H, \quad \text{..........(11.32)}$$

While Eqs. (11.31 & 11.32) are in general valid for paramagnetic substances, we can have useful relations for ions in O_h field (or T_d field) having the crystal field term $^{2s+1}A_1$, $^{2s+1}A_2$, or ^{2s+1}E. We can advantageously use the energy expression (11.25) to find an alternative form of N$\overline{\mu}$.

$$\text{Thus N } \overline{\mu} = \frac{N\sum_m \mu_m e^{-\varepsilon_m/kT}}{\sum_m e^{-\varepsilon_m/kT}} \text{ and since } \mu_m = -\frac{\partial \varepsilon_m}{\partial B} = -\frac{\partial \varepsilon^{(m)}}{\partial B} = 2\beta M_s$$

(cf. Eqs. 11.24 & 11.25) where we have confined ourselves to first order Zeeman energies,

$$N\overline{\mu} = \frac{N\sum_{Ms} 2\beta M_s e^{-2\beta M_s B/KT}}{\sum_{Ms} e^{-2\beta M_s B/kT}} = \frac{N\sum_{Ms} 2\beta M_s\left(l - 2\frac{\beta Ms}{kT}\right)}{\sum_{Ms}\left(l - \frac{2\beta Ms}{kT}\right)}$$

In writing the above expressions, the summation over states, m, been taken over the effective Boltzmann populated levels with different Ms values lying adjacent to the unperturbed level after switching on the external magnetic field. Furthermore, the approximation $M_s\beta \langle\langle\, kT$ has been made. It may also be noted that this summation over M_s is not over the unperturbed excited levels as are implied in a full perturabation correction.

Now since $\sum_{Ms} Ms = 0, \sum_{Ms} 1 = 2S+1$ and $\sum_{Ms} M_s^2 = \frac{1}{3} S(S+1)(S+2)$,

The foregoing expression finally leads to

$$N\ddot{\mu} = \frac{N\beta^2}{3kT} 4S(S+1)B \qquad \text{..........(11.33)}$$

Comparing (11.33) with (11.31), one finally gets for paramagnetic complexes in the foregoing crystal field states, since $\alpha_m \langle\langle\, \mu^2/3kT$ for such molecules.

$$\mu = 2\{S(S+1)\}^{\frac{1}{2}} \text{ in units of } \beta \text{ (cf. Sec. 11.5B(a)} \qquad \text{..........(11.34)}$$

and the paramagnetic susceptibility (cf. Eq. 11.32)

$$\chi_m = \frac{N\beta^2}{3kT} 4S(S+1).\mu_{vac}$$

$$= \frac{N\beta^2}{3kT} 4S(S+1) \text{in esu Gaussion units in which } \mu_{vac} = 1 \qquad \text{..........(11.35)}$$

These relations (11.34) and (11.35) are also valid for d^1 complex in T_d field where the ground state wavefunction is symmetry E. The magnetic moments in such complexes has the "spin-only" value (11.34).

The calculations of magnetic moments and susceptibilities in the $^{2s+1}T_1$ and $^{2s+1}T_2$ in O_h or T_d crystal field states are not easy. We shall handle here the ground state. $^{2s+1}T_{2g}$, or d^1 complex in O_h environment. Employing total magnetic perturbation energy values, comprising spin-orbit and external magnetic field contributions (Eg. 11.27) in the six populated levels $^2T_{2g}$ state the general relation,

$$N\,\ddot{\mu} = \frac{N\sum_m \mu_m e^{-\varepsilon_m/kT}}{\sum_m e^{-\varepsilon_m/kT}}, \text{ where } \mu m = -\frac{\partial \varepsilon_m}{\partial B}$$

yields, under the condition $\beta B \langle\langle\ kT$, the following numerator and the denominator[13]

$$\text{Numerator} = N\left[\frac{8}{3}\frac{\beta^2 B}{\zeta}e^{\frac{\zeta}{2kT}} + \left(\beta - \frac{8}{3}\frac{\beta^2 B}{\zeta}\right)e^{-\frac{(\zeta+\beta B)}{kT}} + \frac{8}{3}\frac{\beta^2 B}{\zeta}e^{\frac{\zeta}{2kT}} + \left(-\beta - \frac{8}{3}\frac{\beta^2 B}{\zeta}\right)e^{\frac{(-\zeta+\beta B)}{kT}}\right] \quad \text{........(11.36)}$$

$$\text{and the denominator} = \left(e^{\frac{\zeta}{2kT}} + e^{\frac{\zeta}{2kT}} + e^{\frac{\zeta}{2kT}} + e^{-\frac{\zeta}{kT}} + e^{\frac{\zeta}{2kT}} + e^{-\frac{\zeta}{kT}}\right) \quad \text{........(11.37)}$$

Using (11.36) and (11.37) in the expression for $N\mu$ and after some laborious algebraic manipulation, one gets finally,

$$N\overline{\mu} = \frac{N\beta^2 B}{3kT}\frac{(3x - 8 + 8e^{\frac{3x}{2}})}{x\left(1 + 2e^{\frac{3x}{2}}\right)} \quad \text{........(11.38)}$$

where $x = \frac{\zeta}{kT}$ and

$$\frac{N\ddot{\mu}}{H} = \chi_M = \frac{N\beta^2}{3kT}\frac{(3x - 8 + 8e^{\frac{3x}{2}})}{x\left(1 + 2e^{\frac{3x}{2}}\right)}\cdot\mu_{vac} \quad \text{........11.39}$$

$\mu_{vac} = 1$ (CGS Gaussian unit)

The calculations of magnetic moments and susceptibilities of higher d^n octahedral complex in their $^{2s+1}T_{1g}$ or $^{2s+1}T_{2g}$ states are done by having recourse to P^n isomorphism. The full treatment is complicated. Only the result is stated here.

$$\chi_M = \frac{\sum_J (2J+1)\chi_J e^{-\frac{vJ(J+1)}{2kT}}}{\sum_J (2J+1)e^{-\frac{vJ(J+1)}{2kT}}} \quad \text{..........(11.40)}$$

$$\text{where } \chi_J = \frac{N\beta^2}{3KT}\left\{A + \frac{CkT}{\zeta}\right\} \quad \text{..........(11.41)}$$

The values of ν, A and C of (11.40) and (11.41) are recorded for some of the octahedral d^n ion complexes in their ${}^{2s+1}T_{1g}$ or ${}^{2s+1}T_{2g}$ states for different values of J in the corresponding Russell Saunders' states of the isomorphic p^n configuration of the ions.[14]

Before we turn to the table (next page), we should not lose sight of a basic stipulation underlying the simplified formulations of the effect of an external magnetic field on splitting of some parent levels. All our simple-looking formulations, dealt with previously, in respect of free atoms (chapters 3.5), are valid subject to the condition, viz. $M_s\beta << kT$ holding good. This stipulation infects also our present treatment of complexes of d^n ions, described by their crystal field states. For free ions, there exist treatments of cases like $Ms\beta \sim kT$ or even $Ms\beta > kT$. Our formulations, therefore, tend to be less and less valid with decrease and further decrease in temperature.

Table $T_{11.1}$

Values of v, A and C [Eqs. 11.40 and 11.41] of d^n –octahedral complexes corresponding to J values in p^n isomophim.

Configuration	Crystal field state	J	ν	A	C
$d_1(t_{2g}^1)$	${}^2T_{2g}$	1/2	$-\zeta$	3	-8
		3/2	$-\zeta$	0	4
$d^2(t_{2g}^2)$	${}^3T_{1g}$	0	$-\zeta/2$	0	-72
		1	$-\zeta/2$	1/2	9
		2	$-\zeta/2$	3/2	9
$d^2({}^3F)$	${}^3T_{1g}$	0	$-3\zeta/4$	0	-196/3
		1	$-3\zeta/4$	1/8	49/6
		2	$-3\zeta/4$	3/8	49/6
$d^4(t_{2g}^4)$	${}^3T_{1g}$	0	$\zeta/2$	0	72
		1	$\zeta/2$	1/2	-9
		2	$\zeta/2$	3/2	-9
$d^5(t_{2g}^5)$	${}^3T_{2g}$	1/2	ζ	3	8
		3/2	ζ	0	-4
$d^6(t_{2g}^4e_g^2)$	${}^5T_{2g}$	1	$\zeta/4$	49/2	54
		2	$\zeta/4$	27/2	-10
		3	$\zeta/4$	12	-16
$d^7(t_{2g}^5e_g^2)$	${}^4T_{1g}$	1/2	$\zeta/3$	12	60
		3/2	$\zeta/3$	27/5	-264/25
		5/2	$\zeta/3$	28/5	-324/25

11.9 Diamagnetic and Temperature Independent Paramagnetic Susceptibilities

The dependence of paramagnetic χ_M as an inverse function of temperature is evident both from Eqs. (11.32, 11.38) which is also an incidental confirmation of the Curie Law. However, α_m of Eq. (11.32) of which a (restricted) form is provided there, viz. $-2\sum_n^{\prime}\frac{\langle m|\mu|n\rangle\langle n|\mu|m\rangle}{\varepsilon_m^0-\varepsilon_n^0}$, shows a temperature-independent contribution to total paramagnetic susceptibility arising from interaction of the given state with the excited states under the main magnetic perturbation operator, $\frac{\beta}{\hbar}(L_z+2S_z)B$. This is a second order effect and has magnitude of the order $\sim+10^{-6}$ per gm-mole of the substance.

For systems with total spin S=0, the full contribution to magnetic susceptibility comes from the second order effect of both $\beta/\hbar(L_z+2S_z)B$ and $\frac{e^2}{8mc^2}\sum_j\left(x_j^2+y_j^2\right)B^2$ of the perturbation operator of Eq. (2.37). The second order energy correction due to magnetic perturbation is

$$\varepsilon_m^{(2)}=\frac{e^2B^2}{8mc^2}\left\langle m\left|\sum_j\left(x_j^2+y_j^2\right)\right|m\right\rangle+\left(\frac{\beta}{\hbar}\right)^2\sum_n^{\prime}\frac{\langle m|l_z|n\rangle\langle n|l_z|m\rangle}{\varepsilon_m^0-\varepsilon_n^0}$$

This, when employed to find the ensemble average $\bar{\mu}$ as in sec. 11.8, enables us finally to equate

$$\alpha_{zm}=-\frac{e^2}{4mc^2}\left\langle m\left|\sum_j\left(x_j^2+y_j^2\right)\right|m\right\rangle\mp2\left(\frac{\beta}{\hbar}\right)^2\sum_n^{\prime}\frac{\langle m|l_z|n\rangle\langle n|l_z|m\rangle}{\varepsilon_m^0-\varepsilon_m^0}\qquad\text{..........(11.42)}$$

On averaging, for freely tumbling molecules, the above over equivalent x, y and z directions, the expression for χ_M becomes

$$\chi_M=N\alpha_m=-N\left(\frac{e^2}{6mc^2}\right)\langle r^2\rangle\mu_{vac}+N\frac{e^2}{6m^2c^2}\sum_n^{\prime}\frac{l_{mn}^{\prime}\,l_{nm}^{\prime}}{\varepsilon_n^0-\varepsilon_m^0}\mu_{vac}\qquad\text{..........(11.43)}$$

$$=\chi_M^d+\chi_M^P\ (\mu_{vac}=1\text{ Gaussian unit})$$

where $l_{mn}^{\prime}$ etc. represents the values of the corresponding integrals. The diamagnetic susceptibility, χ_M^d, is a always negative (~ - 10^{-6}) and temperature independent. As pointed out, χ_M^P part of paramagnetic susceptibility is also independent of temperature and positive, wherever it exists. The last qualifying phrase is occasioned by the fact that in atomic systems χ_M^P vanishes.

As magnetic moments are derived from experimental measurements of susceptibilities, we conclude our discussion by providing a table in which experimentally observed and the theoretically calculated "spin-only" magnetic moments of a number of transition metal ions are listed". For a system with 1,2,3 ... unpaired electron(s), the "spin-only" values of magnetic moments μ_s, (cf. Eq. 11.34) are

Unpaired electrons	μ_s (in units of β)	Unpaired electrons	μ_s (in units of β
1	1.73	4	4.9
2	2.83	5	5.92
3	3.87	6	6.93
		7	7.94

Table $T_{11.2}$

Magnetic Moments of Transition Metal Ions (in units of β)

Ion	**d^n system**	**μ (observed)**	**μ_s (calculated for the number of unpaired electrons commensurate with the ground spectral state)**
V^{4+}	d^1	1.7 -1.8	1.73(1)
V^{3+}	d^2	2.6-2.8	2.83(2)
Ni^{2+}	d^8	2.8-4.0	2.83(2)
Cr^{3+}	d^3	~3.8	3.87(3)
Co^{3+}	d^6	~5.4	4.90(4)
Mn^{2+}	d^5	~5.9	5.92(5)
Cu^{2+}	d^9	1.7-2.2	1.73(1)

The slight departures of the calculated μ_s from experimental μ are to be accounted for by contributions (positive or negative) from the not-fully quenched orbital moments, spin-orbit interactions and the temperature independent paramagnetism. The partially quenched paramagnetic orbital moment may act, depending on systems, either in the same or in opposition to spin paramagnetism. The figures within the parentheses in the last column indicate the number of unpaired electrons considered in calculating μ_s.

11.10 Worked out Examples

We have already mentioned about quenching of orbital moments. A complex ion having unpaired electrons may have its orbital moments fully or partially quenched. In an O_h or T_d environment when the t_{2g} level is half-filled or fully so, a complete orbital quenching in t_{2g} orbital is expected. This is also true for half-filled or completely filled e_g orbital. Thus $t_{2g}^3, t_{2g}^3e_g^2, t_{2g}^3e_g^4, t_{2g}^6e_g^2, t_{2g}^6e_g^4, e_g^2, e_g^4$ configurations have their orbital moments fully suppressed and their magnetic moments are made up of the "spin-only" values. In cases, where deviations from the "spin-only" values are observed in such configurations, the extra contribution comes from magnetic moments associated either with the spin-orbit coupling of ground and upper state (Sec.8.8) or with the external field-induced mixing of ground and unpopulated state (cf. TIP Sec. 11.9 and also Sec. 8.8).

For the other "not-fully filled" configurations, the orbital moments are partially quenched. The remaining fraction of the unquenched moment interacts with the spin-moments of the relevant electrons much like in the manner of spin-orbit interaction. This contributes some value, though small, to the "spin-only" value. This contribution may be eitherway, i.e., in a positive or a negative sense. Finally, there may also occur TIP contribution for such systems.

Example I.

Praseodymium ion, $Pr^{3+}$with a configuration $4f^2\ 5s^2p^6$, present in an O_h site symmetry, has a measured value of 3.47 BM for its magnetic moment. How far is it supported by calculation?

Since Pr^{3+} is an f^2 system, its free-ion ground state is $^3H_4$ with the spins parallel, the f-shell being less than half-filted As it is present in O_h surrounding, its orbital moment is not to be expected to be fully quenched. Hence its magnetic moment is apparently made up of contribution coming mainly from spin moments and partly from the unquenced orbital moments.

In such an event,

$$\mu_{eff} = \left\{2\sqrt{s(s+1)} + C\right\} = (2.83 + C)BM$$

which leaves C to assume a value (3.47-2.83), i.e., 0.64 BM resulting from unquenched orbital moment. It is a little bit too high for the usual unquenched moments.

Another possibility is that, the f-electrons of Pr^{3+} central ion are effectively shielded by the 5s and p-electrons from the distortion effect of the ligands of the complex. It will leave Pr^{3+} to effectively behave as a free-ion in $^3H_4$ state having a magnetic moment mainly contributed by spin-orbit interaction. Thus

$\mu_{eff} = g\sqrt{J(J+1)} + BM$, where J = 4 and

g (Lande g-factor) $= 1+\{J(J+1)-L(L+1)+S(S+1)\}/2J(J+1)$

$= 4/5(L=5, S=1)$

$\mu_{eff} = 4.47\times4/5 = 3.57$ BM which is rather close to the measured value.

The second possibility appears more plausible and, in fact, most of the rare earth complex ions behave as if, they are in free-ion states.

Example 2.

Ti^{3+}, a d^1 ion present in O_h site symmetry, has a coupled ground state due to spin-orbit interaction contributing a value of $4\zeta/10Dq$ towards the splitting factor g (see example 1 - Chap 8). Find its magnetic susceptibility and moment $\left[\zeta = 100\ cm^{-1}\ and\ Dq = 20000cm^{-1}\ T = 300^o K\right]$

Since the central ion is an unpaired spin system, S = 1/2, there will be a spin contribution to magnetic moment with a g value = 2. Furthermore, there is some modification of g value by an additive term $4\zeta/10Dq$. Hence the net g value is $(2+4\zeta/10Dq) = 2(1+2\zeta/10Dq)$

$$\mu_{eff} = \beta g\sqrt{s(s+1)} = \beta.2\sqrt{s(s+1)}.(1+2\zeta/10dq)$$

$$=1.73(1+2\zeta/10Dq)\beta M$$

$= 1.75\ \beta\ M$ on substituting the value of $\zeta/10Dq$.

Molar susceptibility $\chi_M = \dfrac{N\beta^2(1.75)^2}{3KT} = 1.27\times10^{-3}$ erg/gauss^2 .mole, on putting in the values of N, p, k and T. Incidentally it simplifies calculations of this nature by noting that $N\beta^2/3K \approx 1/8$.

Example 3.

Compare the experimentally obtained magnetic moments of the following d^n complexes with those calculated by you and offer comments, if necessary.

(a) $K_4Fe(CN)_6$ (strong crystal field): μ_{exp} = very small

(b) Ferrous orthophenanthroline bromide, $[Fe(Phen)_4]Br_2$

(weak field and strong field) μ_{exp} = 5.2BM

(c) Co^{2+} complex (strong O_h field) μ_{exp} = 1.7~1.9 BM

a) $K_4Fe(CN)_6$ - Its central ion Fe^{2+} is a d^6 ion. In a strong O_h field, the electrons are accommodated in t_{2g} orbital with the configuration t_{2g}^6. All the spins are paired and its symmetry is $^1A_{1g}$. Its orbital moment is, therefore, fully quenched. There is no possibility of spin-orbit contribution to the magnetic moment. Hence, its value is nil.

The existence of small magnetic moment, as found experimentally, may be due to TIP.

(b) $[Fe(Phen)_4]Br_2$ - If we take it to be of O_h symmetry its six d-electrons, primarily coupled as a ^5D Russell Saunders ground state, will occupy the split levels of T_{2g} and E_B symmetries in weak field. Tanabe Sugano diagram (Fig.10.9) shows that in both strong and weak fields the lower state symmetry is $^5T_{2g}$. The magnetic susceptibility and moment can be calculated by using the "proxy" free ion state ^5P_J (J = 3, 2, 1) in the usual $(t_2d)^n$ - p^n isomorphism scheme.

Picking A and C values from Table $T_{11\text{-}1}$, and substituting these in Eqs. (11.40 & 11.41) with T = 300°K and $kT/\zeta \approx 0.6$, χ comes out to be 1.32×10^{-2} ergs/gauss^2.mole.

Since $\chi = N\mu_{eff}^2/3kT,\quad \mu_{eff} = 5.64BM$

We thus find quite an appreciable departure from the experimental value. It is, therefore, quite feasible that the symmetry of the complex may not be just O_h, but is somewhat distorted from it.

(c) In the Co^{2+} octahedral complex, the central Co^{2+} ion has 7 d-electrons, accommodated as $t_{2g}^6e_g^1$ in the strong field with one unpaired spin. Thus with S = 1/2, its "spin-only" value for magnetic moment becomes

$\mu_{eff} = \beta g\sqrt{S(S+1)} = 1.73BM$ since g = 2. The small discrepancies from this calculated value found experimentally in different O_h complexes (strong field) may originate from unquenched orbital moment and/or TIP contributions.

We conclude with the observation that the theoretical basics of the atomic and the crystal field states, though somewhat involved and demanding, are nevertheless refreshingly elegant and intellectually delightful. Once begun, it will always induce one to go more deeply into it till he emerges at the other end, hopefully, satisfied but enquiring for more.

Exercises

1. Show that there can be no spin-orbit interaction for crystal-field states of A_1 and A_2 symmetries.
2. For T_1 or T_2 crystal field states, spin-orbit perturbation splitting is feasible. Justify.
3. In an external magnetic field, there is no orbit-moment contribution to the Zeeman levels of a non degenerate crystal field state. Establish the validity of this

statement.

4. Find the symmetry of the pair of wave functions given by Eq.(1 1.9b).
5. What are the qualitative reasons for quenching of orbital moments? Why should a d^8 metal ion in O_h complex in its ground state have its orbital moment fully quenched in ideal situation?

References

1. A. Abragan and M.H.L. Pryce: *Proc. Roy. Soc,* A206, 164, 1951; also J. S. Griffith : *Trans. Far. Soc,* 54, 1109, 1958.

2. W. Moffitt and C. J. Ballhausen : *Ann. Rev. Phys Chem.,* 7, 107, 1956.

3. J. S. Griffith : *Rec. Trav. Chim.,* 75, 676, 1956.

4. S. C. Rakshit: *Molecular Symmetry Groups and Chemistry* (New Book Stall, Calcutta)

5,6 C. J. Ballhausen : *Introduction to Ligand Field Theory* (McGraw-Hill).

7. E. U. Condon and G. H. Shortley : *Theory of Atomic Spectra* (Cambridge Univ. Press).

8,12. P. W. Atkins : *Molecular Quantum Mechanics* (Oxford University Press).

9. Hellmann and R. P. Feynman : See P. W.Atkins, *Molecular Quantum Mechanics* (Oxford Univ. Press); also *Phys. Rev.,* 56, 340, 1939.

10, 13,14. R. M. Golding : *Applied Wave Mechanics,* (Addison Wesley).

11. H. Margenau and G. M. Murphy, *Mathematics of Physics and Chemistry* (D. van Nostrand).

Subject index

Subject index